Leveraging Distortions

Leveraging Distortions

Explanation, Idealization, and Universality in Science

Collin Rice

The MIT Press
Cambridge, Massachusetts
London, England

© 2021 Massachusetts Institute of Technology

All rights reserved. No part of this book may be reproduced in any form by any electronic or mechanical means (including photocopying, recording, or information storage and retrieval) without permission in writing from the publisher.

The MIT Press would like to thank the anonymous peer reviewers who provided comments on drafts of this book. The generous work of academic experts is essential for establishing the authority and quality of our publications. We acknowledge with gratitude the contributions of these otherwise uncredited readers.

This book was set in Stone Serif and Stone Sans by Westchester Publishing Services, Danbury, CT. Printed and bound in the United States of America.

Library of Congress Cataloging-in-Publication Data

Names: Rice, Collin, author.
Title: Leveraging distortions : explanation, idealization, and universality in science / Collin Rice.
Description: Cambridge, Massachusetts : The MIT Press, [2021] | Includes bibliographical references and index.
Identifiers: LCCN 2020040789 | ISBN 9780262542616 (paperback)
Subjects: LCSH: Causation. | Science--Philosophy.
Classification: LCC Q175.32.C38 R53 2021 | DDC 501--dc23
LC record available at https://lccn.loc.gov/2020040789

10 9 8 7 6 5 4 3 2 1

For Laura and Willa

Contents

Acknowledgments   xi

1 **Introduction**   1

    1.1 Pervasive Distortion Is Central to Science   8
    1.2 Responding to Pervasive Distortion   12
    1.3 Plan of the Book   15

2 **Causal Explanation, Accurate Representation, and the Decompositional Strategy**   21

    2.1 The Role of Causation in Explanation   21
    2.2 Accurate Representation of Relevant Features   29
    2.3 Decomposition of Scientific Representations into Accurate and Inaccurate Parts   34
    2.4 Motivation for an Alternative Approach   37

3 **Many Explanations Are Noncausal**   41

    3.1 Optimality Explanations   42
    3.1.1 The Core Components of an Optimality Explanation   42
    3.1.2 A Frequency-Independent Example: Parker's Dung Flies   45
    3.1.3 Explaining General Patterns: Equilibrium Sex Ratios   47
    3.1.4 Equilibrium Explanation   48
    3.1.5 Essential Idealizations in Optimality Explanations   51
    3.1.6 Synchronic Mathematical Representation of Noncausal Features   55
    3.2 Statistical Explanations   60
    3.2.1 Galton's Autonomous Statistical Explanations   60
    3.2.2 Natural Selection and Autonomous Statistical Explanations   63
    3.2.3 Distinguishing Two Kinds of Natural Selection Explanations   67
    3.2.4 Essential Idealizations in Statistical Explanations   69
    3.2.5 Statistical Dependence and the Role of Irrelevance   71

3.3 Minimal Model Explanations 73
3.3.1 Essential Idealizations in Minimal Model Explanations 81
3.3.2 The Role of Irrelevance Information in Minimal Model Explanations 82
3.4 A Way Forward 86

4 A Counterfactual Account of Explanation 89

4.1 Three Requirements for an Account of Explanation 90
4.2 A Counterfactual Account of Explanation 93
4.3 Unifying Causal and Noncausal Explanations 111
4.4 Connecting Explanation and Understanding 117
4.5 Accounting for the Positive and Ineliminable Contributions of Idealizations 120
4.6 Comparison with Other Counterfactual Accounts of Explanation 124
4.6.1 The Relevance of Irrelevance 124
4.6.2 Extracting Modal Information from Idealized Models 128

5 Models Don't Decompose That Way: The Holistic Distortion View of Idealized Models 133

5.1 Against the Decompositional Strategy 135
5.1.1 Many Scientific Models Don't Decompose That Way 135
5.1.2 Many Models Distort Difference-Making Features 140
5.2 An Alternative Approach: The Holistic Distortion View 145
5.3 Conclusion 150

6 Using Universality to Justify the Use of Holistic Distortions to Explain 153

6.1 Appealing to Universality Classes to Justify the Use of Idealized Models to Explain 154
6.2 Parameters of Universality 162
6.3 Explicit Appeals to Universality to Justify the Use of Idealized Models to Explain 164
6.3.1 Modeling Melting Ice Ponds 165
6.3.2 Modeling Bacterial Growth 169
6.3.3 Gaussian Universality 173
6.3.4 The Universality of the Tracy-Widom Distribution 176
6.4 Universality and Other Concepts of Stability 180
6.5 Conclusion 184

7 Multiscale Modeling and Universality 187

7.1 The Problem of Inconsistent Models 190
7.2 Universality and the Problem of Inconsistent Models 197

Contents ix

    7.3 Multiscale Modeling and the Tyranny of Scales   203
    7.4 Tackling the Tyranny of Scales   206
    7.4.1 Scale Dependence, Scale Separation, and Universality   207
    7.4.2 Scale Invariance, Power Laws, and Universality   211
    7.4.3 Modeling across Scales: Renormalization, Homogenization, and Universality   217
    7.5 Philosophical Lessons: Practical Constraints, Pluralism, and Relative Autonomy   224

8   Understanding, Realism, and the Progress of Science   233

    8.1 Understanding without Explanation   234
    8.1.1 Investigating Necessity Claims   235
    8.1.2 Modeling Hypothetical Scenarios   237
    8.1.3 Exploring Possibility Space   240
    8.1.4 Summing Up   243
    8.2 An Account of Understanding   247
    8.3 Understanding Realism   259
    8.4 Understanding and Scientific Progress   264
    8.5 The Role of Diversity in Promoting Understanding   269
    8.6 Conclusion   273

9   Leveraging Holistic Distortions: The Positive Contributions of Idealizations to Explanation and Understanding   275

    9.1 Some Other Questions about Explanation   275
    9.2 Holistic Distortion Is Necessary for the Modeling Techniques Used to Explain   283
    9.3 Holistic Distortions Often Provide Better Explanations   286
    9.4 Holistic Distortions Improve Understanding   288
    9.5 Intertwined Idealizations and Holistic Distortion   291
    9.6 Universal Patterns and the Epistemic Goals of Science   293
    9.7 Realism about Epistemic Products   296
    9.8 Conclusion   297

Notes   301
References   315
Index   337

Acknowledgments

Many people have been extremely helpful and supportive during my writing of this book. I am particularly indebted to my mentors and advisors who have counseled me along the way. André Ariew fostered my love of philosophy of science and has opened innumerable doors for me. Similarly, Bob Batterman has offered me continual guidance, support, and inspiration. Both André and Bob have also offered countless feedback, comments, and conversations that have shaped the ideas and arguments presented here. Michael Weisberg has also been incredibly supportive of me over the years and sparked my interest in the issues surrounding scientific modeling. Several other friends, colleagues, and students have helped me improve my thinking, arguments, and understanding of these issues. I would particularly like to thank Yasha Rohwer, Alisa Bokulich, Julia Bursten, Kareem Khalifa, Angela Potochnik, Denis Walsh, Michael Strevens, Catherine Elgin, Christopher Pincock, James Woodward, Kyle Stanford, Kyra Hoerr, and several anonymous reviewers for comments and feedback on previous versions of these ideas, arguments, and chapters. Their feedback has been invaluable for improving the final version of the book. I am also indebted to several scientists who have helped me to better understand the practice of science, the methods of their particular fields, and their own work. In particular, I'd like to thank Kristen Whalen, Kathryn Daniel, Piper Sledge, Greg Davis, and Erica Graham. I'd also like to thank Erica for help constructing several of the figures.

I presented much of this material at a variety of places, and I am grateful to those audiences for their criticisms and feedback, especially the audiences at the 2016 and 2018 meetings of the Philosophy of Science Association. A number of institutions also supported my work financially along the way. In particular, I am indebted to support and feedback from various members of the Center for Philosophy of Science at the University of Pittsburgh, the

Logic and Philosophy of Science Department at the University of California, Irvine, and the School of Philosophy, Psychology and Language Sciences at the University of Edinburgh. Bryn Mawr College has also provided me with various kinds of research support throughout the project.

This book contains revised text from several of my previously published articles. A revised version of "Moving beyond Causes: Optimality Models and Scientific Explanation," published in *Noûs*, forms much of section 3.1. The ideas and arguments involved in section 3.2 were first presented in "Autonomous Statistical Explanations and Natural Selection," published with André Ariew and Yasha Rohwer in the *British Journal for the Philosophy of Science*. Section 3.3 is based on the ideas presented in the paper "Minimal Model Explanations," published with Robert Batterman in *Philosophy of Science*. A revised version of "Models Don't Decompose That Way: A Holistic View of Idealized Models," published in *The British Journal for the Philosophy of Science*, forms the basis of chapter 5. My papers "Idealized Models, Holistic Distortions, and Universality," published in *Synthese*, and "Universality and Modeling Limiting Behaviors," published in *Philosophy of Science*, contribute to chapter 6. Finally, "Factive Scientific Understanding without Accurate Representation," published in *Biology and Philosophy*, and "Understanding Realism," published in *Synthese*, contribute to chapter 8. I am thankful to each of these publishers for allowing me to reuse this material.

Finally, I want to thank my best friend and partner, Laura, who has continually supported me throughout my completion of this project. I will be forever grateful for her strength, love, and support.

1 Introduction

A fundamental rule of logic is that in order for an argument to provide good reasons for its conclusion, its premises must be true. This book is about how the practice of science repeatedly, pervasively, and deliberately violates this principle. However, I will argue throughout this book that rather than being a shortcoming of science, scientists are extremely adept at strategically using drastic distortions of the world in order to discover truths that would otherwise be inaccessible to us.[1] In other words, I will argue that one of the most important achievements of science is the discovery of numerous ways to *strategically leverage pervasively distorted representations* in order to discover the truths we are interested in.

Philosophers love truth. Indeed, truth enters at the ground floor of most philosophical accounts of knowledge, understanding, and explanation. Typically, this is via some sort of truth or accuracy requirement for each of these epistemic achievements. Relatedly, it is often claimed that scientific models and theories only provide genuine explanations and understanding when they are accurate with respect to the relevant features of the phenomenon of interest. Scientists, too, typically maintain that scientific methods are a reliable way to develop models and theories that are getting progressively more accurate concerning the nature of reality. As a result, among philosophers and scientists alike, it is typically assumed that the best route to the epistemic fruits of science (e.g., explanation and understanding) is via the construction of models and theories that *accurately describe the relevant parts of reality*. As Peter Godfrey-Smith describes this realist position, "One actual and reasonable aim of science is to give us accurate descriptions (and other representations) of what reality is like" (Godfrey-Smith 2003, 176). Science is certainly the greatest epistemic achievement ever produced by human beings. Without science, we would lack explanations and understanding of planetary motion, chemical reactions, phase transitions, evolution, human and animal minds, social changes, and countless other phenomena. If explaining and understanding these phenomena require true or accurate representation, then the science that produces them must (at least aim to) be true or accurate—at least with respect to the contextually salient, difference-making, or otherwise important aspects of natural phenomena.

Following this generally realist approach, most philosophical accounts of scientific explanation involve some sort of truth or accuracy requirement for the set of claims that do the explaining; that is, the statements appealed to in the explanation must be true (Craver 2007; Hempel 1965; Kaplan and Craver 2011; Salmon 1984; Strevens 2008; Woodward 2003). More specifically, given that most accounts of explanation are explicitly causal, most philosophers have claimed that a necessary condition for a scientific theory or model to be used to explain something is that it be true or accurate with respect to the explanatorily relevant causes of the explanandum. As Michael Strevens puts it, "No causal account . . . allows nonveridical models to explain" (Strevens 2008, 297). Mechanistic accounts of explanation also agree that "good explanations accurately describe the causal structure of the world" (Craver 2007, 61). While several causal accounts have emphasized the role of idealization in achieving the aims of science (Potochnik 2017; Strevens 2008; Weisberg 2007a, 2013), even those accounts that allow for liberal idealization maintain that scientific representations that are used to explain ought to accurately represent the relevant causes of interest (Potochnik 2017, 157).

In addition, most philosophical accounts of scientific understanding involve factive requirements for the body of information included in an agent's understanding (Grimm 2006; Khalifa 2017; Kvanvig 2003; Mizrahi 2012; Strevens 2013, 2017). That is, genuine understanding requires that at least most of the information grasped be true or accurate.[2] In fact, in some cases, these accuracy requirements for explanation and understanding are combined by views that claim that "scientific understanding is the state produced, and only produced, by grasping a true explanation" (Trout 2007, 585–586; see also Strevens 2008, 2013). According to these views, genuine understanding can be achieved only by grasping an explanation that is true.

In line with these accounts of explanation and understanding, most philosophical accounts of scientific modeling have required models that provide explanations or produce understanding to meet certain accurate representation requirements; for example, the model must accurately represent the difference-making, contextually salient, or otherwise relevant causes or mechanisms responsible for the phenomenon (Craver 2006; Kaplan and Craver 2011; Potochnik 2017; Strevens 2008; Weisberg 2007a, 2013; Woodward 2003). In general, the idea is that the accurate representation of the relevant causes of the phenomenon is what enables us to explain or understand why that phenomenon occurred.

Throughout this book, I will refer to this collection of views as *the standard approach* or *the standard view*. This standard approach claims that the epistemic aims of science are best achieved by constructing scientific representations that are accurate with respect to the relevant, focal, difference-making, or otherwise contextually salient causes of the phenomenon. Referring to this collection of views as the standard view is not meant to imply that there is a complete consensus in the field regarding these issues. Indeed, there are numerous authors (e.g., Batterman 2002b, 2010; Bokulich 2008, 2011, 2012; Elgin 2017; Lange 2013a, 2013b; Morrison 2009, 2015; Potochnik 2017; Walsh 2007, 2010) whose departures from this standard view have inspired many of the views I defend in this book.[3] Still, despite this dissent, the tenets of the standard view are common to the majority of the most influential accounts of explanation, idealization, and modeling in the philosophical literature: for example, the mechanistic and causal views defended by Craver (2006), Kaplan (2011), Potochnik (2017), Strevens (2008), Weisberg (2013), and Woodward (2003), among others. These influential views all adopt causal (or causal-mechanical) accounts of explanation, agree that there is a special set of causes that must be accurately represented in order to explain, and suggest that idealizations within models that explain should be restricted to causes that are not difference makers, are negligible, nonfocal, or otherwise not of interest.[4]

Scientists frequently discuss how highly idealized models aid in their explanations and understanding of various phenomena. For example, Bohr's collective model of the nucleus is useful "in explaining nuclear fission. It is also useful for understanding a large class of nuclear reactions" (Halliday, Resnick, and Walker 2011, 1185). Indeed, a brief look at any science textbook or journal reveals a wide range of cases in which science relies heavily on the use of models that are always abstract and idealized to some degree.[5] Several philosophers have suggested that we reserve the term "abstraction" for the omission of features or details from a representation of a target system or systems; that is, abstraction leaves things out (Godfrey Smith 2009; Jones 2005). Idealization, in contrast, is a process that directly introduces some degree of distortion of the features of the target systems; that is, idealizations attribute features that are (typically known to be) inaccurate. In practice, however, this distinction is not nearly so clean (or even possible), because the abstraction of certain features will almost always introduce distortions with respect to other features. For example, leaving out environmental features distorts how genetics contributes to development (Longino 2013). For this reason, among others, my focus throughout this book will be on the role of *distortion* in science, which is typically but not exclusively introduced by the use of idealization.

Relatedly, while my discussion will tend to focus on the use of models in science, I think most of the claims that I make regarding models can be easily extended to theories as well.[6] While I will not be defending any particular account of the relationship between models and theories, I am sympathetic to several accounts that would warrant this extended application of my conclusions. For example, if theories are just collections of models (Giere 1988, 2006), then my conclusions about highly idealized models will likewise apply to theories. Alternatively, according to Nancy Cartwright (1983) and Margaret Morrison (2015), models play a mediating role between the claims of a theory and the application of that theory to real cases. Similarly, Ronald Giere (2006) argues that theories are sets of principles that provide the basis for constructing more specific models. If these views are correct, then models and theories will often include similar assumptions, abstractions, and idealizations. Consequently, although there are various ways of conceiving of the relationship between models and theories, I think the arguments presented throughout this book can be applied to both types of scientific representation. The important point, for my purposes, is that science continuously presents us with a vast array of representations that are, both purposively and unintentionally, drastically distorted descriptions of their target systems.

The widespread and essential use of distortion in science raises a puzzling question that has been at the heart of the idealization and modeling literature over the last several decades: if the epistemic aims of science are best achieved by the accurate (or true) representation of reality, then how can drastic distortion contribute to science's epistemic aims? More specifically, as a result of philosophers' focus on accurate representation of relevant causes, the central question of this book is: how can highly idealized models *that directly distort difference-making and contextually salient causes* be used to provide explanations and understanding of natural phenomena. In many ways, these questions are just more specific versions of one of the oldest and most central questions raised by the discipline of philosophy (Appiah 2017): what is the relationship between the external mind-independent reality and our human representations of that reality? While much of the past philosophical discussion focused on how (or whether) our mental representations reflect or distort reality, the rapidly growing philosophical literature on scientific modeling and idealization asks similar questions regarding the representations used within scientific theorizing. In light of this, we require philosophical accounts of science that enable us to address the fundamental tension between the accurate representation requirements suggested for the epistemic successes of science and the widespread use of pervasively inaccurate representations in scientific practice. The representations used to accomplish the epistemic achievements of science have always been, and will continue to be, pervasive and drastic distortions of reality. We need a philosophy of science that can make sense of how this can be so.

To this end, the main goal of *Leveraging Distortions* is to provide an alternative philosophical approach to science that focuses on the positive and ineliminable contributions that the drastic distortion of causally and contextually relevant features makes to the epistemic achievements of science. That is, my alternative view will be built around the strategic leveraging of essential distortions *of relevant features* in order to discover explanations and understanding that would otherwise be inaccessible. While several other philosophers (e.g., McMullin 1985; Potochnik 2017; Strevens 2008; Weisberg 2007a, 2013; Wimsatt 2007) have emphasized the role of idealization in science, rather than focusing on eliminating, isolating, sidelining, or quarantining the role of idealization to the distortion of irrelevant or nonsalient features, my approach puts the positive epistemic contributions of pervasive misrepresentation of relevant features front and center.

As a result, I follow Angela Potochnik (2017) and Catherine Elgin (2017) in arguing that the epistemic aims of science can often be better achieved by rather drastic departures from the truth (i.e., idealizations are "felicitous falsehoods" [Elgin 2007, 39]).[7] While I agree with these authors that idealization plays a much more significant role than other philosophers have suggested, we take this recognition in quite different directions. For example, whereas Potochnik's central thesis is that science does not aim at truth and Elgin argues that scientific understanding must be nonfactive, I will argue that science still aims at factive (i.e., true) explanations and understanding—those epistemic aims are just achieved via scientific representations that drastically distort the relevant features of the phenomenon of interest.[8] Moreover, where Potochnik focuses on causal explanations and understanding, I offer noncausal accounts of explanation and understanding.

However, the most crucial difference to note at this point is that, while both Potochnik and Elgin allow for the distortion of some difference-making causes, their views still require that the salient, or contextually relevant, features of interest be accurately represented by scientific models that are used to explain or understand. For example, Potochnik's account of explanation requires that "posits central to representing a focal causal pattern in some phenomenon must accurately represent the causal factors contributing to this pattern" (Potochnik 2017, 157). This parallels the standard view's suggestion that idealized models explain only when they accurately describe the relevant causes and restrict their distortions to features that are nonfocal, negligible, or otherwise not of interest. Similarly, Elgin's view relies heavily on the idea that models that produce understanding must *exemplify* the salient features of interest: "Idealizations are fictions expressly designed to highlight subtle or obscure matters of fact. They do so by exemplifying features they share with the facts" (Elgin 2007, 39). The use of highly idealized models to produce scientific understanding is then justified "because the models are approximately true, or because they diverge from truth in irrelevant respects, or because the range of cases for which they are not true is a range of cases we do not care about" (Elgin 2017, 261). This, again, is fairly close to the standard view's claim that idealized models can be used to explain and understand only when they accurately describe the relevant features of the system and their idealizations are restricted to the distortion of irrelevant, nonfocal, or negligible features. In contrast with both of these views, I will argue that the epistemic aims of science are often best achieved by scientific models that directly and drastically distort precisely those features that are known to make a difference, that are salient, and that are of interest to the current research program.

In sum, I will argue that the pervasive and drastic distortion of contextually salient difference-making features routinely makes epistemic contributions that are essential to the explanations and understanding produced by science. That is, I will argue that pervasive distortions of the relevant features of real systems are ineliminable from science's epistemic successes in the sense that, without those distortions, our best scientific models and theories would be unable to provide the explanations and understanding that scientists have given for many puzzling phenomena. However, despite their being produced via pervasively distorted representations, I will argue that these epistemic achievements of science can still be factive. By surveying case studies of idealized modeling from various sciences, we will see that scientific modelers are extremely adept at identifying diagnostic features of similar problems and then using idealizing assumptions to bring existing mathematical and theoretical modeling tools to bear on those problems. Indeed, the strategic use of the pervasive distortion of relevant features is one of the most widely employed and fruitful tools used in scientific theorizing.

What's more, scientific modelers typically justify the use of these idealized modeling techniques by appealing to features other than accurate representation. Specifically, scientific modelers often justify their introduction of these distortions by appealing to the assumptions required to use various modeling techniques and to the universality (i.e., stability) of various patterns across systems that are causally heterogeneous. Doing so enables scientific modelers to employ highly distorted models to explain and understand various patterns that occur across various real, possible, and model systems that are heterogeneous in most of their causes or mechanisms. This suggests that investigating the stability of universal patterns across causally heterogeneous systems is just as important to science as producing representations that accurately represent the causal relationships that led to particular events.

Along the way, I hope to break philosophers of science's heavy reliance on the accurate representation of relevant causes in order to understand the epistemic successes of science.[9] In many places, I will frame my views by critiquing the ways that other philosophers have addressed these questions, but I make no attempt to provide an exhaustive survey of all the available alternatives. Indeed, given the dominance of causal and mechanistic approaches, doing so would be nearly impossible! Consequently, I will discuss other approaches only to the extent that comparison with them helps to illuminate the ideas and arguments I aim to defend. My primary aim will be to argue that philosophers' narrow focus on the concepts of causation, accurate representation, and decomposition (into accurate and inaccurate parts) ought to be replaced by an alternative approach based on the concepts of modal information, holistic distortion, and universality. Such a shift will enable us to see that the pervasive distortion of causally and contextually relevant features is, and ought to be, an ineliminable and epistemically fruitful tool of scientific practice. However, before embarking on this path, it will be important to see in a bit more detail just why such an alternative account of science is needed.

1.1 Pervasive Distortion Is Central to Science

As I noted above, there is a puzzle at the heart of scientific practice: by introducing idealizing assumptions—which typically are known to be inaccurate—scientists frequently claim to have provided explanations and understanding of various natural phenomena. One of the main reasons for this widespread use of idealization is that the complexity of the world, in combination with our cognitive limitations, leads to many intertwined reasons to idealize (Potochnik 2017; Wimsatt 2007). For example, Michael Weisberg (2007a, 2013) distinguishes three kinds of idealization: Galilean, minimalist, and multiple models. According to Weisberg, what distinguishes these kinds of idealizations are the reasons that motivate their introduction and the "representational ideals" they ultimately aim at (Weisberg 2007a, 639). For instance, Galilean idealizations are introduced for reasons of computational tractability. However, given that tractability limitations will presumably be overcome at some point in the future, "Galilean idealization takes place with the expectation of future deidealization and more accurate representation" (Weisberg 2007a, 642). In other words, as Ernan McMullin originally argued, Galilean idealization takes place under the assumption that "models can be made more specific by eliminating simplifying assumptions and 'deidealization,' as it were" (McMullin 1985, 261).

Because these idealizations ought to be removed as science advances, if most idealizations in science were Galilean, then the widespread use of idealization wouldn't be so worrisome. That is, if most idealized representations were simply way stations on the way to more accurate models that did the real epistemic work, the use of idealizations in science could easily be justified by showing that they ultimately lead to accurate representations or true theories (McMullin 1985; Wimsatt 2007). However, throughout this book, I will argue that many of the idealizations in science—even those that aid in computational tractability—are not Galilean because they cannot be eliminated from the scientific representation in any principled way without also destroying the explanation or understanding achieved (Batterman 2002b, 2009; Batterman and Rice 2014; Morrison 2009, 2015; Rice 2013, 2017, 2018; Wayne 2011). In short, most of the idealizations used in science simply cannot in principle be eliminated in the way that accounts of Galilean idealization suggest. Often, this is because the idealizations are *deeply entrenched* in our best scientific models and theories, such that we have no way of knowing what a similar representation without those idealizing assumptions would look like. In short, not only is deidealization rare in practice, but also, in most instances, it is a philosopher's pipe dream.

More problematic for the realist's claim that science aims for accurate representation are what Weisberg calls minimalist idealizations. Minimalist idealizations aim at the construction of models that include only the core causal (or difference-making) factors that gave rise to the phenomenon of interest (Weisberg 2007a, 643–645). According to Weisberg, the reason for using minimalist idealization is tied to the goal of providing scientific explanations because—following the standard approach—most accounts suggest that explanations ought to accurately represent the difference-making causal factors and leave out (or idealize away) causal factors that are irrelevant (Craver 2007; Strevens 2008; Weisberg 2007a, 2013; Woodward 2003). Moreover, according to these minimalist accounts, these idealized models provide better explanations than more accurate models because accurately describing those features would misleadingly suggest that they are relevant to the phenomenon. Consequently, minimalist idealizations should not be removed as science progresses. Due to this permanence, minimalist idealization presents a bit more of a challenge to realists' claims that science aims at truth or accurate representation (Weisberg 2007a).

However, as we will see in later chapters, the problem of idealization is far more potent than accounts of minimalist idealization suggest because many of the explanations provided by science appeal to highly idealized models that *drastically distort relevant features as well* (Batterman and Rice 2014; Bokulich 2012; Potochnik 2017; Rice 2013, 2017, 2018). This is because in many cases, the only (or at least the best) explanation of a phenomenon requires idealizations that directly distort difference-making and contextually salient causes of the phenomenon (Batterman and Rice 2014; Potochnik 2017; Rice 2013, 2017, 2018). For example, optimality models in biology drastically distort the difference-making processes of natural selection that are of interest to adaptationists (Rice 2013, 2017, 2018). Models of fluids pervasively distort the causal processes that physicists know make a difference to phase transitions (Batterman and Rice 2014; Rice 2017). Modelers studying human behavior knowingly distort difference-making features of genes and the environment (Longino 2013). And the list goes on. Consequently, in addition to being ineliminable (i.e., non-Galilean), many of the idealizations used in scientific explanations cannot be restricted to the distortion of irrelevant features in the way that minimalist accounts of idealization suggest.

In addition, there are many essential idealizations in science that are used for purposes other than explanation. For instance, what Yasha Rohwer and I have elsewhere called "hypothetical pattern idealizations" are used in the construction of hypothetical scenarios that aim at the production of understanding (Rohwer and Rice 2013). These idealized models do not aim to explain the target phenomenon, but instead aim to investigate necessity or possibility claims. As a quick example, the Hawk-Dove game is a highly idealized game-theoretic model that investigates the possibility of individual-level selection producing restraint in combat. The model describes a population that is infinitely large, randomly pairs players, involves asexual reproduction, allows only symmetric interactions, and stipulates that all interactions involve exactly two players, that the payoffs for players are constant across iterations of the game, and that there is perfect correlation between winning the game and fitness (Maynard Smith 1982). Although biologists are well aware that such a system is impossible and drastically distorts how natural selection actually produced the phenomenon, the Hawk-Dove game enables us to understand that restraint in combat could *possibly* be the product of individual-level selection. However, as with many of the models used to explain, these hypothetical models make these epistemic contributions without accurately representing the difference-making or contextually salient causes of any real-world target systems.

What's more, while philosophical discussions of idealization have focused on the most egregious false assumptions (e.g., infinite populations, infinite numbers of particles, or perfectly rational agents), science is also filled with many more *garden-variety idealizations*: subtle distortions that are so common that, taken together, they add up to rather drastic departures from the truth (Cartwright 1983). This includes the use of discrete variables for continuous features (e.g., Bursten 2018a) or the use of continuous variables for discrete features. It also includes the smoothing of curves and the simplification of equations in order to make their solutions easier to calculate (Elgin 2017). As I mentioned above, also important here are the distortions of various features introduced by the abstraction of other features (Cartwright 1983; Walsh, Ariew, and Matthen 2002). Philosophers tend to ignore these kinds of distortion because they often seem easy to correct as science progresses. However, even when this is possible, it is rarely done in actual practice. Furthermore, in many cases, it in fact cannot be done without dismantling the scientific model or theory in question.

As a result, most of the explanations and understanding produced by science derive from models and theories in which myriad subtle distortions are maintained throughout repeated applications, right alongside more drastic idealizing assumptions. Consequently, when they are considered collectively, the various motivations and types of idealization used in scientific theorizing typically result in scientific representations that *pervasively distort most of the features of their target systems*—including those that are of interest and make a difference to the phenomenon. Although most of the cases that I will focus on in this book involve more egregious idealizing assumptions that cannot be removed in principle, it is important to remember that even these subtler distortions are widespread and rarely removed in practice.

To make matters even worse, in addition to pervasive idealizations that are introduced into particular scientific models (and theories), very often in science we find multiple conflicting idealized models being used to study the same phenomenon (Chakravartty 2010; Massimi 2018; Mitchell 2009; Morrison 2011; Odenbaugh 2005; Parker 2006; Rice 2019b; Sklar 1993; Weisberg 2007a, 2013; Wilson 2017; Wimsatt 2007). That is, across many scientific fields—including biology, physics, nanoscience, climate science, and economics—multiple idealized models are often used to investigate the same phenomenon, despite those models often including inconsistent theoretical assumptions about (or representations of) that phenomenon. We can see this clearly in Weisberg's characterization of "multiple models idealization" as "the practice of building multiple related but incompatible models, each of which makes distinct claims about the nature and causal structure giving rise to a phenomenon" (Weisberg 2007a, 645). The challenge is to show how such a situation can result in an improved overall understanding of the target phenomenon by integrating the insights provided by multiple inconsistent models. Furthermore, given that most philosophical accounts have relied on accurate representation of relevant causes, it is unclear how multiple conflicting (and known to be false) representations of the relevant features of a system could produce genuine explanations and understanding. I will refer to this challenge as *the problem of inconsistent models* (Chakravartty 2010; Massimi 2018; Morrison 2011, 2015; Rice 2019b).

There are certainly several other kinds of idealization in science besides those that I've mentioned here; this list is nowhere near exhaustive! The main point that I want to make here—and will argue for extensively throughout this book—is that essential and ineliminable idealization of relevant and irrelevant features is widespread in science. This raises a serious challenge to the standard philosophical approach to science because *we know that our best scientific theories and models are representations that pervasively distort both relevant and irrelevant features because we deliberately made them that way.* In sum, the challenge is to clarify how the widespread use of pervasive distortion can make essential and positive contributions to scientific explanations and understanding.

1.2 Responding to Pervasive Distortion

Within the philosophy of science, there have been two basic responses to science's widespread use of idealized models and theories to explain and understand. The first is to argue that the epistemic results of science depend only on the true (or accurate) parts of our scientific models and theories. That is, many accounts argue that the "premises" of scientific inferences are true (or accurate) with respect to the relevant, difference-making, or otherwise contextually salient features that scientists are interested in (e.g., Craver 2006, 2007; Elgin and Sober 2002; Hempel 1965; Kaplan and Craver 2011; Kitcher 1981; Pincock 2011; Potochnik 2007, 2009b, 2017; Strevens 2004, 2008; Weisberg 2007a, 2013). There are, of course, numerous differences among these views. However, each attempts to show that we are justified in accepting the knowledge, understanding, or explanations provided by science because the inferences used depend on models and theories that accurately describe the relevant, difference-making, or contextually salient features of real-world systems. Indeed, as I will argue in more detail in chapter 2, this decompositional strategy is widespread across a range of philosophical debates about explanation, idealization, modeling, and realism

(Rice 2017). The decompositional strategy tries to preserve the factive epistemic achievements of science by decomposing scientific theories and models into their accurate and inaccurate parts (or assumptions) and showing that the accurate parts are responsible for science's epistemic successes.

The other main response has been to argue that because the inferences of science essentially depend on models and theories that are known to be inaccurate, the epistemic results of those inferences must be nonfactive. That is, because some of the essential "premises" of the inferences are false, the conclusions of those inferences (i.e., the epistemic outputs of science) cannot aim at truth (Elgin 2007, 2017; Potochnik 2017).[10] This kind of response is also advocated by various nonrealists such as van Fraassen (1980) and Stanford (2003). In short, this response argues that we are somehow mistaken about the epistemic successes of science. Science does not in fact achieve factive explanations or understanding of natural phenomena; all that can be hoped for is empirical adequacy or nonfactive understanding.

Throughout this book, I will argue against both of these approaches and propose an alternative. According to my view, the representations used by scientists to explain and understand our world drastically and deliberately distort the relevant features of their target systems—including precisely those features that scientists are interested in and that are known to make a difference to the phenomenon of interest. In short, the "premises" of scientific inferences are known to be inaccurate even with respect to the relevant, difference-making, or otherwise contextually salient features of interest. As I suggested above, scientists often use idealizations in positive and ineliminable ways that pervasively misrepresent the model's target systems, including the relevant (e.g., difference-making or contextually salient) features of the target phenomenon. Moreover, scientists *should* be doing this. Philosophical views that suggest that such uses of idealization cannot contribute in positive ways to the epistemic achievements of science not only inaccurately describe the practice of science, but also harmfully suggest that scientists should not be using many of the most epistemically valuable tools in their arsenal. Often, it is only by pervasively distorting the relevant features of reality that scientists can discover the explanations and understanding we seek.

Despite this pervasive distortion, in contrast with the second approach, I will also argue that the epistemic outputs of science—in particular the explanations and understanding produced by science—are nonetheless factive. That is, I will argue that the "conclusions" that science extracts from its use of highly idealized (and abstract) models are true. The challenge, of course, is to show exactly how science is able to accomplish this seemingly magical feat, given that it violates one of the most basic rules of logical inference. The only way to do this, I suggest, is by looking in detail at the various (typically mathematical) modeling techniques used in scientific practice and the ways that scientists strategically leverage those distorted representations in order to extract the knowledge, understanding, and explanations we desire. Time and time again, scientists demonstrate that they are extremely adept at using drastic distortions of reality to further their (and our) epistemic aims.[11] As a result, even if one could construct a version of science that could proceed without the use of essential and pervasive distortion of relevant features, the history of scientific practice suggests that such a science would be *far less epistemically successful*.

My goal in this book is to develop new philosophical accounts of explanation, idealization, modeling, and understanding that enable us to make sense of how this could be so. Unfortunately, because essential idealization of relevant features fails to sit well with standard philosophical approaches to science, the centrality of these distortions to scientific practice often goes unmentioned in discussions about the methods of science. I have found that this is often also true in discussions among scientists themselves—at least in discussions of how to teach science to students, or how to promote a general public perception of how science works. In both cases, the essential and pervasive role of distortion of relevant features is largely sidelined in favor of characterizing science as a slow march toward more and more accurate representations of the relevant features of the natural world. Unfortunately, the willingness to promote this misleading description of how science works has led to widespread misunderstanding of how science has achieved the epistemic successes it has and how it might be made even more epistemically successful going forward.

Therefore, my goals in this book are both descriptive and normative. If it is to be useful to practicing scientists, the philosophy of science must go beyond merely describing science, and offer ways of improving scientific practice by illustrating how science works when it works well, as well as showing how to avoid erroneous inferences or methods. However, if the normative guidance offered by philosophers is too far divorced from the actual practice of science, then the lessons will be of little use. As a result, our normative philosophical accounts of science need to be closely tied to actual scientific practice.

I will argue that the standard responses to the challenges raised by idealizations are too descriptively inaccurate to offer much in terms of normative guidance to practicing scientists. Only if philosophers develop pictures of science that more accurately reflect the actual practices, methods, and justifications used by scientists will their normative prescriptions be of genuine use to the practice of science. My intended audience, therefore, is both philosophers and scientists (as well as students in these fields). I hope to get philosophers of science to update their views concerning scientific explanation, understanding, idealization, modeling, and realism in ways that scientists will find useful. I also hope that in doing so, scientists might find the ideas and arguments presented in this book genuinely helpful to their practice and to the teaching of that practice to others. In order to accomplish these goals, throughout this book I deliberately adopt a case-based, bottom-up methodology that begins with the detailed analysis of numerous case studies from scientific practice, from which I will construct my philosophical accounts of explanation, idealization, modeling, and understanding.

1.3 Plan of the Book

After grouping several prominent accounts of explanation, idealization, and modeling under the standard approach in chapter 2, I will begin by arguing that—contrary to most accounts of explanation—many explanations in science are *noncausal*. I do this in chapter 3 by providing an in-depth analysis of three types of noncausal explanations from across the sciences: optimization explanations, statistical explanations, and minimal model explanations. Drawing on features extracted from these examples, I present in chapter 4 an alternative view of explanation that focuses on providing information about both counterfactual dependence *and* counterfactual independence. The main advantage of this counterfactual approach is that it can capture the wide variety of noncausal explanations found across the sciences and show what they have in common with the numerous causal explanations provided in science. In addition, the account provides a direct connection between the discovery of explanations and the production of understanding. Finally, I will argue that the counterfactual approach provides a framework that is better able to accommodate the positive contributions made by essential and ineliminable idealizations to many scientific explanations.

After arguing for this counterfactual account of explanation, I move to show how essential idealizations routinely enable scientific modelers to produce explanations (and understanding) without accurately representing the relevant features of their target systems. Instead of attempting to decompose scientific representations into their accurate and inaccurate parts, I argue in chapter 5 for an alternative, *holistic distortion view* of idealized models. According to this view, the use of idealization is so pervasive that philosophical accounts of science ought to characterize scientific models as pervasive (rather than partial) distortions of their target systems. This isn't to say that models always distort every single feature of their target systems. Rather, according to the holistic distortion view, the distortions introduced by widespread idealization (and abstraction) in science are typically so pervasive that we are unable to identify which features of the target systems are being accurately represented and which are distorted. As a result, we need philosophical accounts of how science can be epistemically justified that do not require scientists to know how to decompose idealized models (or theories) into their accurate and inaccurate parts. The holistic distortion view contributes to this project by changing the way that we conceive of the representations provided by idealized models and how scientists' appeals to those representations ought to be justified. In particular, the holistic distortion view focuses our attention on the specific ways in which pervasively distorted models are leveraged by scientists to discover explanations and understanding that would otherwise be inaccessible.

In chapter 6 I develop this holistic distortion view further by showing how physicists' concept of a universality class can clarify how holistically distorted models can be justifiably used in the construction of scientific explanations and understanding. A *universality class* is a group of systems that display some universal (i.e., stable) behaviors despite (sometimes drastic) differences in their physical features (Batterman 2002b, 2009, 2010; Morrison 2015, 2018a). As the mathematician Terence Tao explains, "Over the decades, many such universal [patterns] have been found to govern the behavior of wide classes of complex systems, regardless of the components of a system or how they interact with each other" (Tao 2012, 25). By being in the same universality class as their target systems, idealized models—which are conveniently amenable to the various modeling techniques that scientists have on hand—can be justifiably used to discover explanations and understanding of real-world phenomena without having to accurately represented the causes or mechanisms responsible for those phenomena. What's more, by looking at a variety of examples from scientific modeling, I argue that this account better aligns with the justifications provided by scientists themselves for using highly idealized models to explain complex phenomena.

Furthermore, in chapter 7 I argue that universality classes also help us understand some of the ways that scientific modelers can overcome various challenges that arise in cases of *multiscale modeling* (e.g., the problem of inconsistent models and the tyranny of scales). Scientific modelers are often interested in phenomena whose relevant features span an incredibly wide range of spatial and temporal scales. The problem is that most scientific modeling tools have been developed to capture features at only one (or a few) scales of a system (Batterman 2013; Green and Batterman 2017). The concept of universality shows how multiscale modelers are able to overcome some of these challenges by investigating the scale-dependent, scale-invariant, and interscale relationships that are required to model complex phenomena with multiple conflicting models. Moreover, this investigation into various multiscale modeling techniques has important implications for philosophical debates concerning modeling, reduction, and emergence.

In addition to providing explanations, another primary epistemic aim of science is the production of human understanding (de Regt 2017; Elgin 2017; Khalifa 2017; Potochnik 2017). In chapter 8 I first argue that holistically distorted models can produce factive scientific understanding via routes other than explanation. I then use these cases to develop a factive account of scientific understanding that is compatible with science's widespread use of holistically distorted models. After we have seen the various ways that holistically distorted models can positively contribute to the epistemic achievements of science, we can begin to make the case for adopting a more nuanced form of scientific realism. This form of realism must acknowledge that our best scientific theories and models are pervasively inaccurate representations. However, according to the view that I will call *understanding realism*, these inaccurate scientific representations can still be used to produce a factive corpus of understanding of natural phenomena. I argue that this can be accomplished by moving our focus away from the accuracy of scientific models and theories themselves, and instead evaluating the factivity of the epistemic achievements produced by scientists' strategic leveraging of those distorted representations. Furthermore, my account shows how past models and theories can contribute to our current understanding of phenomena and how the epistemic aims of science are directly improved by increasing the diversity of researchers, methods, and modeling approaches in science.

In combination, these accounts of explanation, idealization, modeling, and understanding will allow us to better characterize the wide variety of ways that scientists leverage holistic distortions in order to better explain and understand our world. Putting these ideas together into an overall picture of scientific theorizing, in chapter 9 I first show how the ideas and arguments from earlier chapters provide answers to several important questions regarding scientific explanation that are largely ignored by traditional accounts (Potochnik 2018). In addition, I argue that these views clarify the ways that holistically distorted models provide better explanations and improve our overall understanding when compared to less distorted scientific representations. Finally, these views focus our attention on the essential role played by universal patterns in science, rather than on uncovering the causes of the phenomenon.

In summary, the central arguments of the book make the case for replacing each of the components of the standard approach with the following components of my leveraging holistic distortions approach:

	Replace: Standard approach	With: Leveraging distortions approach
Chapters 3 and 4	Causal explanation	Counterfactual explanation
Chapter 5	Decomposition into accurate/inaccurate parts	Holistic distortion
Chapters 6 and 7	Accurate representation	Universality classes
Chapter 8	Realism about models and theories	Realism about understanding

More specifically, I will argue that a counterfactual account of explanation ought to *subsume* causal approaches to explanation. In contrast, my holistic distortion view is meant to *replace* accounts that focus on accurate representation of relevant features. Thus some parts of the standard view can be subsumed, but others will have to be replaced. Doing so will enable us to develop philosophical accounts of science that explicitly recognize the essential and ineliminable use of pervasive distortion of relevant features *as a virtue rather than an obstacle* to the epistemic accomplishments of science.

Science is the greatest epistemic tool we have, but we need better accounts of how it does and should produce explanations and understanding via idealized models and theories. However, in order to provide alternative answers to these questions, we need to rethink many of our foundational philosophical assumptions and focus our attention on the variety of ways that holistically distorted models are strategically leveraged by scientists to explain and understand our world.

2 Causal Explanation, Accurate Representation, and the Decompositional Strategy

In order to motivate the views that I defend later in the book, in this chapter I briefly survey several of the most prominent philosophical accounts of scientific explanation, idealization, and modeling. While there are certainly myriad differences among these views, I argue that they can be grouped under what I call the *standard approach* or the *standard view* by showing that all of them adopt three main tenets: (1) the essential role of causation in explanation; (2) the aim of accurately representing relevant (e.g., difference-making or contextually salient) causes; and (3) the accommodation of idealization by decomposing scientific representations into their accurate and inaccurate parts. The following chapters of the book argue that each of these assumptions ought to be rejected, and suggest alternative accounts of explanation, idealization, modeling, and understanding that ought to replace (or subsume) them.

In the following section I briefly survey the explanation literature to show that the vast majority of extant accounts of explanation have been explicitly causal. Next, in section 2.2, I argue that these influential accounts have also required the scientific representations that are used to explain to accurately represent the relevant causes of the phenomenon. Then, in section 2.3, I show that most of these accounts have responded to science's use of idealizations by attempting to decompose scientific models (and theories) into their accurate and inaccurate parts. After laying out these central claims of the standard view, section 2.4 provides some initial motivation for thinking that philosophical accounts of science ought to move beyond these claims.

2.1 The Role of Causation in Explanation

It is difficult to overstate the importance of explanation to science. Indeed, there is a fundamental distinction between knowledge *that* something occurred and knowledge about *why* something occurred. Knowledge of the latter kind is *explanatory* knowledge, and it produces scientific understanding of our world. Philosophers have long been interested in the concept of explanation because explaining why things happen is central to both our daily lives and scientific practice (Salmon 1984). For example, we are each continuously generating (and testing) possible explanations for why other people behave as they do, why we feel sick, why our appliances have stopped working, why our tomato plant died, and so on. In addition, one of the primary goals of scientific inquiry is to provide explanations of natural phenomena. For example, we may want to explain why various fluids undergo phase transitions, why salt dissolves in water, why gas prices rose over the last decade, or why polar bears have thick fur. By providing the right kind of information, we can explain why these events happen. In fact, Ernest Nagel once argued that: "It is the desire for explanations ... that generates science; and it is the organization and classification of knowledge on the basis of explanation that is the distinctive goal sciences" (Nagel 1961, 4). At the very least, explanation is typically seen as one of the distinctive aims of scientific inquiry. The concept of explanation is also crucial to science education: "Over the years, evidence has accumulated indicating that getting students to articulate not just *that* something happens, but *why* something happens, is exceedingly valuable for a variety of reasons" (Cooper 2015, 1274).[1] Furthermore, discovering explanations is the most common way that science produces understanding of the phenomena we observe. I will address the growing debate concerning whether or not understanding can be produced without going through explanation later on in the book (e.g., see Khalifa 2012, 2013; Lipton 2009; Rice 2016; Rohwer and Rice 2016; Strevens 2008, 2013). For now, I will focus on explanation because it is one of the primary epistemic achievements of scientific inquiry.

Just as it is difficult to overstate the importance of explanation to science, it is also difficult to overstate just how dominant the causal (or causal-mechanical) paradigm has been in the philosophical literature on explanation. A major reason for this is that most philosophers have taken for granted that the only (or at least the best) solution to various problems with Carl Hempel's *deductive-nomological* (DN) model is to adopt a causal account of explanation.

Hempel aimed to characterize one kind of scientific explanation as a particular form of deductive argument (Hempel 1965; Hempel and Oppenheim 1948). The explanatory argument has the explanandum as its conclusion.

The premises of the argument, the explanans, must include antecedent conditions and at least one lawlike generalization (and the law must be vital to the validity of the argument). When the explanandum can be logically deduced from the explanans, the argument is said to subsume the event to be explained under the laws of the explanans.[2] Combining laws with the antecedent conditions is then supposed to show us why the event to be explained was *expected* to occur (Hempel 1965, 337). If these structural conditions are satisfied, the argument is a potential explanation. If the antecedent conditions and laws of the explanans are also *true*, then the argument is an actual explanation. Hempel's account is the starting point for the vast amount of philosophical work that constitutes the modern discussion of scientific explanation.

Despite its influence, the DN model has been widely rejected, largely due to the presentation of numerous counterexamples. The most famous counterexample is the flagpole's shadow (Bromberger 1966). A vertical flagpole of a certain height casts a shadow of a certain length on the ground. Now, from facts about the pole's height and the sun's elevation, in conjunction with laws concerning the propagation of light, we can deduce the length of the shadow. Conversely, however, given facts about the shadow's length and the sun's elevation, in conjunction with the same laws, we can deduce the height of the flagpole. However, few will agree that the length of the shadow can be used to explain the height of the flagpole. The difference appears to be due to the fact that the flagpole causes its shadow, and can therefore explain its length, whereas the shadow does not cause the flagpole and thus cannot explain its height. The DN model fails to capture this asymmetry because deduction is symmetric. This is typically referred to as the *problem of asymmetry* (Salmon 1984).

A second set of counterexamples focuses on showing that Hempel's account allows explanations to cite features (or laws) that are explanatorily irrelevant. One of these counterexamples is the hexed-salt case. In this case, one might try to explain the dissolving of table salt in water by citing the fact that it was hexed and the (true) law-like generalization that "All hexed salt dissolves in water." This explanation meets the criteria of the DN model. The problem, of course, is that the correct explanation for the salt dissolving in water is that all salt is soluble in water (i.e., the chemical properties of salt explain the event). The salt's being hexed is completely irrelevant. This has come to be called the *problem of relevance* (Salmon 1984; Strevens 2008).

Finally, Hempel's DN model struggles to handle the numerous instances in science in which explanations are provided without referencing laws of nature. For example, many explanations in biology, economics, and social science are provided by models or generalizations that are limited in scope and have many exceptions (Mitchell 2009; Potochnik 2017). In fact, there is some debate about whether these sciences have laws at all (Beatty 1995; Sober 1997). Regardless of whether there are laws in these "special sciences," it is simply a fact that many of the explanations provided by these sciences cite models or generalizations that appear to lack most of the features required for being considered a law of nature. Moreover, even in sciences were laws are more easily found (e.g., physics), many explanations are provided without referencing laws.

One way to respond to these problems is to make causation central to explanation. Causation is asymmetric in that, on most accounts of causation, a cause must precede its effects. Therefore, if A causes B, B does not cause A. In addition, the reason that the chemical structure of salt, and not its being hexed, is explanatorily relevant is that those properties (but not the hexing) cause the salt to dissolve. Finally, causal explanations need not cite laws of nature, but instead focus on causal dependencies of varying levels of generality (Bechtel and Abrahamsen 2005; Potochnik 2017; Strevens 2008; Woodward 2003). As a result, most philosophical accounts of explanation have explicitly adopted a causal approach after briefly mentioning some of these problems for Hempel's account.

According to the causal approach, what explains an event is the event's causal history. More specifically, causal accounts of explanation suggest that the context of the explanation will determine a set of relevant factors in the event's causal history and the explanans ought to describe those causal factors and how they relate to the explanandum. In order to identify these causal relations, Wesley Salmon's (1984, 1989) influential account distinguishes between causal processes and pseudoprocesses. Causal processes are able to transmit marks; pseudoprocesses cannot. A process is capable of transmitting a mark if, once a mark has been introduced at one spatiotemporal location, it will persist to other spatiotemporal locations even in the absence of any further interaction. For example, if you introduce a mark into a stationary light beam by placing a blue lens at one location, that modification will persist from that point on without further intervention. In contrast, if a light beam from a lighthouse is rotating, a mark introduced at one location will not be passed on as the beam rotates away from the lens. Therefore, the stationary beam is a genuine causal process; the rotating beam is a pseudoprocess. This distinction is important, Salmon argued, because only genuine causal processes are able to transmit *causal influences*, which connect events that occur at different times and places. In particular, for Salmon, a causal explanation "provides knowledge of the mechanisms of *production* and *propagation* of structure in the world" (Salmon 1978, 701). As Salmon summarizes his account: "causal processes, causal interactions, and causal laws provide the mechanisms by which the world works; to understand why certain things happen, we need to see how they are produced by these mechanisms" (Salmon 1984, 132). Thus, according to Salmon's causal-mechanical (CM) model, an explanation must trace (at least some portion of) the spatiotemporally continuous causal processes that produced the event to be explained.

Another example of the causal approach is James Woodward's interventionist account (Woodward 2003, 2010, 2018). In contrast with Salmon's view, Woodward's account does not require the causal relationships cited in the explanans to be spatiotemporally continuous causal processes. According to Woodward's account, c is a cause of e just in case the following interventionist counterfactual is true: if c had not occurred due to an intervention, and had all other variables been held constant, e would not have occurred (Woodward 2003, 203). Furthermore, on Woodward's account, causal relations connect causal *properties* represented by a set of variables in a causal graph. To say that there is a causal relationship between variables X and Y is to say that intervening on property X, while holding the other causal contributions to Y fixed, would change (the probability distribution of) property Y. Finally, in order for these interventions to be a genuine test of a causal relationship, there must be a possible intervention on X that is "surgical" in such a way that any change in Y will occur only "through" the change in X (Woodward 2003, 14).

To explain a phenomenon, on Woodward's account, is to cite these difference-making causes that produced the phenomenon. According to Woodward, this is explanatory because evaluating the outcomes of such interventions enables us to answer a range of "What if things had been different?" questions (Woodward 2003, 11). More generally, manipulationist or interventionist accounts of explanation claim that a property X is causally (and therefore explanatorily) relevant to the explanandum Y if and only if there is some ideal intervention on X that changes the probability of Y.[3] Furthermore, while he admits that there may be noncausal explanations that do not tell us about the results of interventions (as I will discuss in the next chapter), Woodward tells us explicitly that: "My view is that the sorts of counterfactuals that matter for purposes of causation and explanation are [those] that describe how the value of one variable would change under interventions that change the value of another" (Woodward 2003, 15).

Another causal account of explanation is defended by Michael Strevens, who explicitly adopts the causal approach after citing the problem of asymmetry and the problem of relevance for Hempel's DN model (Strevens 2008, chap. 2). On Strevens's view, explanations are defined by a metaphysical dependence relation and an explanatory relevance relation. For Strevens, the metaphysical dependence relation specifies what the causes of an event are. Given that the notion of causation is itself a controversial philosophical topic, on Strevens's account, the causes are just those factors that satisfy one's preferred metaphysical view of causal influence. Once the causes of the event have been identified, the explanatory relevance relation specifies *which* of those causes ought to be included in the explanation. For Strevens, like Woodward, the explanatory relevance relation is a relation of difference-making. However, instead of appealing to interventionist counterfactuals, Stevens's view appeals to a veridical, deterministic, and atomic causal model for event e. Such a model is veridical because the statements in the model are true or accurate with respect to the causal processes that produced the event. It is deterministic because the statements in the model must entail that event e occurs. And it is atomic because no intermediate steps in the causal process are spelled out explicitly within the model. Finally, the model is causal because the statements in the model *causally* entail that e occurs, such that the model's derivation of e "mirrors a part of the causal process by which e was produced" (Strevens 2008, 72). Using these models, according to Strevens's view, "If a factor c cannot be removed from a veridical, deterministic, atomic causal model for an event e without invalidating the entailment of e, then c is a difference-maker for e" (Strevens 2008, 87). Once this criterion for difference-making has been applied, on Strevens's kairetic account, to explain an event is to provide a veridical causal model that includes only the causes that make a difference to the explanandum. As Strevens puts it, "Every element in a standalone explanation of an event e, . . . is a difference-maker for e" (Strevens 2008, 121).

More recently, Angela Potochnik (2015, 2017) has offered a causal account of explanation that focuses on *causal patterns*. Like other causal accounts, Potochnik's use of causal dependence relations is partially motivated by various counterexamples that show the asymmetry of explanation (Potochnik 2017, 135). Moreover, on Potochnik's account, the features cited in causal pattern explanations qualify as causal because they meet Woodward's manipulationist requirements for causal relationships (Potochnik 2017, 136). What makes these causal *pattern* explanations is that an explanation must also identify the scope of the causal relationships cited in the explanans (Potochnik 2017, 136). Thus, on Potochnik's view, explanations (1) show how the phenomenon to be explained causally depends on one or more properties of the world, and (2) indicate the scope of those causal dependencies (Potochnik 2017, 139). In addition, Potochnik's account focuses on the causes that are of interest to the current research program (Potochnik 2017, 150–152). This allows difference-making causes that are not of interest to be omitted or idealized (Potochnik 2017, 140). Despite allowing for the elimination of some difference-makers, according to Potochnik's account, what distinguishes explanations from nonexplanations is that they accurately describe the causal factors that contribute to the pattern of interest to the current research program (Potochnik 2017, 157).

Another prominent example of the focus on causation in explanation comes from various mechanistic accounts of explanation (Bechtel and Richardson 1993; Craver 2006, 2007; Craver and Darden 2013; Glennan 2002, 2017; Kaplan and Craver 2011; Machamer, Darden, and Craver 2000). Mechanistic accounts claim that to explain an event is to describe the components of the causal mechanisms—that is, the parts, their organization, and their individual causal contributions—that produced (or constituted) the explanandum. As David Kaplan and Carl Craver put it, providing a mechanistic explanation "will involve describing the underlying component parts, their relevant properties and activities, and how they are organized together causally, spatially, temporally and hierarchically" (Kaplan and Craver 2011, 605). Indeed, Craver's mechanistic view claims that "complete explanations capture all of the relevant causal relations among the components in a mechanism" (Craver 2007, 61–62). The reason why this is important, according to Craver, is that "when one possesses explanations of this sort, one knows how to intervene into the mechanism in order to produce regular changes in the phenomenon" (Craver 2007, 160).

Another prominent account of mechanistic explanation is defended by Bill Bechtel (Bechtel 2015; Bechtel and Abrahamsen 2005, 2010). According to Bechtel, mechanistic explanations aim to decompose mechanisms into their parts and operations and show that, when appropriately organized, these components can generate the phenomenon of interest (Bechtel 2015, 84). However, on Bechtel's view, "since explanation is itself an epistemic activity, what figures in it are not the mechanisms in the world, but representations of them" (Bechtel and Abrahamsen 2005, 425). Thus, the relevant notion for Bechtel's account of explanation is that "a model of a mechanism describes or portrays what are taken to be its relevant component parts and operations, the organization of the parts and operations into a system, and the means by which operations are orchestrated so as to produce the phenomenon" (Bechtel and Abrahamsen 2005, 425). In short, modeling how the parts of the mechanism causally produce the phenomenon is what provides the mechanistic explanation.

More recently, Stuart Glennan has also argued that mechanistic explanations proceed by "showing how the organized activities and interactions of some set of entities cause and constitute the phenomenon to be explained" (Glennan 2017, 223). However, Glennan goes further, suggesting that mechanistic explanations describe the processes that are responsible for causal relationships; that is, mechanistic explanations show *how* the causes produced the explanandum (Glennan 2017, 226). As a result, his account directly ties mechanistic explanation to describing (perhaps in more detail) how the entities and activities of the mechanism caused (or constituted) the explanandum. Indeed, the general goal of mechanistic explanations is to describe the relevant (i.e., difference-making) components and interactions of the mechanisms that causally produce or constitute the explanandum.

In sum, the most prominent approaches to scientific explanation claim that an explanation must describe the relevant causes of the explanandum (Craver 2006; Hitchcock and Woodward 2003; Kaplan and Craver 2011; Lewis 1986; Potochnik 2015, 2017; Railton 1981; Salmon 1984; Strevens 2004, 2008; Woodward 2003). Although these accounts differ over which causal factors must be included in an explanation, they all agree that causal relationships are the key to scientific explanation. Indeed, most philosophical accounts of how explanation works are explicitly causal.

2.2 Accurate Representation of Relevant Features

In addition to the focus on causation's role in explanation, since Hempel, it has been widely accepted that a necessary condition for something to count as a scientific explanation is that it be (at least in some sense) true (Hempel 1965, 248). A version of this requirement clearly exists in contemporary causal and mechanistic theories of explanation. Indeed, according to these theories, to explain an event is to *accurately* describe, or represent, the difference-making (or otherwise contextually salient) causes or mechanisms that give rise to (or constitute) the explanandum. However, as discussed in chapter 1, it is also widely accepted that many scientific explanations are provided by highly idealized models (Batterman 2002a, 2010; Bokulich 2011, 2012; Cartwright 1983; Morrison 2015; Potochnik 2017; Rice 2013, 2018; Strevens 2004, 2008; Weisberg 2007a, 2007b, 2013). This has led defenders of causal and mechanistic accounts of explanation to provide various accounts of how models that include idealizations are still able to explain—and perhaps can do so better than their nonidealized counterparts. The vast majority of these accounts have appealed to one, or typically both, of the following features: (1) the accuracy of the model with respect to the relevant (i.e., difference-making, focal, or contextually salient) causes of the explanandum; and (2) the irrelevance, insignificance, or nonsalience of the features distorted by the idealizations.

For example, according to most mechanistic accounts, an idealized model will only explain if it provides an accurate representation of the relevant features of the causal mechanisms that produced (or constituted) the explanandum (Craver 2006; Kaplan and Craver 2011). As Craver succinctly puts it, "good explanations accurately describe the causal structure of the world" (Craver 2007, 61). Kaplan and Craver's model-to-mechanism-mapping (3M) requirement makes this accurate representation requirement explicit:

> (3M) A model of a target phenomenon explains that phenomenon to the extent that (a) the variables in the model correspond to identifiable components, activities, and organizational features of the target mechanism that produces, maintains, or underlies the phenomenon, and (b) the (perhaps mathematical) dependencies posited among these (perhaps mathematical) variables in the model correspond to causal relations among the components of the target mechanism. (Kaplan and Craver 2011, 611)

Indeed, according to almost all mechanistic accounts, "the goal is to describe correctly enough (to model or mirror more or less accurately) the relevant aspects of the mechanisms under investigation" (Craver and Darden 2013, 94).[4] Furthermore, "An explanatory model suffers, for example, if it includes irrelevant parts that are not in the mechanism, irrelevant properties that play no causal role, or irrelevant activities that are sterile in the mechanism" (Kaplan and Craver 2011, 607). In short, successful mechanistic models will accurately represent the relevant components and interactions of the target mechanism and leave out irrelevant features.

Although Bechtel's account focuses on accurately representing the causal contributions of the components of mechanisms, he does acknowledge that mechanistic strategies of decomposition and localization involve "simplifying assumptions which may turn out to be false" (Bechtel 2015, 92). Specifically, Bechtel discusses how idealizations play a role in specifying the characteristic scale and boundaries of the mechanism (Bechtel 2015, 85). However, he then suggests that the reason why modelers should be explicit about these assumptions is that "if, in imposing a temporal scale, and in imposing a boundary around the parts and operations of a mechanism, researchers are aware of what they are potentially ignoring, then, when it becomes important, they can extend the scale or boundaries and consider additional aspects of a mechanism" (Bechtel 2015, 92). In other words, when scientists discover that the features ignored by imposing idealized mechanistic boundaries are important, the mechanistic strategy suggests that they should work to remove those idealizations by considering additional aspects of the mechanism that were previously left out. Ideally, then, idealizations should be restricted to external features of the mechanism that are not important in the production of the phenomenon. Indeed, the ultimate goal of these mechanistic modeling strategies is to construct models that "accurately describe relevant aspects of the mechanisms [that are] operative in the world" (Bechtel and Abrahamsen 2005, 425).

In addition, most causal accounts require models that explain to provide an accurate representation of the difference-making, significant, or contextually salient causal factors that produced the explanandum (Elgin and Sober 2002; Potochnik 2015, 2017; Strevens 2004, 2008; Weisberg 2007a, 2013; Woodward 2003). For example, Mehmet Elgin and Elliott Sober argue that optimality models in biology can still explain because their idealizations make little difference to the predicted outcome:

> A causal model contains an idealization when it correctly describes some of the causal factors at work, but falsely assumes that other factors that affect the

outcome are absent. The idealizations in a causal model are *harmless* if correcting them wouldn't make much difference in the predicted value of the effect variable. Harmless idealizations can be explanatory. (Elgin and Sober 2002, 448)

That is, Elgin and Sober argue that idealized models can still explain because they only distort features that are largely irrelevant (or insignificant) to the occurrence of the explanandum. We can see the irrelevance of those features by replacing the idealizations with correct assumptions and showing that it makes little difference to the predictions made by the model.[5] According to Elgin and Sober, demonstrating that removing the idealizations makes little difference to the outcome shows that the idealizations are "harmless." What's more, they argue that optimality models are still able to explain because they correctly describe some of the difference-making causal factors at work (e.g., the role of natural selection in bringing about the explanandum).

As another example, on Strevens's view, when idealized models explain, the following occurs:

> The content of an idealized model, then, can be divided into two parts. The first part contains the difference-makers for the explanatory target. . . . The second part is all idealization; its overt claims are false but its role is to point to parts of the actual world that do not make a difference to the explanatory target. The overlap between an idealized model and reality . . . is a standalone set of difference-makers for the target. (Strevens 2008, 318)

In other words, idealized models explain by accurately representing a set of difference-making causes and restricting the use of idealization to the distortion of irrelevant causal factors.

In fact, as Weisberg (2007a) explains, several minimalist accounts of idealization follow Strevens in claiming that a model that explains is one that "accurately captures the core causal factors" because "the key to explanation is a special set of explanatorily privileged causal factors. Minimalist idealization is what isolates these causes and thus plays a crucial role for explanation" (Weisberg 2007a, 643, 645). According to these views, a *canonical* explanation is a causal model that accurately represents all (or at least most) of the difference-making causes. Idealizations can then be introduced in order to emphasize that other features are irrelevant to the explanandum. For example, in the case of Boyle's law, Weisberg tells us:

> Theorists often introduce the assumption that gas molecules do not collide with each other. This assumption is false; collisions do occur in low-pressure gases. However, low-pressure gases behave as if there were no collisions. This means

that the collisions make no difference to the phenomenon and are not included in the canonical explanation. Theorists' explicit introduction of the no-collision assumption is a way of asserting that collisions are actually irrelevant and make no difference. Even with this added, irrelevant fact, the model is still minimalist because it accurately captures the core causal factors. (Weisberg 2007a, 643)

In other words, according to these accounts, when idealized models are able to explain, it is because the model accurately represents the difference-making causes of the explanandum, and the idealized parts of the model are justified by distorting only features that are irrelevant.

As we saw in the previous section, Potochnik's causal account is a bit more flexible, in that it allows idealizations to distort some difference-making causes. In contrast with minimalist accounts of idealization, on her account, "significant causal factors that are not central to the research program can still be set aside" (Potochnik 2015, 1178). However, her account still requires models that explain to accurately represent the causal factors that had a significant impact on the probability of the outcome and are of interest to the current research program. In particular, "posits central to representing a focal causal pattern in some phenomenon must accurately represent the causal factors contributing to this pattern. . . . Idealizations, in contrast, must . . . represent as-if [such that] . . . none of its neglected features interferes dramatically with that pattern" (Potochnik 2017, 157). Therefore, while the set of important causal factors is determined somewhat differently, there is still a particular set of causal factors that need to be accurately represented in order for an idealized model to explain. Specifically, models that explain ought to accurately represent the causal factors that contribute to the causal pattern of interest.[6]

Before moving on, it is worth addressing a possible reply on behalf of some causal views of explanation. In particular, while several defenders of the standard view (e.g., Craver 2007; Strevens 2008) have put things in terms of "mirroring" the causal relations in the system, other accounts (e.g., Elgin and Sober 2002; Potochnik 2015) sometimes put things in terms of "capturing" the causes of interest.[7] In addition, the view of similarity that Weisberg (2013) defends seems to allow for some distortion (in terms of dissimilarities) regarding difference-making features. In other words, causal accounts might attempt to allow for some distortion (or omission) of causally relevant factors by only requiring the model to "adequately capture" or somehow "take into account" the relevant causes.

Although this would weaken the accurate representation requirement somewhat, the focus of these accounts is still on accurately describing (to some degree), capturing, or taking into account the most important causal factors (e.g., those causes of interest that have a major impact on the probability of the outcome). That is, the aim is still to provide a relatively accurate representation of the relevant causes of the phenomenon—even if that representation somewhat distorts those causes. As a result, although different causal views use slightly different criteria to determine which causal factors are relevant to the explanation and how accurately those causes need to be described, they still identify a set of relevant causal factors that idealized models must accurately capture in order to explain. Consequently, I think the arguments presented throughout this book will show that requiring explanations to "mirror," "accurately describe," "adequately capture," or "take into account" a set of relevant causes are all problematic. Specifically, none of these claims will be able to accommodate cases in which the explanation *drastically distorts* or *completely ignores* most (if not all) of the difference-making (or contextually salient) causes of the explanandum. However, because these claims are somewhat different and are defended by different authors, I will try to be explicit about how the individual arguments that follow apply to each of them.

In addition to the role of accurately describing causes for purposes of explanation, philosophers of science have long recognized a strong connection between explanation and understanding (de Regt 2009b; Friedman 1974; Grimm 2008; Salmon 1984; Strevens 2013). For example, Salmon writes: "understanding results from our ability to fashion scientific explanations" (Salmon 1984, 259). The link between explanation and understanding is also echoed by epistemologists: "Understanding why some fact obtains . . . seems to us to be knowing propositions that state an explanation of the fact" (Conee and Feldman 2011, 316). While I argue later in this book that understanding can be obtained without having an explanation (Lipton 2009; Rohwer and Rice 2013, 2016), the fact that much of our scientific understanding comes from providing explanations means that scientific understanding is typically thought to require some kind of accurate representation as well. Indeed, like the intuition that knowing a proposition requires that the proposition be true, most epistemological accounts of understanding maintain that genuine understanding must be constituted by accurate information (Grimm 2006; Khalifa 2012, 2013; Kvanvig 2003, 2009; Mizrahi 2012; Rice 2016; Strevens

2013). In fact, some accounts suggest that scientific understanding can be achieved only by grasping a *correct* (or true) explanation (Strevens 2013; Trout 2002).

When accounts of scientific understanding allow falsehoods to play a role in understanding, the false claims are often restricted to peripheral or nonessential propositions within one's understanding. For example, Kvanvig (2003) distinguishes between central propositions and peripheral propositions within one's understanding. He then argues that all of the central propositions of one's understanding must be true, but a few false beliefs about peripheral propositions do not undermine one's understanding. That is, similar to the accounts of how models explain surveyed here, on Kvanvig's view of understanding, idealizations ought to be restricted to irrelevant or peripheral parts of our understanding.

In sum, according to most accounts, idealizations are justified in models that explain and produce understanding by showing that they only distort causes, mechanisms, or features that are irrelevant, insignificant, or not of interest—that is, their distortions do not get in the way of the accurate representation of the relevant causal mechanisms, difference-makers, or contextually salient causes. Consequently, each of these accounts suggest that a necessary condition for scientific models to provide explanations or understanding is that the idealized model must accurately represent the relevant causes or causal mechanisms that produced the explanandum.

2.3 Decomposition of Scientific Representations into Accurate and Inaccurate Parts

The restriction of idealizations to the distortion of irrelevant, nonfocal, or nonsalient features within models that explain is one example of a far more general strategy used across numerous debates within the philosophy of science—what I will refer to as the *decompositional strategy*. This strategy involves the following three assumptions (see Rice 2017 for more details):

1. *Target decomposition assumption*: The real-world system is decomposable, such that the contributions of the features that are relevant to the occurrence of the target phenomenon (e.g., the difference-makers) can be isolated from the contributions of features that are irrelevant (or are largely insignificant) to the target phenomenon.[8]

2. *Model decomposition assumption*: The scientific model is decomposable, such that the contributions of its accurate parts can be isolated from the contributions of its inaccurate (i.e., idealized or abstracted) parts.
3. *Mapping assumption*: When successful, the accurate parts of the model can be mapped onto the relevant, important, or contextually salient parts of the real-world system and the inaccurate parts of the model only distort the irrelevant, negligible, or uninteresting parts of the real-world system.

Adopting these three assumptions allows one to claim that the accurate parts of the scientific representation are what "do the real work," while the inaccurate parts are justified by distorting only what is irrelevant, insignificant, or otherwise not of interest. This decompositional strategy is central to a wide variety of philosophical accounts of how to model complex systems, how models explain, how idealizations contribute to model explanations, robustness analysis, and how idealized modeling is compatible with scientific realism (Rice 2017). While these accounts are committed to the decompositional strategy to different degrees and for different reasons, I argue that they are all equally committed to a general approach to accommodating idealizations that requires some form of these three assumptions.[9]

Why have philosophers continually focused on decomposing scientific representations into their accurate and inaccurate parts? The answer, I think, is that the debate over scientific realism has been cast as a debate about whether the theories and models used in science are accurate representations of the parts of reality that enable us to explain and understand our observations. This has led most philosophers to assume that the epistemic achievements of science are best achieved by models or theories that accurately describe the relevant parts of reality. Consequently, because philosophers have started to acknowledge that many of our best scientific representations are inaccurate due to idealization (or just because they have been later disconfirmed), they have sought to show that those inaccuracies can be tolerated by showing that they are inessential to, or separable from, the epistemic successes of science. Indeed, the focus on accurately representing only the relevant causes has enabled philosophers to argue that distortions of irrelevant causes can be made consistent with a generally realist approach. Therefore, according to the standard view, the contributions made by idealizations can be separated from the contributions made by the accurate parts of science that describe relevant features. As a result, the widespread use of idealizations can be made consistent with the claim that the epistemic aims of science are best accomplished by accurate representation of relevant features.

This motivation is made explicit by several philosophers. For example, Kaplan and Craver suggest that the common use of idealization and abstraction "should not lead one to dispense with the idea that models can more or less accurately represent features of the mechanism in the case at hand. . . . These practices of abstraction and idealization sit comfortably with the realist objectives of a mechanistic science" (Kaplan and Craver 2011, 610). In other words, the realist objectives of mechanistic modeling allow for idealization and abstraction as long as the model accurately represents the relevant features of the causal mechanism of interest. Similarly, while other mechanists have emphasized the virtues of abstracting away from the details of mechanisms in certain contexts (Becthel 2015, 2017), even in those cases, "the model aims to track those features of the system that make a difference to the behavior being explained" (Levy and Bechtel 2012, 256).

In addition, as we saw in the previous section, several accounts of idealized modeling focus on separating out the accurate representation of difference-making or contextually salient causal factors from the idealization of causes that do not make a difference or are not of interest. For example, as Weisberg notes, minimalist accounts of idealization are compatible with scientific realism because the representation of the important causal factors "must be accurate" (Weisberg 2007, 658). Indeed, as Strevens argues, one of the key upshots of this kind of view is that the "factors distorted by idealized models are details that do not matter to the explanatory target—they are explanatory irrelevancies. The distortions of the idealized model are thus mitigated" (Strevens 2008, 315). Following this line of reasoning, according to most causal accounts of explanation, we can maintain a kind of realism despite the role of idealizations in science because scientific models that explain and produce understanding are accurate with respect to the relevant, difference-making, or otherwise contextually salient causes of the phenomenon of interest (Potochnik 2017; Salmon 1984; Strevens 2008; Weisberg 2007a, 2013; Woodward 2003).

More generally, many defenses of scientific realism have used a version of the decompositional strategy in order to suggest that scientific models and theories are *partially* accurate representations. These are sometimes referred to as selective confirmation approaches to defending realism (Kitcher 1993; Leplin 1997; Peters 2014; Stanford 2003). A recent example of this is Christopher Pincock's defense of realism, which seeks to "draw a distinction between the different parts of a scientific model and what these parts represent. For a model with parts P_1 through P_n, we can have a good reason to think that P_1 accurately represents some aspect A_1 of the system even when we lack a reason to think that the remaining parts accurately represent other aspects of the system" (Pincock 2011, 21). In particular, Pincock contends that the realist can respond to the use of idealization by arguing that scientific models are partially accurate with respect to certain aspects of the system that are relevant for explanation. As a result, he argues that "there is reason for optimism that a segmented approach to the content of scientific models can help us to reconstruct the scientific knowledge that the success of science suggests we have" (Pincock 2011, 31). Similarly, Arnon Levy argues that "while model descriptions are typically idealized, hence not true of their targets simpliciter, they are nevertheless *partly* true, at least when successful" (Levy 2015, 792). Levy develops this idea by suggesting that models can be divided into their true parts and their false parts. The general idea is that we can maintain a version of scientific realism because scientific representations are often accurate representations of the relevant features of real-world systems, even if they distort other (presumably irrelevant) features.

This brief (and nonexhaustive) survey of accounts makes it clear that many of the most influential philosophical accounts of scientific explanation, idealization, and modeling have assumed three main tenets: (1) the essential role of causation in explanation, (2) the requirement that explanations and understanding be provided via the accurate representation of relevant causes, and (3) the accommodation of idealization by decomposing scientific representations into their accurate and inaccurate parts.

2.4 Motivation for an Alternative Approach

While this standard approach has provided one way to justify the use of idealized models to achieve the epistemic aims of science, I contend that it cannot be anywhere close to the whole story. For one thing, as I will argue in chapter 3, many of the explanations we find in scientific practice are noncausal explanations (Ariew, Rice, and Rohwer 2015; Batterman 2002b; Batterman and Rice 2014; Bokulich 2011, 2012; Huneman 2010; Pincock 2012;

Rice 2013, 2017, 2018; Walsh 2010, 2015; Walsh, Lewens, and Ariew 2002). However, unlike the counterexamples to Hempel's DN account, these counterexamples have yet to generate a clear alternative account of scientific explanation. My goal in the next chapter is to analyze three classes of counterexamples to causal approaches to explanation in order to extract several features of noncausal explanations that this new account of explanation ought to include. I will then argue for a counterfactual account of explanation (and later, of understanding) that can capture these features and identifies what noncausal explanations have in common with the numerous causal explanations we find in science.

Another serious problem with the standard approach is that pervasive idealizations often make positive epistemic contributions by directly distorting relevant causes of the target phenomenon. As the examples analyzed throughout this book will make clear, there are many scientific models that are used to explain and understand complex phenomena whose idealizations drastically distort difference-making causes and mechanisms—including those that are of interest to the current research program. These cases also show that most idealized models that are used by scientists to explain and understand cannot be easily decomposed into their accurate and inaccurate parts. Instead, the models used to explain and understand in scientific practice routinely use modeling frameworks that *pervasively distort* the features of their target system(s).

In response, I will argue for an account of idealized modeling that characterizes models as holistically distorted representations of their target systems—that is, the models distort both relevant and irrelevant features. This account enables us to move beyond the mistaken assumptions of the decompositional strategy and prioritizes the central and positive contributions made by idealizations that distort relevant features to explanations and understanding. Moreover, the account is applicable even when scientists are not in the position to know which features of their models are accurate and which are inaccurate. Because this is an extremely common situation in actual scientific practice, this holistic distortion view provides more useful normative guidance for the practice of scientific modeling.

Having provided alternative accounts of explanation, modeling, and idealization, the later chapters of the book develop an alternative account of *how to justify* the use of holistically distorted models to explain and understand real-world phenomena. Rather than focusing on accurate representation relations, my account uses universality classes (Batterman 2002a, 2002b, 2005, 2010; Morrison 2002, 2004, 2009, 2015) and various mathematical modeling techniques to license explanatory inferences that enable scientists to extract the modal information required to explain and understand various phenomena. Putting these various accounts together will ultimately allow me to defend a version of scientific realism concerning the epistemic products of science that is compatible with the widespread use of pervasive idealizations that distort relevant features.

In short, while the standard approach is useful in some cases, it is woefully inadequate as a complete account of the epistemic contributions made by idealizations to scientific explanations and understanding. The epistemic contributions made by idealizations go well beyond their roles as tools for simplifying, isolating, or focusing our attention on the accurate representation of the relevant causes of interest. In fact, drastically distorting relevant causes is often the best (or even the only) way to access the information required to explain and understand the phenomenon of interest. As a result, we require alternative accounts of explanation, understanding, idealization, and modeling that can accommodate the plethora of *noncausal* and *pervasively distorted* representations that scientists use to explain and understand natural phenomena.

3 Many Explanations Are Noncausal

The first issue with the standard approach is that, while science contains many causal explanations, it also contains many explanations in which explicit appeals to noncausal features do the real explanatory work. These phenomena certainly *depend* on the features cited in the explanation, but there is no relationship of causal dependence between the features cited in the explanans and the explanandum. The point isn't that these phenomena have no causes—they certainly do—but that these phenomena can also be explained by citing noncausal features of the system that account for why the phenomena occurred. In this chapter I survey three classes of noncausal explanations as a means of extracting a set of features that an account of noncausal explanations ought to capture.

Given the wide array of different causal and mechanistic accounts, some of the arguments provided here will apply to certain causal views, but not to others. Rather than surveying every available causal or mechanistic view of explanation, my goal will be to show that these cases fail to meet the general requirements provided by several of the most influential causal accounts of explanation. In particular, contra Kaplan and Craver (2011), these models fail to satisfy the model-to-mechanism-mapping requirement; contra Strevens (2008) and Weisberg (2007a, 2013), they do not accurately represent the difference-making (or core) causes of the explanandum; contra Potochnik (2017), they do not accurately represent the causal factors that contribute to the patterns of interest to the research program; and contra Woodward (2003), they do not appeal to causal factors that could in principle be manipulated independent of other causes of the explanandum. In what follows, I try to be explicit about which lines of argument apply to which causal accounts.

In the following section I discuss optimality explanations. Then, in section 3.2, I analyze statistical explanations, and in section 3.3 I discuss minimal model explanations. Finally, in section 3.4 I extract a set of features from these cases that our revised account of explanation should account for in order to accommodate these noncausal explanations.

3.1 Optimality Explanations

An important example of idealized modeling in science is the use of optimality models (Orzack and Sober 2001; Potochnik 2007, 2009a, 2009b; Rice 2012, 2013; Sober 2000).[1] Optimization models serve as an excellent case study because they are widely used across biology (Maynard Smith 1978, 1982; Orzack and Sober 2001; Sober 2000), physics (Hartmann and Rieger 2002), economics (Pindyck and Rubinfeld 2009), cognitive science (Carruthers 2006; Churchland 2013), and chemical engineering (Corsano et al. 2009). In this section I will argue that the way that biologists typically use optimality models to provide explanations of phenotypic traits conflicts with the requirements of various causal approaches to explanation (Rice 2013).[2]

3.1.1 The Core Components of an Optimality Explanation

Optimality models are distinguished by their use of *optimization theory*, whose goal is to identify which values of some *control variables* will optimize the value of some *design variables* in light of some *constraints and tradeoffs* (Beatty 1980; Maynard Smith 1978; Parker and Maynard Smith 1990; Seger and Stubblefield 1996). An optimality model specifies a constrained set of possible strategies known as the *strategy set*. The design variables to be optimized constitute the model's *currency*.[3] An optimality model also specifies what it means to optimize these design variables (e.g., should a design variable be maximized or minimized?). This is referred to as the model's *optimization criterion* (Parker and Maynard Smith 1990).

Once the strategy set and optimization criterion have been identified, an optimality model describes an *objective function*, which connects each possible strategy to values of the design variables to be optimized.[4] These equations build in various context-specific design constraints and trade-offs among the quantities represented within the model. Indeed, what distinguishes optimization models from other kinds of mathematical models is the fact that the solution to the design problem is a result of navigating various constraints and trade-offs among the control and design variables (Seger and Stubblefield 1996, 94). Once these components of the optimality model are specified, one can deduce which of the available strategies will yield the optimal values of the design variables. This strategy is then deemed the *optimal strategy*. In short, by mathematically representing the important constraints and trade-offs, an optimality model can demonstrate why a particular strategy is the best available solution. Here is a quick example to illustrate how this works.

Lapwings search for food by moving a few paces before pausing to look for their prey, insects (Parker and Maynard Smith 1990). The available strategies here are the possible distances traveled between each scan of the environment. The benefit of moving farther away is that the lapwing will be less likely to scan terrain that it has already inspected (and thus more likely to find prey). However, once it has moved beyond the diameter of its visual field, moving farther does not help because all of the new ground will already be outside the previously inspected area. Furthermore, although moving increases the lapwing's chances of finding new prey, each step also *costs energy*. In this model, curve B represents the benefit of adopting strategy x and curve C represents the cost of adopting strategy x. The benefit is the *average* energy value of prey items obtained after moving distance x, and the cost is the *average* energy utilized in moving distance x. Assuming that $B(x)$ will increase (perfectly) asymptotically and $C(x)$ will increase (perfectly) linearly, we get the optimization model shown in figure 3.1.

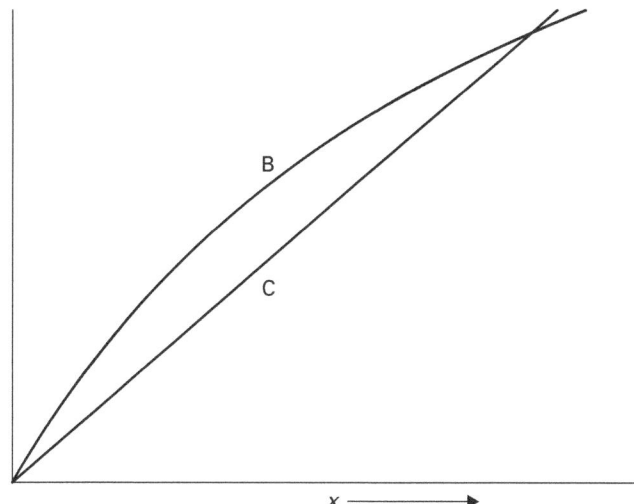

Figure 3.1
Parker and Maynard Smith's model of the costs and benefits of lapwing foraging strategies (Parker and Maynard Smith 1990, 28).

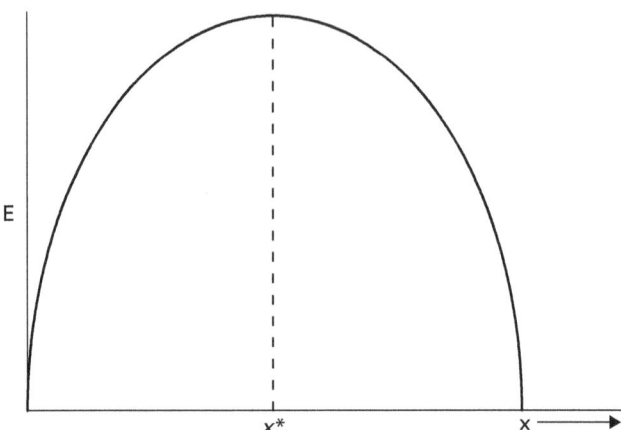

Figure 3.2
The model representing the average energy gain for various lapwing foraging strategies. The average energy intake is maximized by the strategy x^* (Parker and Maynard Smith 1990, 28).

The proposed optimization criterion of this model is the average net energy gain per move, E, where $E(x) = B(x) - C(x)$ (figure 3.2).

Therefore, the optimal strategy, x^*, maximizes $E(x)$ by best navigating this cost-benefit trade-off. This occurs when $dE(x)/dx = 0$ and $d^2E(x)/dx^2 < 0$.

Now that we understand what an optimality model is, we can turn to the question of how an optimality model might be used to provide an explanation. In biology, optimality models usually provide a species of *equilibrium explanation* (Sober 1983).[5] In the simplest cases, we can identify which of the available strategies will maximize some currency (e.g., fitness). In these frequency-independent cases, it is assumed that the system will tend to increase the model's currency, thereby making the strategy that maximizes the model's currency an equilibrium point. However, sometimes the optimal strategy will depend on what strategies other individuals in the population are playing (or the population as a whole).[6] In these frequency-dependent cases, the goal is typically to use game-theoretic techniques to show that the optimal strategy (or distribution of strategies) is evolutionarily stable (Lewontin 1961; Maynard Smith 1982).[7] In either case, in order to provide an explanation, a biological optimality model must (1) describe the salient constraints and trade-offs involved in the design problem and (2) make some optimization assumptions to show why the optimal strategy is expected to evolve (and be maintained).

3.1.2 A Frequency-Independent Example: Parker's Dung Flies

A widely cited example is G. A. Parker's use of an optimality model to explain why dung flies (*Scatophaga stercoraria*) copulate for 36 minutes on average (Parker 1978; Sober 2000). Parker's model is an instance of a general class of biological optimality models that stem from the prey and patch choice models used in foraging theory (Krebs and Davies 1993; Pyke 1984; Stephens and Krebs 1986).[8] These foraging models analyze the trade-off between energy intake from the current patch or food item and the lost opportunity to perform other foraging tasks. Parker's model uses this kind of cost-benefit analysis to calculate the optimal time for dung flies to invest in copulating during their mating cycles.

First, Parker observed that female dung flies typically mate with multiple males. He then discovered by experimentation that when this occurs, the second male fertilizes far more eggs (80 percent) than the first (20 percent). Consequently, after copulating, a male spends some time guarding the female before searching for other mates. The total behavioral cycle time is given by summing search time, copulation time, and guard time. Parker then observed that the average time spent searching plus guarding was 156 minutes. Therefore, the total cycle will last $156+c$ minutes, where c is the amount of time spent copulating. Different values of c constitute the model's strategy set.

Parker's model assumes that increasing c increases the average number of eggs fertilized. However, there is an important trade-off: time spent copulating is time that cannot be spent on other parts of the mating cycle. In addition, Parker's observations show that additional copulation time brings diminishing returns in terms of the average number of eggs fertilized. These constraints and trade-offs are captured by a mathematical curve that represents average fertilization as a function of copulation time (figure 3.3).

According to Parker's optimization criterion, the optimal value for c is the value that maximizes the *rate* of eggs fertilized across several iterations of the behavioral cycle. Consequently, the optimal strategy occurs at the point where a line that passes through the origin and intersects the asymptotic curve where its slope is steepest (line A–B in figure 3.3) intersects the curve. This optimal strategy occurs when c is equal to 41 minutes, which is fairly close—given the statistical standards of the field—to the observed average value of 36 minutes. Given this predictive accuracy, as well as the fact that Parker's model was based on detailed empirical observations, the model is often thought to have captured the salient constraints and

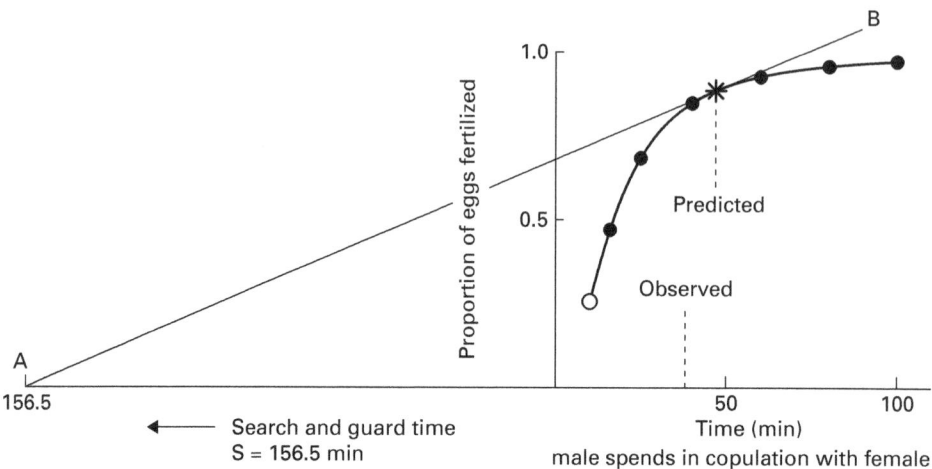

Figure 3.3
Parker's optimality model, used to investigate the copulation time of dung flies (Parker 1978; Sober 2000).

trade-offs involved in the selection of this phenotypic trait (Sober 2000, 137–138).[9]

However, in order for Parker's model to provide an explanation of the observed trait value, several additional optimization assumptions—many of which are idealizations—are required. First, we must assume that a strategy's fitness is strictly increasing with the increased (average) rate of egg fertilization. In addition, Parker's model assumes that various other factors, such as the search and guard times, were held fixed while the copulation time evolved (Sober 2000, 138). Furthermore, we require assumptions that eliminate the influence of other evolutionary factors (e.g., drift or genetic constraints). For example, in order to eliminate drift, the optimality model makes the idealizing assumption that the population is infinite. Other idealizations, such as assuming that phenotypes are inherited perfectly by offspring, result in the elimination of other evolutionary factors. In combination, these idealized optimization assumptions entail that the optimal strategy identified by the model is the equilibrium point of the evolving system. In this way, Parker's optimality model can be used to provide an equilibrium explanation for why dung flies copulate for approximately 36 minutes on average.

3.1.3 Explaining General Patterns: Equilibrium Sex Ratios

In addition to explaining system-specific phenomena, optimality models are frequently used to explain highly general patterns. Although there are numerous examples to choose from, for my purposes, it is sufficient to consider R. A. Fisher's (1930) widely cited model, which is used to explain why the sex ratio is 1:1 across many different species.

Roughly, Fisher's model claims that if one sex is more common in the population, there will be a fitness payoff to parents who produce the minority sex because their children will have more mating opportunities. To illustrate this, suppose that male births are less common than female births. A newborn male thus has better mating prospects than a newborn female and is, therefore, expected to have more offspring on average. Therefore, parents who produce males tend to have more grandchildren on average. Consequently, male births will become more common in the population. However, as the population approaches a 1:1 ratio, this fitness advantage fades away. The exact same reasoning applies if we assume that female births are less common. The only state in which selection will not favor the production of the minority sex is when the sex ratio is 1:1. Therefore, a 1:1 sex ratio is the equilibrium state of the evolving population.

Fisher's model relies on a key trade-off that economists refer to as the *substitution cost*. In this case, the substitution cost tells us how many sons can be produced if one less daughter is produced. In the model, this trade-off is perfectly linear—that is, males and females cost the same amount of resources to produce. This is why Fisher's model predicts a 1:1 sex ratio. Because males and females cost the same amount of resources to produce, producing the minority sex will yield higher fitness payoffs for the same investment of resources (Fisher 1930, 142–143). Eric Charnov (1982, 28–29) shows that generalizing Fisher's model leads to the conclusion that the equilibrium ratio of a population, r, can be calculated using the following formula:

$$r = C_2 / (C_1 + C_2)$$

where C_1 is the average resource cost of one son and C_2 is the average resource cost of one daughter. This trade-off between using resources to produce sons and daughters is the key to explaining the ubiquity of the 1:1 sex ratio across various populations that realize this trade-off in very different ways.

For Fisher's optimality model to provide an explanation, we must assume that natural selection will optimize the model's criterion (e.g., mating opportunities per unit of resource investment). Additionally, the model assumes that other evolutionary factors (e.g., drift) will not deter the population from reaching the equilibrium favored by natural selection. In order to derive this result, these optimality models typically make the idealizing assumptions that organisms reproduce asexually, mate randomly, have equal access to resources for investing in offspring, the population has an infinite size, and so on.[10] These assumptions allow the model to show why a 1:1 sex ratio is the equilibrium point of many evolving populations.

The two examples provided here are insufficient to provide an account of all optimality modeling in science (for additional examples, see Rice 2012, 2016, 2017, 2018). Still, in light of these representative instances, we can identify three key features common to most biological optimality explanations. First, these models usually provide a type of equilibrium explanation. Second, these models are typically highly idealized. Third, most of the explanatory work in these models is done by synchronic mathematical representation of structural constraints and trade-offs between the control and design variables involved in the optimization problem. In what follows, I argue that each of these key features of optimality explanations move the model in precisely the opposite direction suggested by the standard approach by eliminating or drastically distorting the relevant causes of the system.

3.1.4 Equilibrium Explanation

An equilibrium explanation tells us that a wide range of potential causal trajectories would all yield the same result, but it does *not* tell us which of these is the actual causal trajectory of the target system(s) (Sober 1983). One way that a causal account might try to understand the explanatory power of this kind of equilibrium explanation is to note that it still provides some minimal information about the actual causes of the explanandum. For example, Strevens suggests that an equilibrium explanation of a ball's arriving at the bottom of a basin describes the actual causal mechanism that produced the phenomenon "at an extremely abstract level, so abstract that the description is satisfied by every process by which the ball might have reached the bottom of the basin" (Strevens 2008, 268).[11] Similarly, Levy and Bechtel (2012) point out that several techniques for modeling complex systems (e.g., graph theory) abstract away from the specific details of the mechanisms and instead aim to "represent parts and operations in actual mechanisms" such that the description "applies to any mechanism in which those value obtain" (Levy and Bechtel 2012, 247).

I maintain that grounding a biological optimality model's explanatory power in this minimal information about the actual causes is a mistake. For one thing, it suggests that a model that provided additional information about the actual causes would provide a superior explanation. Indeed, Strevens claims that models in ecology and economics that "black-box" the details of the dynamics of causal mechanisms responsible for the explanandum are at best partial explanations (Strevens 2008, 159–162).[12] While Levy and Bechtel (2012) explicitly reject this idea, other mechanistic accounts of explanation have favored explanations that accurately represent the "real components, activities, and organizational features of the mechanism that in fact produced the phenomenon" (Craver 2006, 361). Although Craver and Kaplan (2020) have recently argued that only including more relevant details of the actual mechanism makes a model more explanatory (Craver and Kaplan 2020, 303), the focus is still on accurately representing features of the actual causal mechanisms that produced (or constitute) the explanandum.

In contrast, I maintain that there are several reasons for preferring the kind of equilibrium explanations provided by optimality models, by virtue of the fact that they do not cite the actual causes or mechanisms of real systems—even at a more abstract level (Batterman 2002a; Garfinkel 1981; Pincock 2007, 2011; Weslake 2010). One reason is that citing the actual causal history or mechanisms of the system would mean that the optimality model would be unable to apply to other systems which are subject to similar constraints and trade-offs, but which are heterogeneous in their causes and mechanisms. While the features of those systems' mechanisms are certainly *causally* relevant to particular occurrences of the explanandum, they are not necessary for the equilibrium result to occur. As Sober puts it, "Where causal explanation shows how the event to be explained was in fact produced, equilibrium explanation shows how the event would have occurred regardless of which of a variety of causal scenarios actually transpired" (Sober 1983, 202). Many philosophers have recognized the explanatory value of this kind of modal information (Garfinkel 1981; Jackson and Pettit 1992; Woodward 2003). Although the actual causal history of these populations includes many difference-making causal factors, the optimality model explains without referencing those causes. Instead, the optimality model provides us with an important piece of explanatory information: *the actual causal history of the system is not important for understanding why the explanandum occurred because several very different causal histories would have led to the same outcome.*

Of course, several causal and mechanistic accounts have emphasized the value of generality as well (Bechtel and Abrahamsen 2005; Glennan 2002; Potochnik 2017; Strevens 2008). However, almost all of these accounts have assumed that "generality is a virtue because it points to a common underlying causal structure" (Levy and Bechtel 2012, 259). What is important to notice here is that often the generality of the explanation isn't captured by ignoring some causal details in order to focus on a more abstract causal pattern; instead, it is provided by abstracting away from *all* the actual causes of the phenomenon—regardless of their level of detail. For example, a crucial aspect of Fisher's explanation is that it does not matter what causes or mechanisms produce the substitution cost; what counts is that a 1:1 substitution cost will lead to a 1:1 sex ratio within the population. By moving away from the actual causes or mechanisms of the target systems, the optimality model is able to capture a wider range of possible systems that are extremely heterogeneous in their difference-making causes. As I will argue in more detail later in this chapter, many scientific explanations focus on stable patterns across different systems *without* assuming that the stability is the result of an underlying common causal structure (Rice manuscript).

Providing an equilibrium explanation is the first step away from attempting to describe the actual causes of the phenomenon. This line of argument targets causal accounts that focus on including the relevant details of the actual causal mechanisms that produce (or constitute) the phenomenon (e.g., Craver 2007; Kaplan and Craver 2011; Salmon 1984). It is important to note, however, that the features that make an optimality explanation noncausal go well beyond merely noting that the explanation provided is an equilibrium explanation. Indeed, as I will argue, even if we grant that some equilibrium explanations can provide a kind of causal explanation (Potochnik 2017: Strevens 2008), many optimality models will still provide noncausal explanations because the features cited as responsible for the equilibrium point of the system are not causes of the explanandum.

3.1.5 Essential Idealizations in Optimality Explanations

Even though optimality explanations do not reference the system's actual initial conditions, causal trajectory, or relevant details of the causal mechanisms that produced the phenomenon, they may still provide a different type of causal explanation by accurately representing some of the causal factors that gave rise to the explanandum. For example, Potochnik argues that when optimality models provide explanations, "the model accurately represents the selection dynamics, and . . . there are no other important evolutionary factors to represent (for the target evolutionary outcome in question)" (Potochnik 2009b, 186).[13] This, of course, parallels the view presented in Potochnik (2017), arguing that an explanation must accurately represent the factors contributing to the causal pattern of interest. This interpretation of optimality explanations is similar to Elgin and Sober (2002)'s claim that these models correctly describe the causal factors involved in selection despite distorting other negligible evolutionary factors.

However, even if optimality models can be characterized as attempting to describe a more limited set of causal factors, they will almost always fail to meet the veridical standards required by causal approaches to explanation. This is because optimality explanations are almost always so highly idealized that they fail to accurately represent the contextually salient difference-making causes of their target systems (Rice 2012, 2013). In particular, these models fail to accurately represent the difference-making causal relationships involved in the selection of the phenotypic trait. Yet it is precisely these causes that are of interest to the adaptationist research program in which these models are formulated.

To begin with, the mathematical representations of constraints and trade-offs in biological optimality models are often inaccurate when compared to the causes acting within the populations of interest. For example, in Parker's model, the assumption is that average fertilization rate increases with increased time spent copulating, according to a perfectly asymptotic curve. In addition, many game-theoretic models assume that a conflict is symmetric or that payoffs are constant across iterations of the game so they can use certain mathematical operations that are required to deduce the equilibrium state (for some examples, see Hammerstein and Selten 1994). But these smooth curves and constant parameters will almost always be inaccurate when compared with the actual causes and mechanisms acting within the model's target systems. While some of these distortions may be compatible with attempting to describe the actual causes or mechanisms operating in the system (Burnston 2017, 2019; Potochnik 2017), in many cases the mathematical equations used will drastically distort the causal relationships that actually produced the phenomenon of interest.

Next, the strategy sets of most optimality models do not accurately represent the set of strategies actually causally interacting within the system.[14] For instance, although a sex ratio model may assume that the strategy set includes all the probabilities that a birth will be a male, this does not mean that the target population contained individuals that were playing this range of strategies. Rather, the model assumes that the optimal strategy will evolve regardless of which particular distribution of strategies was actually causally interacting within the population's history (so long as this set includes the optimal strategy). The actual strategies, however, are causally relevant to the evolutionary process that occurred. For example, their removal from a veridical causal model of the system would result in the model failing to produce the explanandum.

In addition, a biological model's optimization assumptions are almost always inaccurate when compared with a population's causes—even if we consider causes at the level of the population. For instance, in foraging models, it is often assumed that natural selection will maximize average energy intake. However, this is only one thing that might influence the survival and reproduction of organisms in the population. In fact, the various causal mechanisms (or, more generally, natural selection) acting in the population often will not optimize the model's criterion. The goal of a biological optimality explanation, however, is not to accurately describe the causal dynamics that led to the evolution of a trait. Rather, biological optimality models identify optimal strategies (or states) that are only outcomes or end states of an evolving system that approximates their optimization assumptions in the long run.[15] Therefore, optimality models are frequently used to explain why a system has evolved the optimal strategy, regardless of whether their optimization assumptions are true of the causal processes acting in the system at any point along the way to that strategy.

Next, many biological optimality models idealize the way in which phenotypic strategies are inherited. For example, many models assume that strategies are inherited perfectly by offspring (Parker and Maynard Smith 1990, 30). These idealizations are introduced because it is assumed that changing these inheritance assumptions will have no effect on the occurrence of the phenomenon. However, the causes underlying these inheritance idealizations are important difference-makers in the explanandum's evolutionary history in at least one sense—namely, that without some causal process of inheritance, nothing would have evolved at all. Nevertheless, an accurate representation of these difference-making causes is not required for the optimality model to explain the target phenomenon.

Another idealization is the assumption that environmental pressures are constant. This idealization is required for the optimal strategy to be an obtainable equilibrium state (i.e., the population needs time to arrive at the predicted optima before changes in environmental conditions can alter the equilibrium point of the system). However, the causal selection pressures in a population are never constant in this way. Therefore, this idealization grossly distorts the causal relationships that actually produced the selection of the trait.

Finally, most biological optimality models assume that the population being modeled is infinite (or effectively infinite). However, population size *does* make a difference to every evolutionary process (i.e., drift is a statistical fact of every real-world biological population). Assuming an infinite population size has the effect of eliminating drift (i.e., statistical error) from the model by utilizing various laws of large numbers. By incorporating the idealization of infinite population size, the optimal trait according to natural selection is what we expect to evolve. However, once again, this idealization drastically distorts the causal mechanisms (or processes) operating within the model's target populations.[16]

Considered together, these idealizations entail that biological optimality models usually provide little if any accurate information about the actual causes, or causal mechanisms, within the model's target systems—even those causes that played a crucial role in the selection of a trait. In the end, the highly idealized optimality model represents mathematical relationships between constraints, trade-offs, and the system's equilibrium point that do not accurately describe any causal relationships or processes in the target system. Put in a different way, optimality models fail the kind of mapping requirement involved in most causal theories that requires that the "dependencies . . . among variables in the model correspond to the . . . causal relations among the components of the [system]" (Kaplan and Craver 2011, 611).

The key point here is to recognize that rather than distorting irrelevant causal factors so that we can focus on those causes that make a difference (Strevens 2008) or focus on the causal factors of interest to the research program (Potochnik 2017), I argue that these idealizations, considered collectively, move us away from attempting to accurately represent causes *at all* and toward a representation of relationships that do not mirror or accurately describe any of the causal relationships within the model's target systems—even if we consider causes at some macro-level of the system. These models do not aim to accurately represent some modular part of the causal process that led to the phenotypic trait. Instead, these models focus on an entirely different set of dependence relationships between the structural constraints and trade-offs in the systems of interest and the explanandum.

What's more, these idealizations play essential roles within these optimality explanations. In particular, these explanations often require that certain idealizations be introduced in order to employ certain mathematical techniques used to derive the target explanandum. Without these idealizations, the features represented in the mathematical model are insufficient for deriving the explanandum. Therefore, for many optimality explanations, it is unclear how the various idealizations could be removed from the model without consequently eliminating the explanation being offered (i.e., the idealizations are ineliminable).

By moving away from attempting to provide an accurate representation of causes, optimality models are able to provide an explanation that applies to systems in which these causes are different. That is, by introducing various idealizations that distort or eliminate the difference-making causes of particular systems, the models are able to provide an explanation that captures a wider range of possible systems. This is not intended to suggest that an increase in generality always makes optimality models objectively better than less general models. In science, we want generality sometimes, but we want detail at other times (Sober 1999). However, in many cases, our interests will dictate that the explanandum is best explained by a more general optimality explanation. In addition, this kind of explanation is precisely what is required in order to explain highly general patterns that range over systems that are heterogeneous in most of their causal features.

In addition, these idealizations make essential contributions to the explanation provided by an optimality model because they show why most, if not all, of a system's causal factors, mechanisms, or variables are not important for understanding why the explanandum occurred. Although many of these details are relevant to the veridical causal explanation (i.e., they make important causal contributions to the occurrence of the explanandum in real systems), they are not required in order for the optimality model to explain the phenomenon. This is because extremely different physical causes would have been sufficient, so long as the constraints, trade-offs, and optimization assumptions of the optimality model are satisfied. This modal (i.e., counterfactual) information about the irrelevance of these causes is key to explaining many of the repeatable patterns that we observe because it shows us why extremely heterogeneous causal systems will nevertheless display similar behavior.

3.1.6 Synchronic Mathematical Representation of Noncausal Features

A biological optimality model explains by showing how the equilibrium point of the system counterfactually depends on a set of constraints and trade-offs—that is, the model shows how changing the constraints or trade-offs of the system results in a change in the phenotypic strategy that is expected to evolve. Given this, one might argue that these relationships are causal relationships after all because we can see how changes to these structural features would change the equilibrium point of the system. Consequently, optimality models might be understood as providing some kind of nonveridical causal explanation. In response, I will argue that—independent of their ability to accurately represent causes—when we look closer, the key dependence relationships within an optimality explanation are best interpreted as noncausal relationships of dependence.

To begin, it is often difficult to see how the trade-offs represented within optimality models can be understood as causal relationships. For instance, in many biological optimality models, average energy intake and average predation risk will exhibit a trade-off that is vital to the explanation of the observed phenotype, but it is unclear how we ought to understand the metaphysical claim that average energy intake is a cause of average predation risk (or vice versa). Moreover, within an optimality explanation, it is really the trade-offs *between* variables that do the key explanatory work. As Charnov claims, "The trade-offs themselves are the fundamental objects of evolutionary interest, at least with respect to stabilizing or equilibrium selection" (Charnov 1989, 115). This is precisely why biologists have provided detailed analyses of how "evolutionary outcomes . . . depend on the shape and position of the tradeoff curves constraining the course of evolution" (de Mazancourt and Dieckmann 2004, 769). In other words, it is the overall shape of the relationships between variables, rather than particular changes in the variables themselves, that do the explaining. However, it would be rather puzzling to claim, for example, that the population-level trade-off between average energy intake and average predation risk causes anything. In short, it is extremely difficult to see how our metaphysical intuitions about causes can be codified in the case of the trade-offs that are central to optimality explanations.

Another reason to think that the constraints and trade-offs of optimality explanations do not represent causal relationships is that the optimality model's mathematical representation of them does not reference any processes or events that unfold prior to the explanandum. A central feature of causal representations is that they are essentially *diachronic*—that is, there is a temporal dimension to the representation that captures changes over time (Pincock 2012). This diachronic component is especially prominent in process and mechanistic theories of causal explanation (Craver 2006; Kaplan and Craver 2011; Machamer, Darden, and Craver 2000; Salmon 1984). This temporal component also exists in causal accounts that attempt to use the temporal asymmetry of causation to solve the problem of asymmetry (e.g., Potochnik 2017; Strevens 2008; Woodward 2003). The explanation provided by an optimality model, in contrast, merely identifies the optimal strategy by showing that the model's currency is optimized by a particular strategy, given the constraints and trade-offs synchronically represented within the model. For instance, in Parker's model, the optimal strategy (i.e., the point at which the average rate of fertilization is maximized) is simply represented as the value of c at which the slope of a line that intersects the curve and passes through the origin is maximized. Nowhere does the model describe a causal process (or causal trajectory) that unfolds prior to the explanandum. Instead, the model merely identifies $c = 41$ minutes as the optimal strategy. Furthermore, none of the points along these mathematical curves needs to be instantiated in order for the model to explain the outcome. This is because, contrary to the standard way that causes bring about their effects, *optimality explanations do not provide a dynamical account of the processes or events that led to the explanandum*. Instead, these models identify optimal strategies by using synchronic representations of structural features of the system and mathematically deriving the optimal strategy. So, although temporal dynamics are the main focus of causal explanations, this is precisely the kind of information that is almost entirely absent from an optimality explanation.

Finally, we can consider modularity and interventions. Interventionist accounts of causal explanation typically require that causes be *modular*, such that they can be manipulated independent of other causes within the system (Hausman and Woodward 1999; Woodward 1997, 2003, 2010). Woodward (2003) makes this modularity assumption explicit. He says that "it should be possible to alter or disrupt each of the equations in the system (i.e., by altering the operation of the mechanism with which it is associated) without altering or disrupting the others" (Woodward 2003, 328). More recently, however, instead of thinking of modularity as a necessary condition for all causal models, Woodward has suggested that modularity is merely a plausible default assumption for purposes of causal modeling. However, he still maintains that "among all of the observationally equivalent representations, we should prefer the one that is modular because it will be the one that correctly and fully represents causal relationships and mechanisms" (Woodward 2003, 332). This is because "if there is to be any systematic connection between causal claims and claims about what will happen under manipulations, one needs something like modularity" (Hausman and Woodward 2004, 158). In short, even if we drop modularity as a necessary condition for a system of causal equations, being able to isolate one cause while keeping the others fixed is essential for applying the interventionist approach to interpreting causal claims. Such modularity is also crucial to most mechanistic accounts (e.g., Craver 2007), given that mechanistic explanations also focus on isolating the causal contributions of specific parts of the system that could (at least in principle) be independently manipulated (Levy and Bechtel 2012, 244).

Yet, in evolving biological systems, the trade-offs between (what are often statistical) higher-level properties of the system usually depend on the fitness of various individuals, and these variables in turn depend on a complex and integrated network of causal contributions to individual fitnesses. Moreover, in many cases, these higher-level properties arise from complex systems whose dynamics are chaotic, nonlinear, or involve feedback loops that make modularity assumptions problematic (Mitchell 2008, 2012). Indeed, as Cartwright notes, in many cases, "the causal laws are harnessed together and cannot be changed singly" (Cartwright 2004, 811). While causal and mechanistic models can sometimes be used to investigate these kinds of complex dynamics (Bechtel and Abrahamsen 2010), none of these complex causal relations are described or modeled anywhere in the optimization explanation. Furthermore, given the causal entanglement and complex integration of evolving biological systems, it is unlikely that one would (even in principle) be able to intervene in such a way that changed only a particular trade-off's influence on the target phenomenon.

For example, as biologist Graham Pyke notes concerning one optimal foraging model, "It is hard . . . to imagine or implement an experiment in which patch residence time is artificially varied and fitness consequences determined" (Pyke 2019, 113).[17] In the example given here, the key trade-off in Parker's model is that time spent copulating is time that cannot be spent on other parts of the behavioral cycle. Intervening on this trade-off would presumably require altering not only the causes that impinge on each individual dung fly, but also the basic principle that time spent on one task cannot be spent on others. Precisely what this kind of (in principle) intervention would even look like is unclear. Similarly, while some authors have suggested that the key trade-off in Fisher's model could be subject to empirical intervention (e.g., Chirimuuta 2018, 856), the proposed interventions are always achieved through manipulation of the organisms or environment such that the cost of producing male and female offspring is not equal. However, as before, this kind of intervention in the causal interactions between species and their environments does not surgically intervene on the trade-off itself; instead, it manipulates the various causal interactions that underlie particular instantiations of the trade-off (none of which are cited in Fisher's original explanation).[18] Thus, it is extremely difficult to see how we could, even in principle, manipulate the influence of these trade-offs on the equilibrium point of the population independent of other causal factors.

Potochnik has more recently argued that an optimality model of the foraging behaviors of redshank sandpipers describes a manipulable causal pattern because "if large worms had been historically more difficult to find (an intervention), the bird's preference, or at least degree of preference, would have been different" (Potochnik 2017, 137). The first issue is that this intervention would not change the observed explanandum that "the birds spent most time feeding where prey density was highest and profitability was greatest" (Goss-Custard 1977, 26). Changing which patch has the highest density of food, or which types of food are available in various patches (e.g., large worms), would not change the fitness of this overall foraging strategy—indeed, the birds would still prefer patches with the highest density and profitability of prey.[19]

The main problem, however, is that Potochnik's discussion leaves out the key relationships that make the sandpiper case an instance of optimality modeling. If all we needed to model was a causal relationship between changes in food availability and changes in foraging behavior, there would be no need to employ optimization theory. However, as Goss-Custard goes on to discuss, there is often an important trade-off in these cases because "the tendency of redshank to congregate in the most profitable feeding areas is counteracted by the tendency to avoid areas of high bird density" (Goss-Custard 1977, 29). In other words, while foraging in high-density areas increases the amount of food available, foraging in those areas comes with the fitness cost of having to compete with other birds foraging in those same patches. This trade-off is what explains why, in the second population, the birds did not distribute themselves only according to profitability, but rather spread out more to avoid overpopulating the most profitable patches. Consequently, the intervention proposed by Potochnik does not intervene on the structural features cited in the explanans of the optimality explanation (i.e., the constraints or trade-offs).

This shows that just because a model—or its background assumptions—may include some causal information about the outcomes of certain interventions does not mean that the explanation provided by the model is causal. Simply providing some correct information about the result of interventions is insufficient if that causal relationship is largely incidental to the explanation. Noncausal explanations do not require that *no* causal information can be gleaned from the model. What makes the explanation noncausal is that the features of the system that do the real explanatory work—in this case the structural trade-offs between various control and design variables—are not causes of the explanandum.

Most importantly, however, optimality models are able to provide satisfactory explanations without adding any claims about interventions or causation. Nowhere in the description of Parker's explanation did we require any mention of these concepts. As a result, it appears that an optimality model need not tell us how things would have been different under an intervention in order to explain why we observe the explanandum. More recently, Woodward has acknowledged this as one form of noncausal explanation, in which the explanation answers a range of "What if things had been different?" questions, but it does not do so by providing answers to questions about what would happen in response to interventions (Woodward 2018, 122). Although claims about interventions may be important for testing (or interpreting) causal claims and may be a crucial component of causal explanations, they are not required to establish that the *explanation* is sufficient—that is, the explanatory claim and the causal claim are independent of one another.

In sum, not only is it difficult to see how the relationships represented within optimality models can be made to square with our metaphysical intuitions about causes, but also several of the key features of causal explanations are absent from an optimality model's representation of these structural features. Therefore, I conclude that the key dependence relationships cited within an optimality explanation are best interpreted as *noncausal* relationships. Optimality models primarily focus on noncausal counterfactual dependence relations between structural constraints and trade-offs of the overall system and the system's equilibrium point. In addition, by moving away from causes, an optimality model will often (1) apply to more possible systems, (2) provide counterfactual information about why a phenomenon would occur in other causally heterogeneous systems, and (3) highlight the noncausal structural features essential to understanding why the explanandum occurred. Put differently, moving away from causes is often what enables optimality models to provide the kind of explanation we seek.[20]

3.2 Statistical Explanations

A second set of noncausal explanations is the widespread use of statistical models to explain across the sciences (Ariew, Rice, and Rohwer 2015; Ariew, Rohwer, and Rice 2017; Lange 2013b; Matthen and Ariew 2009; Walsh, Lewens, and Ariew 2002; Walsh 2007, 2010). These explanations often involve making deductions from laws that appeal only to the statistical features of the population (e.g., its mean and variance). As with optimality explanations, we will see that these explanations appeal to noncausal relationships of counterfactual dependence. Moreover, we will again see that the idealization of difference-making causes plays an essential role in the construction of the mathematical models appealed to in the explanation. In addition, statistical explanations (once again) illustrate the important role of irrelevance information in explaining patterns across causally heterogeneous systems.

3.2.1 Galton's Autonomous Statistical Explanations

As a first example, we can consider Francis Galton's (1886) use of statistical laws (e.g., regression) to explain various patterns of inheritance (Ariew, Rice, and Rohwer 2015; Ariew, Rohwer, and Rice 2017; Lange 2013b). Galton's example is historically significant because his pioneering work "represents the most important step in perhaps the single major breakthrough in statistics in the last half of the nineteenth century" (Stigler 1990, 281).[21] In particular, Galton was one of the first to recognize that laws that focus on statistical dependencies could be sufficient *on their own* to explain why an event occurred, without referencing the causes operating within the population (see Ariew, Rohwer, and Rice 2017 or Hacking 1990 for more details).

Galton was interested in what he called the "reversion to mediocrity" observed in many instances of inheritance. The puzzle is that, although parents with exceptional characteristics tend to have children with less exceptional characteristics, the statistical distribution of the population (i.e., its mean and variance) tends to be stable across generations (Hacking 1990, 182–184). This is the case even when selection favors certain traits over others. As Galton puts it, "The processes of heredity are found to be so wonderfully balanced, and their equilibrium so stable, that they concur in maintaining a perfect statistical resemblance" (Galton 1877, 1). Galton's target explanandum, then, is the stability of the population's statistical distribution despite character variation in and selection of some of those variants. Thus, Galton's question is, "How is it, that although each individual does not as a rule leave his like behind him, yet successive generations resemble each other with great exactitude in all their general features?" (Galton, 1877, 2). Instead of trying to identify the underlying causes that gave rise to this phenomenon, Galton focused on identifying the statistical properties of the overall population that maintained this stability across generations (Sober 1980, 370). In particular, "Galton's discovering of the standard deviation gave him the mathematical machinery to begin treating variability as obeying its own laws, as something other than an idiosyncratic artefact" (Sober 1980, 368). In other words, Galton's explanation appeals to the explanandum's dependence on statistical features of prior generations of the overall population and various laws (e.g., the law of deviation) that describe how those features relate to the statistical features of later generations. Galton's key insight was arguing that this is sufficient to provide the explanation on its own, without uncovering the causes that give rise to that variation or the causal processes that maintain its stability in any particular case.

In order to perform the statistical derivations used within these explanations, Galton needed to construct an idealized statistical model of the system that would admit of a mathematical treatment. In order to accomplish this, Galton assumed that the traits of interest were numerous and independent of one another. These idealizing assumptions allowed him to represent the system as a (perfect) Normal distribution, which enabled him to employ important mathematical modeling tools. As Galton puts it:

> Since the characteristics of all plants and animals tend to conform to the law of deviation, let us suppose a typical case, in which the conformity shall be exact, and which shall admit of discussion as a mathematical problem, and find what the laws of heredity must then be to enable successive generations to maintain statistical identity. (Galton 1877, 4)

Importantly, this move to a mathematical model required a fair amount of idealization: real-world traits are rarely perfectly distributed according to an ideal Normal curve. In fact, only in an infinite population with complete independence of factors will the conformity be exact. Still, while real-world populations typically only "conform to" or "approximate" a Normal curve, we know the conditions that make such conformity highly likely: the sample size must be relatively large, and the samples must be largely independent of one another.[22] Making these assumptions about his sample population enabled Galton to construct an idealized model of the population to which various statistical laws could be applied.

Applying those statistical laws allowed Galton to derive that "if a population is Normally distributed, it can be deuced that in a second generation there will be a Normal distribution of about the same mean and dispersion" (Hacking 1990, 186). Galton's breakthrough is recognizing that although these idealized statistical models and laws fail to reference the underlying causes of the system, "the typical [statistical] laws are those which most neatly express what takes place in nature generally; they many never be exactly correct in any one case, but at the same time they will always be approximately true and *are always serviceable for explanation*" (Galton 1877, 17). In other words, Galton recognizes that making certain assumptions so as to construct an idealized model of the population, and then applying laws that appealed only to the statistical properties of the population, could be sufficient for explaining the various patterns observed across generations.

Galton's explanations show that many statistical results can be explained independent of the details of the causes or components of the system. As he says, "The law of deviation is purely numerical: it does not regard the fact whether the objects treated of are pellets in an apparatus like [the quincunx], or shots at a target, or games of chance, or any other of the numerous groups of occurrences to which it is or may be applied" (Galton 1877, 7). Thus, it isn't just the fact that these explanations abstract away from "lower-level" causal details that makes them noncausal. Many causal explanations do the same thing. What makes these explanations noncausal is that the statistical generalizations appealed to in these explanations are independent of the particular components or causes operating in the system (at any scale).

3.2.2 Natural Selection and Autonomous Statistical Explanations

Galton's statistical explanations are an important historical case because his work on regression would later provide the foundation for the development of the statistical theory of population genetics in the 1920s (Ariew, Rohwer, and Rice 2017; Stigler 2010). Indeed, Fisher's modeling of biological populations shares many of the assumptions used by Galton and other statistical modelers (Fisher 1930). This is because population genetics was largely developed by treating evolutionary patterns as irreducibly statistical phenomena. In line with this interpretation, Morrison (2002, 2004, 2015) argues that Fisher's contribution to the development of the genetical theory of selection requires modeling biological populations in ways that are analogous to the statistical treatment of gases:

> Random mating as well as the independence of the different factors were also assumed. Finally, and perhaps most importantly, he assumed that the factors were sufficiently numerous so that some small quantities could be neglected; in other words, large numbers of genes were treated in a way similar to large numbers of molecules and atoms in statistical mechanics. As a result, Fisher was able to calculate statistical averages that applied to populations of genes in a way analogous to calculating the behaviour of molecules that constitute a gas. (Morrison 2004, 1197)

By making these idealizing assumptions, Fisher was able to replace the causal complexity of actual biological populations with a proxy statistical model that focused on only a few parameters. In other words, Fisher's approach to modeling of biological populations succeeded "only by invoking a very *unrealistic* . . . model of a population (Morrison 2015, 24). However, as is the case with Galton, Fisher's use of these idealizing assumptions enabled him to apply mathematical derivations that would not otherwise be applicable. For example, assuming that the genes in the population are sufficiently numerous and that factors within the population are statistically independent enabled Fisher to apply the central limit theorem in order to model biological populations as a normal distribution of the fitnesses of trait types

(Morrison 2015, 44; Plutynski 2004, 1209). More generally, these idealizing assumptions enabled Fisher to use a statistical modeling framework in order to model biological populations as a distribution of trait types, in which the mean value was the expected outcome due to natural selection and the variance of the population was statistical error due to genetic drift.

Modern population geneticists continue to adopt Fisher's statistical modeling framework in order to provide explanations of evolutionary change across generations (Walsh 2015; Walsh, Lewens, and Ariew 2002). As a simple example, suppose that we want to explain evolutionary changes in a population with two genotypic trait types: T_1 and T_2. However, all we know about these trait types is that T_1 has a reproductive rate of 0.8 and T_2 has a reproductive rate of 0.7. These reproductive rates represent *trait* fitness; that is, they are averages of the number of offspring (with those trait types) had by individuals with those trait types. This information does not tell us which of the individuals with traits T_1 and T_2 will survive and reproduce, nor does it tell us the environmental conditions or other features of the population that cause these averages. Moreover, these trait fitnesses are not components within the population, nor are they equivalent to the fitness of individual members within the population. Instead, they are population- (or trait-type-) level averages of the reproductive outputs of populations of individuals. Still, although this kind of model fails to provide this information about the causal interactions between organisms and their environment, the model can still be used to explain why T_1 eventually becomes the dominant trait in the population simply by citing the differences between the fitness of these trait types (Walsh, Lewens, and Ariew 2002, 462). The point is that many of the models used in evolutionary biology explain changes in trait frequency distributions by citing statistical features of the evolving population without referencing the various causal interactions operating within the population.

In this example, it is the mean reproductive rate of the different trait types that is doing the real explanatory work, but a trait type's statistical variance can also be used to explain certain evolutionary outcomes (Walsh 2015). For example, in several papers, Gillespie (1974, 1977) argues that variance of reproductive output and population size ought to be included in calculations of fitness. Specifically, Gillespie argues that the most effective measure of trait fitness is

$$w_i = \mu_i - \sigma^2_i/n$$

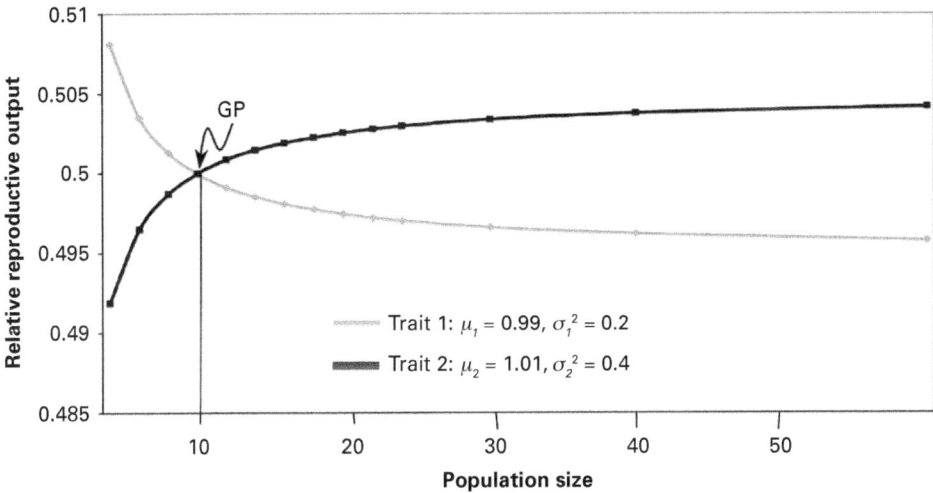

Figure 3.4
Relative reproductive output of two traits as a function of population size. The lower-mean but lower-variance trait outperforms the higher-mean, higher-variance trait when the population size is below the Gillespie Point (GP) (Walsh 2015, 481).

where w_i is the fitness of trait i, μ_i is the mean reproductive output of individuals with trait i, σ^2_i is the variation in reproductive output of individuals with trait i, and n is the population size. The important point, for my purposes here, is that variation in reproductive output affects the fitness consequences for traits within the biological population. Denis Walsh (2015) gives the following example for a population of relatively constant size:

Trait 1: $\mu_1 = 0.99$, $\sigma_1^2 = 0.2$
Trait 2: $\mu_2 = 1.01$, $\sigma_2^2 = 0.4$

In cases where the population size is less than 10 (what is called the *Gillespie Point*, or GP), trait 1 actually has higher fitness than trait 2. When $n > 10$, trait 2 will have higher fitness than trait 1 (figure 3.4).

The important point to notice is that variation in reproductive output can have an impact on evolutionary outcomes by affecting the fitness of trait types. As a result, we see how appeals to fitness differences and natural selection often depend on both the mean and variation of the reproductive outputs of different trait types. However, as before, we need not know about the underlying causes of these means and variances in order to explain why the biological population's trait frequency distribution shifts in later generations.

While I will not be arguing that *all* explanations that appeal to natural selection are noncausal explanations, considering the role of statistical models in evolutionary biology shows that "selection and drift can be explanatory ... without being causes of evolutionary change" (Matthen and Ariew 2009, 16). Although there are also many causal explanations in biology (Woodward 2010), in many cases, evolutionary explanations "appeal to a set of statistical properties of populations, viz. the mean (and variance) of fitnesses between trait types. Explanations of this sort *do not avert to forces*" (Walsh, Lewens, and Ariew 2002, 462). I maintain that this kind of statistical explanation can succeed independent of any description of causes or causal processes—even if there are macro-level causal mechanisms operating in the population. The important point is that, whatever the causes or mechanisms are within a biological population, many statistical models of natural selection (and drift) can provide satisfactory explanations without making any reference to those causes.

The key to understanding these statistical explanations is to recognize that natural selection is a deductive consequence of another statistical feature of populations: variation in the fitnesses of trait types. That is, *any* population that displays variation in the fitnesses of trait types will evolve by changing the population's frequency distribution of trait types. Because of this, biologists can often explain a statistical result by showing how it counterfactually depends on other statistical features of the population (e.g., the distribution of the fitnesses of trait types in previous generations). While the fitnesses of various trait types will certainly have underlying causes that are responsible for the lives and deaths of the individuals within the population, those causal interactions need not be cited (or known) in order to explain why the population evolves in the way it does. Indeed, there are multiple sets of causal interactions that would all give rise to the same distribution of fitness across trait types. This is because fitness describes the probability of a trait type being passed on to future generations, but it does not tell us whether any particular individual with that trait is able to survive and reproduce. In other words, the distribution of trait fitness within a population is a statistical property that is multiply realizable in terms of the causal interactions within the system.

However, the argument here is *not* that these features are noncausal simply because they are multiply realizable. The reason these explanations are noncausal is that, contrary to Shapiro and Sober (2007), the multiply realizable features cited in these statistical explanations are not macro-level causes, nor are they robust higher-level mechanisms (Wimsatt 1994). They are, instead, statistical features of the population (e.g., the mean and variance of the fitness distribution and statistical laws about how all statistical distributions change over time). Therefore, even though there certainly can be multiply realized macro-level causes (and mechanisms) in biological systems, those types of causal features are explicitly *not* referenced by many of the explanations of evolutionary outcomes that appeal to the population's statistical distribution of trait fitnesses.

Consequently, my primary argument against a causal interpretation of these cases does not depend on the claim that natural selection is not a causal process. Instead, it relies on the fact that the statistical models appealed to in these explanations do not aim to accurately represent, mirror, or capture the causal processes involved in natural selection. Instead of appealing to those causal processes, these explanations appeal to statistical features of the population that the explanandum depends on. So, regardless of whether there are causal processes involved in natural selection, these statistical models do not purport to accurately represent those causal processes.

3.2.3 Distinguishing Two Kinds of Natural Selection Explanations

Some critics of the statistical interpretation of natural selection explanations claim that natural selection is a cause of evolution by appealing to manipulationist (or interventionist) criteria for causation (Northcott 2010; Reisman and Forber 2005; Shapiro and Sober 2007; Stephens 2004). The basic idea is that because changes in the fitnesses of various traits results in changes in the later distribution of the population, that the former must be a cause of the latter. In response, Matthen and Ariew (2009) argue that if these manipulation tests are appropriately applied, natural selection often fails to meet the basic requirements of manipulability tests of causation.[23] For example, while Woodwardian manipulations can show how trait differences, such as variation in camouflage, can cause evolutionary change, these same manipulability tests fail to establish causation when we consider the following claim: natural selection causes evolutionary change (Matthen and Ariew 2009). This is because, although we can manipulate the statistical fitness distribution of a population by intervening on the traits and environments within the population, the dependence relationship between the population's statistical fitness distribution and the evolutionary changes in the population are *purely mathematical*. Just because we can mathematically derive how the later distributions of traits will change in response to changes to the population's statistical of trait fitnesses does not mean that the relationship described by those mathematical derivations is causal.

In response, Sober (2011) suggests that mathematical statements such as "*A* is fitter than *B*" can be causally relevant to why "*A* increases in frequency in the population," even if the relationship between these two statements is a priori and mathematical. Sober argues that these a priori statements are causal because they describe how fitness "promotes" *A*'s increasing in frequency. I certainly agree that "*A* is fitter than *B*" can explain why "*A* increases in frequency in the population," but I don't see why we should interpret such a relationship in terms of causal promotion, given that it tells us nothing about any of the factors that causally contribute to the fitness distribution that entails the evolution of the population (Lange and Rosenberg 2011).[24] Now, one could suggest that any invocation of natural selection must result in a causal explanation if we maintain that natural selection is a population-level causal process or mechanism (Millstein 2006). However, without further argument for why this must be so, this would just beg the question against noncausal (e.g., statistical) interpretations of natural selection explanations. More importantly, however, even if we grant that natural selection is a population-level causal process, if all the explanation tells us is that selection played a role, without specifying any details about how the trait was causally selected, then the minimal causal information provided seems completely *incidental* to the explanation provided (Lange and Rosenberg 2011). To use Sober's own terminology, many evolutionary explanations cite only the mathematical consequences of fitness differences without describing the sources of those fitness differences (Sober 1984, 2011).

However, rather than getting deeper into this (largely metaphysical) debate about whether natural selection is a causal mechanism, I think the more important point is about *what is referenced* in the explanations provided by these biological models. Whether or not there is a macro-level cause or a causal mechanism that might be called "natural selection," many of the explanations provided in evolutionary biology make use of statistical models that do not reference those causes, mechanisms, or interventions. Consequently, one can argue that many of the explanations provided by the modern genetical theory of natural selection are *autonomous statistical explanations* without having to demonstrate that there is not a population-level causal process that *could* be cited to explain changes in trait frequency (Ariew, Rice, and Rohwer 2015). The fact that causal explanations for these phenomena also exist—even at the macro-level—is insufficient to undermine the claim that there are also autonomous statistical explanations for those phenomena that succeed without representing causes.

As a result, in order to answer whether a natural selection explanation ought to be characterized as causal or statistical, we have to pay attention to what is being explained in a particular case and how biological modelers propose to explain it. In doing so, what we find is that there are at least two theories of natural selection that are appealed to in explaining evolutionary phenomena. As Marjorie Grene explains: "We must . . . distinguish between 'genetical selection,' which is purely statistical, and Darwinian selection, which is environment-based and causal. They remain two distinct concepts with a common name" (Grene 1961, 31). While a causal theory often provides the correct description of explanations that appeal to Darwin's theory and the interactions between organisms and their environments, a statistical characterization is typically required to understand how the modern genetic theory of natural selection explains large-scale statistical regularities without referencing the population's causes. This is because "in the hands of Fisher, selection becomes irreducibly statistical because the mathematics used to describe it no longer allows it to explain the occurrence of individual traits. It is now understood as a population-level phenomenon explained in terms of changing gene frequencies. In that sense, the mathematics used to formulate an account of how selection operates determines the way it would be understood" (Morrison 2009, 140). The more general point is that there are a wide variety of types of explananda in biology. Some of those explananda will be best accounted for by causal or mechanistic explanations, but many other biological explananda are best explained by statistical explanations that abstract and idealize away from the causes of the system and focus on statistical regularities that are stable across causally heterogeneous systems. I argue that *both* kinds of explanation are routinely offered in evolutionary biology (and many other sciences).

3.2.4 Essential Idealizations in Statistical Explanations

These examples also illustrate that many statistical explanations necessarily depend on the contributions made by idealizations that distort many of the causes operating within the target system—even causal processes that are required for the explanandum to occur (e.g., inheritance). Without these idealizations, the statistical modeling techniques required to provide the explanations of these patterns would be inapplicable (e.g., the derivation of the explanandum would not go through). For example, "Fisher's mathematisation of selection created a new framework in which its operation was understood as an irreducibly statistical phenomenon, a reconceptualisation that emerges in conjunction with the application of specific mathematical . . . techniques" (Morrison 2015, 41). In other words, this idealized reconceptualization of biological populations was necessary for applying the mathematical modeling techniques that Fisher used to explain and understand evolutionary patterns. As Plutynski notes, "As with almost all of Fisher's theorietical work, his assumptions were idealized and crucial to his conclusions" (Plutynski 2004, 1208). Without the assumptions of infinite population size and random mating, many of the statistical modeling techniques used by Fisher and contemporary population genetics would be inapplicable. The role of these idealizations is to enable these modelers to move to a representational framework that distorts many of the causal interactions within the system in order to focus our attention on the statistical features on which various explananda counterfactually depend.[25]

It is also important to note that these idealizations are required for almost every instance of statistical modeling because Gaussians and other widely used statistical distributions are *limiting* distributions; that is, they hold only in limiting cases (Tao 2012; Wilson 2017). As Wilson explains with regard to the mean and variance of Gaussian distributions, "these simple numbers don't mathematically coalesce until we reach an infinite population. . . . The asymptotic maneuver of blowing up a [population] to infinite size acts as a filter that brings out the 'good part' (=dominant behavior) of our accumulated data" (Wilson 2017, 218). What is crucial for statistical modelers is that the real systems approximate the conditions required for these distributions enough to be justified in making the idealizing assumptions required to use various statistical modeling techniques. The justification for employing these statistical modeling techniques, then, is provided by appealing to various limiting theorems (e.g., the central limit theorem tells us that the more events there are, and the more statistically independent those events are, the more closely the distribution will resemble the Gaussian distribution).

Importantly, this use of idealization is different from the kind suggested by the standard approach. These idealizations are not introduced to emphasize the irrelevance of the distorted features or to isolate causal difference-makers. Instead, these idealizations distort many of the causal components and interactions that make a difference to evolutionary outcomes. For example, like optimization models, Fisher's statistical models distort the causal processes of biological populations by representing selection as occurring in a highly idealized population in which drift does not occur, mating is completely random, and the selection pressures on the population are constant. In addition, rather than attempting to isolate macro-level causes, these models focus on identifying statistical parameters that aggregate over all the causes (and causal trajectories) of the population. In short, rather than appealing to the accurate representation of causes and distortion of irrelevant features, these idealized modeling techniques are justified because they enable scientific modelers to use the mathematical modeling tools that have been most fruitful for explaining the kind of explananda they are interested in.

In addition, while these statistical models focus on certain minimal conditions required for the application of statistical modeling techniques, these conditions are only *approximated* by real systems. This raises a potential problem for Marc Lange's account of statistical explanations. Lange argues that "an explanation is [really statistical] if and only if it works by identifying the result being explained as an instance of some characteristically statistical phenomenon such as regression toward the mean" (Lange 2013b, 177–178). The main problem with this suggestion is that it seems to imply that statistical explanations can be provided only when the real system *instantiates* a particular statistical result. However, only infinite systems with completely independent interactions will exactly instantiate the statistical results associated with the central limit theorem and Gaussian distributions. Therefore, Lange's account misses two important features of how statistical explanations are provided in scientific practice: (1) they require certain idealizing assumptions and (2) they are typically justified by approximation and limit theorems (Rice, Rohwer, and Ariew 2018).

3.2.5 Statistical Dependence and the Role of Irrelevance

A possible objection to my analysis of these cases is that these statistical explanations depend on deduction of the explanandum using statistical laws, but the counterexamples to Hempel's deductive-nomological model have shown that mere deduction of the explanandum is insufficient for explanation. In response, I argue that these derivations from statistical models are sometimes sufficient to explain because they provide the required information about how the explanandum counterfactually depends on the features cited in the explanans (Bokulich 2008; Rice 2013; Woodward 2003). For example, a statistical model in evolutionary biology can show us how a large-scale statistical regularity counterfactually depends on the statistical properties of the distribution (Ariew, Rice, and Rohwer 2015; Ariew, Rohwer, and Rice 2017; Matthen and Ariew 2009; Walsh 2007, 2010). The explanation shows how changes in those statistical features—such as the mean and variance of the population's fitness distribution—would change the resulting phenomenon.

As I have argued thus far, what makes these explanations noncausal is that the statistical features cited in the explanation—on which the phenomenon counterfactually depends—are not causes of the explanandum. Features such as the mean of the fitness distribution, the variance of the fitness distribution, and the size of the population do not causally contribute to the production of the explanandum. Nor are these features components, entities, or activities within any causal mechanism within biological populations. Instead, these features are statistical (or mathematical) properties of the overall population that emerge out of extremely complex causal relationships operating between organisms and their environments. While many causes certainly underlie these population-level statistical parameters, they are not referenced by many of the explanations that biologists provide of large-scale statistical patterns. This is because statistical explanations are typically employed in cases where the explanandum is a statistical pattern that ranges over many systems that have *different* causal difference-makers that give rise to their statistical behaviors. While some evolutionary biologists, such as Darwin, are often interested in tracking how the causal interactions of the system give rise to the fitnesses of particular traits, that causal information is not required to explain many of the evolutionary outcomes that we observe in real populations.

As a result, these statistical models also provide extensive counterfactual information about why the explanandum would have occurred, regardless of changes to the causal features within the population. Indeed, as various investigations of statistical relevance have shown, statistical relationships greatly underdetermine causal relationships because several causal relationships can give rise to the same statistical distributions. Consequently, "the statistical explanation can tell us why a particular trajectory at the level of ensembles is highly likely across systems with extremely heterogeneous causal processes" (Ariew, Rice, and Rohwer 2015, 20). In other words, like optimality explanations, statistical explanations show us that within a certain range of possibilities, "the event would have occurred regardless of which of a variety of causal scenarios actually transpired" (Sober, 1983, 202). This counterfactual information about the lack of dependence on the causes of these systems is an essential part of how these statistical models explain the stability of evolutionary patterns across extremely heterogeneous biological populations. Indeed, the great unifying power of these statistical explanations is that they show how certain evolutionary patterns are universal across systems that are extremely heterogeneous in terms of their components and causes.

3.3 Minimal Model Explanations

As a final example of noncausal explanation, scientists often appeal to extremely minimal models within their explanations, despite the fact that these models are thoroughgoing caricatures of real systems—that is, they are highly idealized and distort or ignore most (if not all) of the difference-making causal entities, interactions, and properties of the systems whose behavior they purport to explain (Batterman 2002a, 2002b, 2010; Batterman and Rice 2014; Bokulich 2011, 2012; Hartmann 1998; Pexton 2014). What is crucial in these minimal model explanations is that the explanation of the universality of the patterns we observe is provided only by combining the minimal model with a detailed story about why most of the features of a class of systems are irrelevant to their universal behaviors (Batterman and Rice 2014). That is, these explanations involve two main components: (1) a minimal model and (2) a detailed demonstration of why most of the features of a class of systems are irrelevant to the explanandum. I will argue that both these components make noncausal contributions to minimal model explanations.

As an example, physicists use the *lattice gas automaton* (LGA) model to investigate and explain the behavior of real fluids, despite the fact that the model represents a possible system that is nothing like the real fluids whose behavior they want to explain (figure 3.5).

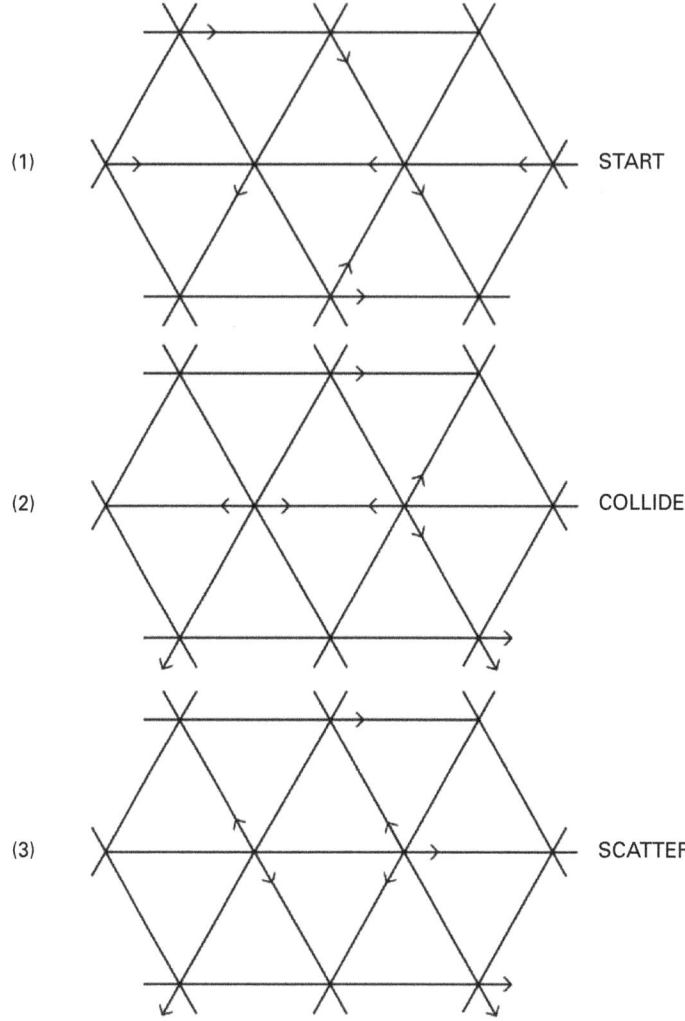

Figure 3.5
The LGA model (Batterman and Rice 2014; Goldenfeld and Kadanoff 1999). (1) shows the initial state of the system, (2) shows how the particles on the lattice collide at various nodes, and (3) shows the resulting scattering following those collisions.

The LGA model is an extremely minimal computational model. It places a set of point particles on a hexagonal lattice and allows them to move in one of six directions. The model then uses the following updating algorithm: Between (1) and (2), in figure 3.5, each particle moves in the direction of its arrow to the nearest neighboring node. If the momentum at that node sums to zero, the particles undergo a collision, resulting in a jump of 60 degrees, as shown in (3).

Somewhat surprisingly, running this simple algorithm many times for many particles is able to reproduce many of the macroscopic patterns of behavior of real fluids (e.g., phase transitions, fluid flow through a pipe, or fluid flow around a barrier, as shown in figure 3.6).[26]

Indeed, physicists have appealed to the LGA model in their explanations of a wide range of patterns of fluid flow, despite the fact that the model is a highly idealized caricature of any real fluid (Batterman and Rice 2014; Goldenfeld and Kadanoff 1999). Unlike real fluids, the LGA model is discrete, is composed of point particles, involves interactions that are restricted to a lattice, and follows only the simple algorithm described previously. In other words, the LGA model provides a pervasively distorted representation of the basic ontological components and interactions of its target systems. In fact, the only minimal features that the model has in common with real fluids are the following (Goldenfeld and Kadanoff 1999, 87):

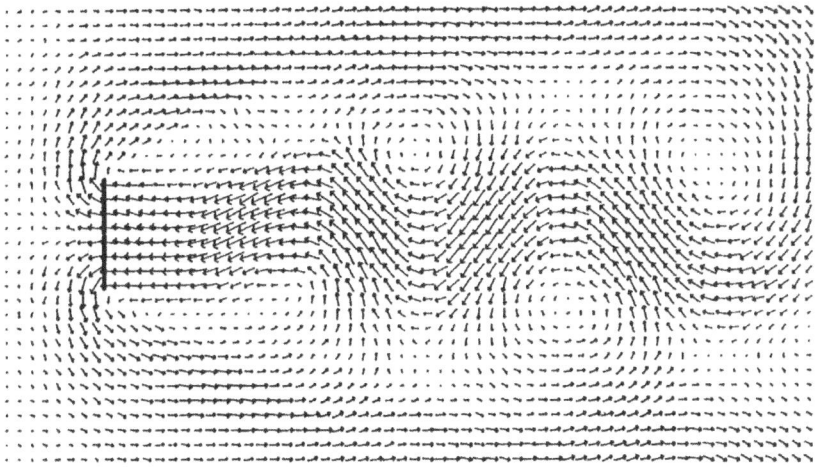

Figure 3.6
Fluid flow past a plate as modeled by the LGA model (D'Humiéres and Lallemand 1986, 334).

Locality: The fluid contains many particles in motion, each of which is influenced only by other particles in its immediate neighborhood.

Conservation: The number of particles and the total momentum is conserved over time.

Symmetry: The fluid is isotropic and rotationally invariant.

These are the only features that are necessary for the occurrence of the behaviors of interest (i.e., these are the features that the explanandum counterfactually depends on). However, I think it is fairly clear that these minimal features are not causes of the explanandum. For one thing, they describe general constraints on the overall system rather than specifying any factors, components, interactions, or mechanisms that bring about the explanandum. For example, the system's being rotationally invariant is surely not a cause of the explanandum, and neither is the conservation of the number of particles. Furthermore, while the patterns of fluid flow counterfactually depend on these features, it is unclear how one would (even in principle) intervene on these features in real fluids without altering other causes within the system—that is, it is difficult to see how interventions on these features could be "surgical." Finally, the description of these features is relatively static (i.e., no temporal causal process or mechanism is described for how these features produce patterns of fluid flow). The only dynamical aspects of the minimal model are involved in the time steps of the computational algorithm, but those algorithms do not accurately describe any of the causal interactions that produce these patterns in actual fluids. In short, these three minimal features do not contribute to the explanation by accurately representing causal factors or mechanisms that produce the phenomenon in any real-world fluid.

Instead of aiming to accurately represent causes, cases like the LGA model focus on modeling macroscale limiting behaviors of the system that are stable across a wide range of real, possible, and model systems. These limiting behaviors are often interesting to scientists precisely because they are independent of the complex causal interactions or mechanisms involved in real systems. As a result, these modelers build highly idealized models that distort (or often just ignore) most of the causal difference-makers of the target systems, but they are still able to display the patterns of interest in the limit (Corwin 2016; Frank 2009; Huneman 2010; Morrison 2015; Rice 2018).

A second feature of these cases is that the models' target systems are typically drastically different in most (if not all) of their causes. Indeed, many patterns are stable across systems as diverse as fluids and magnets (Batterman 2002b), the evolution of extremely diverse biological populations (Ariew, Rice, and Rohwer 2015), and the phase transitions of composite materials and growing arctic melt ponds (Golden 2014; Hohenegger et al. 2012). In these cases, it is extremely difficult to see how each of the systems could be claimed to exhibit the pattern of interest because they all have the same causal difference-makers or the same causal mechanisms—even if the causal mechanisms are described very abstractly at some macro-level. Instead of describing the causes of these systems, the explanations provided in these cases routinely employ noncausal features of the systems that are stable despite changes to the systems' causes.

In addition, even if these minimal features might contain some minimal causal information about the system, simply citing these features that the model has in common with real fluids fails to provide an explanation of the universality (i.e., stability) of the patterns of fluid flow that we observe (Batterman 2002b; Batterman and Rice 2014). That is, while the minimal model plays a role in these explanations, it is not sufficient to provide the explanation of the observed universal patterns on its own. In fact, the explanatory role played by these minimal models has very little to do with the minimal features that are accurately represented by the model (causal or otherwise). This is for two main reasons: (1) we still need to know why we are justified in inferring explanatory relationships in real fluids from the operations of this extremely minimal model and (2) we would like to understand why these patterns are so universal despite drastic differences between the physical features of various fluids (and the features of the LGA model). Merely looking at the features represented by the LGA model fails to answer these further explanatory questions. In order to answer them, a minimal model explanation is only provided by combining the model with various mathematical techniques that demonstrate the irrelevance of most of the features of real, possible and model fluids.

In particular, the LGA model provides a minimal model explanation only in combination with the use of the renormalization group to delimit a universality class. *Renormalization* is a strategy for extracting stable macro-behaviors by eliminating irrelevant degrees of freedom from the mathematical description of the system and increasing the coefficients of relevant parameters. This process is importantly different from simply averaging over the microstates of the system (Morrison 2014, 1148). Instead, the renormalization group transforms the representation of the system in such a way that it identifies which features are relevant and irrelevant—across multiple scales—to the macroscopic behavior of the system, and describes a new, renormalized representation of the system that preserves those macroscale behaviors. Simply averaging over the microstates of the system would fail to tell us which of the system's degrees of freedom (at various scales) are relevant and which are irrelevant to the macroscale behaviors of interest. We don't just want a macroscale description of the system; we want a description of how various smaller-scale details of the system are related to the macroscale behaviors.

In applying the renormalization group, one constructs an abstract space of possible systems described mathematically as a space of Hamiltonians (Batterman and Rice 2014). Each point in this space might represent a real, possible, or model fluid (or solid). For example, one of these points might be a real system of hydrogen atoms, another a possible system of neon atoms, and yet another the model system represented by the LGA model (figure 3.7).

Next, one induces a transformation on this abstract space of systems that has the effect of eliminating various degrees of freedom that are irrelevant to the stable macroscale behaviors of the system. For example, in the block spin transformation, one replaces a collection of spins (from a lattice)

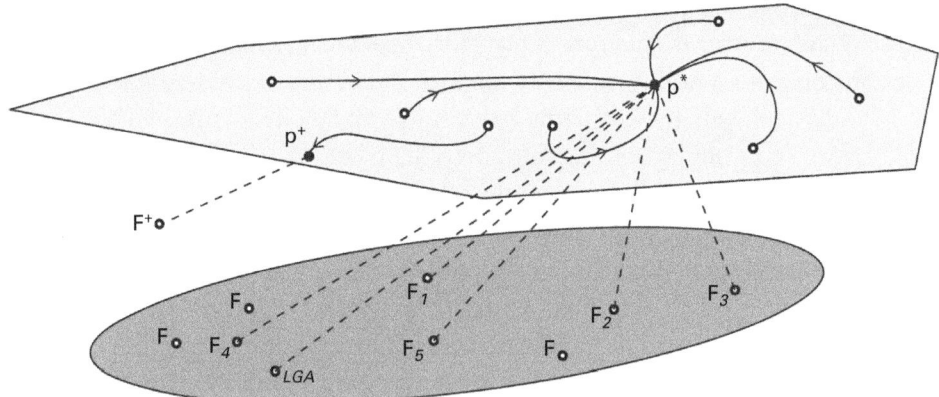

Figure 3.7
Fixed point p* for a universality class that includes real fluids (e.g., F_3), possible fluids (the unnumbered Fs), and the LGA model. The bottom oval represents the systems themselves, whereas the top shape represents the space of abstract mathematical descriptions of the systems (i.e., their Hamiltonians) on which the renormalization group transformation is performed (Batterman and Rice 2014, 363).

with an average spin that captures the interactions among the spins of the original system. Then, by rescaling, one arrives at a new possible (and idealized) system that still exhibits the continuum-scale behaviors of interest, but involves fewer degrees of freedom than the original system. Performing this renormalization transformation repeatedly shows why most of the features of the system are irrelevant because the transformation effectively eliminates degrees of freedom that are irrelevant to the macroscopic universal behavior. Physicists then look for *fixed points* under these transformations, such that further applications of the transformation yield the same system. When multiple systems all flow to the same fixed points under repeated applications of the renormalization transformation, those systems are said to be in the same universality class (see figure 3.7 again). This universality class tells us that, despite myriad differences in their physical features, each of the systems in the class will display the same universal behaviors at macroscales.

One way in which delimiting the universality class plays an essential role in minimal model explanations is by justifying the appeal to the extremely minimal model in our explanation of patterns of fluid flow. This is because the use of the renormalization group is able to demonstrate that the LGA model is within the same universality class as many real-world fluids (Batterman and Rice 2014). As a result, the LGA model will display many of the macroscale universal behaviors that we observe across extremely heterogeneous fluids. As the physicists Goldenfeld and Kadnoff explain, "In our fluid dynamics example, the large-scale structure is independent of a detailed description of the motion on the small scales. We can exploit this kind of 'universality' by designing the most convenient 'minimal model'" (Goldenfeld and Kadnoff 1999, 87).

On this point, Lange objects that, although we rarely observe scientists using real systems to explain the behavior of a model, "if our demonstrating that the model and the target system are in the same universality class were sufficient to allow us to use one to explain the other, then we might just as well use the target system to explain why the model exhibits the given behavior" (Lange 2014, 296). I do not think this kind of asymmetry objection is successful. For one thing, if this kind of explanatory asymmetry is a requirement, it is not a unique problem for accounts of minimal model explanations because accounts that appeal to models being similar to their target systems (Weisberg 2013) or having features in common with their target systems (Kaplan and Craver 2011; Strevens 2008) will face the same challenge. In addition, it seems that scientists have clear, pragmatic reasons for using the minimal model as part of the explanation of the behaviors of real systems (rather than vice versa). First, scientists simply do not have access to a description of all the features of the real systems, but they do have access to the minimal model. Second, the goals of the modelers can establish the desired asymmetry (Giere 2010). As Knuuttila and Merz put it: "The users' intentions create the directionality needed . . . something is being used and/or interpreted as a model of something else" (Knuuttila and Merz 2009, 148–149). Third, in many cases, only the infinite model systems are amenable to the mathematical modeling techniques necessary to explain (e.g., renormalization group transformations require that infinite systems be considered in order to identify fixed points). These considerations are more than enough to account for the explanatory asymmetry that Lange observes in scientific practice.[27]

In addition to showing that the minimal model is within the universality class, the renormalization group's demonstration of the irrelevance of most of the heterogeneous features of real, possible, and model fluids allows physicists to provide an explanation of the universality (or stability) of those patterns. As Robert Batterman explains, repeatedly applying this renormalization technique at progressively larger scales "demonstrates that many of the details that distinguish the physical systems from one another are irrelevant for their universal behavior" (Batterman 2002b, 42). In other words, applications of the renormalization group demonstrate why various features of real systems are counterfactually irrelevant to their macroscale behavior. In addition, "by telling us what (and why) various details are irrelevant for the behavior of interest, this same analysis also identifies those properties that *are* relevant to the universal behavior being investigated" (Batterman 2000, 127). That is, the renormalization group also shows us that the universal phenomenon depends on only a few minimal features of these varied systems. This modal information about which features are relevant and which are irrelevant is the key to explaining the universality of the patterns of behavior that we observe across various real fluids (and magnets).

The explanatory contributions made by the renormalization group to explanations of universality are also noncausal. First, the renormalization group is not a representation of causal relations within the system, but is instead a mathematical transformation on an abstract space of Hamiltonians

(Batterman 2002b; Reutlinger 2014).[28] Second, the features that are the focus of renormalization techniques are features of the overall space of many systems, rather than the features of any particular system within the universality class (Morrison 2018a). For example, the key feature is often the symmetry of the order parameter, but it is not a causal component or mechanism of any real (or model) system. Finally, rather than aiming to describe causal relevance, the goal of the renormalization group is primarily to demonstrate that most of the features of the systems in the universality class are irrelevant to their macroscale behaviors. This demonstration of irrelevance is quite different from aiming to accurately describe difference-making causes.

3.3.1 Essential Idealizations in Minimal Model Explanations

What's more, the use of renormalization methods requires an essential mathematical idealization called the *thermodynamic limit* (Batterman 2002a, 2002b, 2010, 2011; Morrison 2015), which is the limit in which (roughly speaking) the number of particles of the system approaches infinity. In most instances, the thermodynamic limit is important for introducing qualitative differences between the behavior of the system in the limit and the behavior of the system as it approaches the limit (Kadanoff 2013). Without these singularities, the mathematical models fail to display the qualitatively distinctive behaviors that physicists are interested in explaining (e.g., phase transitions). As Morrison explains, "once we move to statistical mechanics, the equations of motion that govern these systems are analytic and hence *do not exhibit singularities*. As a result, there is no basis for explaining phase transitions in [statistical mechanics]. In order to recover the [thermodynamical] explanation, singularities are introduced into the equations, and thus far the *only* way to do this is by assuming the number of particles in the system is infinite" (Morrison 2009, 128; emphasis added). In other words, for a wide range of cases in physics, the thermodynamic limit is an *essential* mathematical idealization for providing the only explanation available.

Contrary to the standard view, physicists do not introduce this idealizing assumption simply as a way of emphasizing the irrelevance of the distorted features. Instead, the thermodynamic limit plays an essential role in the applicability of the renormalization techniques that demonstrate that most of the degrees of freedom of the systems within the universality class are irrelevant to their critical behaviors. Instead of distorting irrelevant causes, this idealization serves as a *necessary* condition for applying the mathematical techniques involved in the explanation of these behaviors (Batterman 2002b, 2010, 2011; Batterman and Rice 2014; Morrison 2009, 2015; Rice 2018). As Batterman puts it, "We have an explanation for the universality of critical phenomena that depends essentially upon infinities and divergences." (Batterman 2011, 1037). In sum, the thermodynamic limit is an idealization that often plays an essential role by allowing for the application of a mathematical technique that extracts the required explanatory information.

Moreover, rather than being an isolated case, as Kadanoff (2013) explains, "the differences among solid, liquid and gas; the distinctions among magnetic materials and between them and nonmagnetic materials; and the differences between normal materials and superfluids are all best understood as distinctions that apply in the limit in which the number of molecules is infinite" (Kadanoff 2013, 143). Without using these limits, physicists would not be able to provide the best (or only) explanation of the behaviors that we observe across different physical systems. Furthermore, the use of the thermodynamic limit in these explanations is similar to the use of singular limits in many other explanations in physics, such as the use of the ray limit in geometrical optics (Batterman 2002b, 2011; Morrison 2015; Pincock 2011). As I will argue in more detail in later chapters, because these idealizations are essential to the overall mathematical frameworks that physicists use to generate these explanations, the distortions that they introduce cannot be isolated to particular features of the system that are irrelevant to the explanandum. Only by distorting difference-making and non-difference-making features can physicists apply the mathematical modeling techniques that are required to provide epistemic access to the desired explanations.

3.3.2 The Role of Irrelevance Information in Minimal Model Explanations

Minimal model explanations are also useful because they very clearly illustrate the essential role of *irrelevance information* in many scientific explanations. This unique explanatory structure is almost universally mischaracterized by extant philosophical accounts of explanation and modeling that focus on the accurate representation of relevant features (Batterman and Rice 2014). This is because, in addition to focusing on noncausal features of the system and using essential idealizations, minimal model explanations focus on importantly different explanatory questions. Indeed, scientists who study universal patterns across extremely diverse systems—such as phase transitions—often desire answers to explanatory questions such as the following (Batterman and Rice 2014, 361):

1. Why are these minimal features necessary for the phenomenon to occur?
2. Why are the remaining heterogeneous details irrelevant to the occurrence of the phenomenon?
3. Why do very different systems have these minimal features in common?

Simply identifying which features are necessary for the occurrence of the explanandum fails to provide answers to these important explanatory questions. For example, simply noting that many systems have certain minimal features in common fails to explain *why* very different systems share those features and why only those features are necessary to display the universal behaviors.

Both Lange (2014) and Povich (2018) have objected to this account of minimal model explanations by suggesting that answering these three questions is unnecessary to explain patterns. However, Batterman and Rice (2014) never claimed that all explanations of patterns will require answers to these questions. Indeed, this is part of what makes minimal model explanations unique: they are not just explanations of patterns, but explanations of *universality* that aim to show *why* certain behaviors are stable across very different real, possible, and minimal model systems. What's more, while not every explanation of a pattern will raise these additional questions, in many cases these are precisely the explanatory questions that physicists would like to answer. As physicist Michael Fisher puts it:

> The traditional approach of theoreticians, going back to the foundations of quantum mechanics, is to run to Schrödinger's equation when confronted by a problem in atomic, molecular or solid state physics! . . . The modern attitude is, rather, that the task of the theorist is to *understand* what is going on and to elucidate which are the crucial features of the problem. For instance, if it is asserted that the exponent depends on the dimensionality, d, and on the symmetry number, n, but on no other factors, then the theorist's job is to explain *why* this is so and subject to what provisos. If one had a large enough computer to solve Schrödinger's equation and the answers came out that way, one would still have *no understanding* of why this was the case! (Fisher 1983, 46–47)

In short, scientists are often interested in precisely these sorts of explanatory questions that go beyond merely identifying a minimal set of common features across a range of systems. Specifically, they want to understand why the pattern of interest depends only on a few relevant features and why the other features of the system are irrelevant to that pattern. Without answering these further questions, there seems to be no understanding of why the patterns are universal across this range of heterogeneous real, possible, and model systems.

What makes minimal model explanations unique is that they directly focus on answering these kinds of explanatory questions by demonstrating that most of the physical features of a class of systems are irrelevant to their universal behaviors (Batterman and Rice 2014). In the case of the LGA model, this is largely accomplished by using the renormalization group to delimit a class of systems that display the universal behaviors from other systems that fail to display those behaviors. Moreover, by demonstrating that most of the features of the systems in that universality class are irrelevant, the renormalization group shows that we could perturb most of the features of those systems without altering their universal behaviors. In addition, delimiting the universality class via renormalization shows why only the minimal features are necessary, why the other features are irrelevant, and why systems with very different features will have those necessary features in common. In short, the demonstration of the irrelevance of most of the features of the systems within the universality class is what enables us to answer these further explanatory questions.

Woodward (2018) criticizes Batterman and Rice (2014) for seeming to suggest that information about irrelevance is sufficient to explain on its own. Woodward is certainly correct that information about irrelevance is insufficient to provide the complete explanation, and perhaps this point should have been made clearer in Batterman and Rice's original paper. Nonetheless, an absolutely crucial feature of minimal model explanations is that the explanation of universal behaviors focuses largely on the role of irrelevance information rather than on identifying the relevant difference-making features. The important point, according to Batterman and Rice, is that "many explanations focus on finding difference-making factors and implicitly assume that factors left out are irrelevant. In minimal model explanations this process is reversed: one proceeds by showing that various factors are irrelevant. The remaining features will then be the relevant ones" (Batterman and Rice 2014, 363). In other words, although irrelevance information cannot explain on its own, Batterman and Rice (2014) argue that the exclusive focus on accurately representing difference-making features has led most accounts of explanation to miss the most important part of the structure of minimal model explanations. The importance of these explanations is that the explanation of universal behaviors focuses largely on demonstrating that most of the features of the systems are irrelevant. So while appealing to the relevant features within the minimal model is required, it is only in combination with the information about irrelevance provided by techniques like renormalization that we get a complete explanation of the universality of those behaviors. Consequently, I agree with Woodward that in this case, "relevance and irrelevance information work together" (Woodward 2018, 135). The crucial difference is that I further argue that information about irrelevance takes center stage and is ineliminable in minimal model explanations (and other explanations of universality). Unfortunately, this focus on providing information about the irrelevance of various features in order to explain universal patterns (and their stability) is largely ignored by extant accounts of explanation and modeling that focus only on difference-makers (Rice manuscript).

Before moving on, it is important to note that this type of minimal model explanation is not restricted just to cases in physics. First, minimal model explanations are also found in biology (Batterman and Rice 2014). I will present a few of these biological cases in chapters 6 and 7. In addition, similar uses of minimal models to explain can be found in neuroscience (Chirimuuta 2014, 2018; Ross 2015). For example, Lauren Ross (2015) discusses a case in which the canonical model is appealed to in the explanation for why a wide range of (real, possible, and model) neural systems all display the same class I excitability behaviors. Like the case described earlier, here the explanandum of interest is "why neurons that differ so drastically in their microstructural details all exhibit the same type of excitability" (Ross 2015, 41). Moreover, neuroscientists want to know why these patterns of behavior are well captured by the extremely minimal canonical model. As Ross explains: "The canonical model and abstraction techniques used in this approach explain why molecularly diverse neural systems all exhibit the same qualitative behavior and why this behavior is captured in the canonical model" (Ross 2015, 41).

This is precisely the kind of explanatory structure that is used in minimal model explanations. Moreover, Ross argues that "as the neural systems that share this behavior consist of differing causal mechanisms . . . a mechanistic model that represented the causal structure of any single neural system would no longer represent the entire class of systems with this behavior"

(Ross 2015, 46). Indeed, she goes on to argue that this kind of explanation is best captured by a minimal model approach based on delimiting universality classes that include the minimal model.

Consequently, the general structure of minimal model explanations is used across many scientific fields, including physics, biology, and neuroscience. These explanations focus on demonstrating the irrelevance of most of the features of a class of real, possible, and model systems in order to explain the stability of the macroscale behaviors and justify appealing to extremely minimal models—that are within the same universality class—to study those macroscale patterns. Moreover, this information about irrelevance continues to play an essential role in the explanation of the stability of these universal behaviors across heterogeneous physical systems—even after we have identified the minimal features that are relevant.

3.4 A Way Forward

In conclusion, many scientific explanations succeed despite the fact that they do not accurately represent the contextually salient difference-making causes of the explanandum. There are certainly several other examples of noncausal explanation that could also be included here, such as mathematical explanations (Baker 2009; Lange 2013a; Pincock 2012; Reutlinger and Saatsi 2018). However, the examples of optimality, statistical, and minimal model explanations provide sufficient support for establishing the ubiquity of noncausal explanations in science. Each of these cases meet Alisa Bokulich's general description of a noncausal explanation: "A non-causal explanation is one where the explanatory model is decoupled from the different possible kinds of causal mechanisms . . . such that the explanans is not a representation (even an idealized one) of any causal process or mechanism" (Bokulich 2018, 149). Put differently, it isn't just that these models directly and deliberately distort the contextually salient difference-making causes of the explanandum. What makes the explanations noncausal is that they do not even aim to provide a representation (accurate or not) of the causes that produced the explanandum. Instead, the explanation is provided by appealing to noncausal features of the system that account for the occurrence of the explanandum. In light of these noncausal explanations, we must move beyond the standard approach's exclusive focus on causal explanation.

The cases surveyed here also reveal some of the features of noncausal explanations that ought to be captured by our revised account of scientific explanation. First, these explanations provide information about how the explanandum counterfactually depends on noncausal features and why many of the features of the system (e.g., its causes) are counterfactually irrelevant to the occurrence of the explanandum. As a result, information about both counterfactual dependence and independence is essential to these explanations. Second, these explanations routinely employ essential idealizations that distort many of the difference-making (or otherwise relevant) features of the models' target systems. In fact, these essential idealizations often distort precisely the features and processes that scientists know are relevant to the occurrence of the explanandum. As a result, these cases also show that we need to expand beyond the standard approach's focus on accurately representing relevant features. Third, the use of these idealized models to explain is typically justified by noting that their distortions are necessary for the mathematical modeling techniques required to extract information about counterfactual dependence and independence. Our revised accounts of explanation, idealization, and modeling ought to incorporate these features.

4 A Counterfactual Account of Explanation

In this chapter I argue for a counterfactual account of scientific explanation that builds on several existing proposals in the literature (Ariew, Rice, and Rohwer 2015; Batterman and Rice 2014; Bokulich 2008, 2011, 2012; Reutlinger 2016; Rice 2012, 2013, 2016, 2018; Saatsi and Pexton 2013; Woodward 2003). While this account is certainly motivated by the kinds of noncausal explanations discussed in chapter 3, I contend that it can subsume several causal and mechanistic explanations as well. According to my counterfactual account, explanations must provide the required *set* of information about the (contextually salient) counterfactual dependence *and* independence relations that hold between the features cited in the explanans and the explanandum. However, given that I embrace a degree of pluralism about scientific explanation, I will not be arguing that absolutely all scientific explanations can be captured by a single counterfactual account—that is, contrary to Reutlinger's (2016) view, I will not be offering a monistic account of necessary and sufficient conditions for all scientific explanations.[1] In fact, elsewhere I have argued that explanation is a cluster concept: there are multiple overlapping subsets of features that are sufficient to explain, but no set of features is necessary for all explanations (Rice and Rohwer 2020). Consequently, there will likely be some counterexamples to the account proposed below. Despite the existence of a few counterexamples, I will argue that a wide range of causal and noncausal explanations can be unified by looking at the counterfactual dependence and independence relations that hold between the explanans and the explanandum and that my counterfactual account has certain advantages over other contemporary (e.g., causal) accounts of explanation. While I advocate a bottom-up approach to philosophical attempts to analyze scientific explanation, part of that approach ought to involve identifying commonalities among specific cases when they are available.[2] To this end, I will argue that the counterfactual account is the best available account for unifying a wide range of causal and noncausal explanations, connecting explanation with the cognitive achievement of understanding, and accounting for the essential role of idealizations in scientific explanations. Consequently, we have strong reasons for adopting and continuing to develop a counterfactual account of explanation.

There is an ever-growing consensus that noncausal features of a system can be used to explain (Ariew, Rice, and Rohwer 2015; Batterman 2002b, 2010; Batterman and Rice 2014; Bokulich 2008, 2011; Lange 2013a, 2013b; Matthen and Ariew 2009; Pincock 2012; Rice 2012, 2013; Saatsi and Pexton 2013; Walsh 2007, 2010; Walsh, Lewens, and Ariew 2002). However, advocates of causal accounts of explanation have provided extended treatments of how causal explanations work. Therefore, it is important to develop an account of noncausal explanations, or of scientific explanation more generally, that avoids the requirement that an explanation accurately represent (or describe) causes. Fortunately, many of the features of our existing accounts of explanation—such as counterfactuals and invariance (Woodward 2003), generality (Kitcher 1981; Potochnik 2017; Strevens 2008), asymmetry (Salmon 1984; Strevens 2008), and expectability (Batterman 2002b; Hempel 1965; Strevens 2008)—are not restricted to a causal interpretation. Therefore, there is already a widely accepted set of features from which this alternative account of explanation can be constructed. In addition, the analysis of various kinds of noncausal explanations given in chapter 3 has revealed several additional features that our alternative account of explanation ought to incorporate. My goal in this chapter is to provide such an account and argue for its superiority to other prominent accounts of explanation.

The following section lays out three requirements for an account of scientific explanation. In section 4.2 I present my counterfactual account of explanation. In sections 4.3, 4.4, and 4.5 I argue that the counterfactual account satisfies this set of requirements better than other prominent views of explanation. And in section 4.6 I compare my account to some other counterfactual accounts in order to highlight some of its key distinguishing features.

4.1 Three Requirements for an Account of Explanation

Explanations provide information about *why* a phenomenon occurred. Therefore, the first requirement for an account of explanation is that it must identify the kind of information provided by explanations that allows them to show why a phenomenon occurred (Bokulich 2012; Craver 2006; Salmon 1984; Woodward 2003). Specifically, explanations show how the explanandum depends on the features cited in the explanans. As a result, our account of explanation ought to tell us what the explanatory dependence relation is that distinguishes explanations from nonexplanations. Furthermore, the kind of dependence relation that we identify as essential to explanation should enable us to see why the wide array of paradigmatic examples found in the explanation literature (e.g., covering-law, causal, mathematical, equilibrium, statistical, and minimal model explanations) can all sometimes qualify as satisfactory explanations. In particular, in light of the cases discussed in chapter 3, the kind of dependence relation our account takes to be explanatory ought to unify various kinds of causal and noncausal scientific explanations.[3]

In addition to specifying an explanatory dependence relation, a satisfactory account of explanation should tell us why discovering information about those dependence relations is cognitively valuable to us. Specifically, a key motivation for discovering explanations is that they allow us to *understand* the world around us. In addition, we often demonstrate our understanding by being able to offer an explanation and our ability to communicate an explanation is often a function of how well we understand what we are trying to explain (Elgin 2017; Morrison 2015). Accordingly, many accounts emphasize this important relationship between explanation and understanding. As Salmon writes: "Perhaps the most important fruit of modern science is the understanding it provides of the world in which we live, and of the phenomena that transpire within it. Such understanding results from our ability to fashion scientific explanations" (Salmon 1984, 259). In addition, Friedman suggests that our theory of explanation ought to "somehow connect explanation and understanding—it should tell us what kind of understanding scientific explanations provide and how they provide it" (Friedman 1974, 14). Kitcher agrees: "A theory of explanation should show us *how* scientific explanation advances our understanding" (Kitcher 1981, 508). While I will argue in chapter 8 that we can acquire scientific understanding from models even when they fail to explain, discovering explanations is certainly one of the primary ways that science produces understanding. Consequently, the second requirement for a satisfactory account of explanation is that it must explicate the relationship between explanations and the cognitive achievement of scientific understanding (Friedman 1974; Kitcher 1981; Potochnik 2017). This connects the metaphysical project concerning which set of facts ought to be included in a scientific explanation with the cognitive components involved in the discovery of explanations. As Walsh explains:

> An explanation serves two functions—metaphysical and cognitive. The metaphysical function (roughly) is to identify a set of conditions in the world—the explanans conditions—such that when they hold the explanandum does too. The cognitive function is that of enhancing our understanding of the occurrence of the explanandum. An explanation successfully discharges its cognitive function when it elucidates the relation between explanans and explanandum in such a way that knowing that the explanans condition holds is sufficient for us to understand the occurrence of the explanandum. (Walsh 2015, 471)

In other words, it is the grasping of the conditions cited in the explanans and their relation to the explanandum that enables us to understand why the explanandum occurred. As a result, the requirement that an account of explanation show us how explanations produce understanding connects the metaphysical aspects of explanation with the epistemic features of explanations involved in asking and answering particular why questions of interest to humans (Potochnik 2017; van Fraassen 1980).

Finally, an account of explanation should also tell us how scientific practice produces explanations. In particular, this requires showing how practices like highly idealized modeling can provide information about explanatory dependence relations that enables us to understand natural phenomena. Satisfying this requirement is crucial for ensuring that our philosophical account of explanation provides genuine normative guidance for human scientists' attempts to explain and understand various phenomena. Here, I agree with Glennan that "while it may be true that we sometimes speak of one thing in the world explaining another, or of explanations being out there to be discovered, the act of scientific explanation is always an act of describing or representing those things. Moreover, for an explanation to succeed, it must be understood, and theories of scientific explanation should explore the kinds of representations that are successful in promoting understanding" (Glennan 2017, 222). In other words, a theory of explanation must show us how the representations actually used by scientists (i.e., idealized theories and models) enable them to explain and understand phenomena. Furthermore, as I argued in chapter 3, idealizations

A Counterfactual Account of Explanation

often make positive contributions to, and play ineliminable roles within, scientific explanations (Batterman 2002b, 2010; Bokulich 2012; Rice 2013; Strevens 2008; Wayne 2011; Weisberg 2007a). As a result, an account of scientific explanation must provide a satisfactory analysis of the essential contributions made by ineliminable idealizations to the explanations provided by scientific models (Bokulich 2011, 2012). More specifically, the account ought to show why replacing many idealizations with abstractions or more realistic assumptions would result in a worse—or perhaps no—explanation. This is precisely what makes the idealizations ineliminable and illustrates their positive contributions to the epistemic aims of science.

In sum, a satisfactory account of explanation ought to do the following:

1. Identify the kind of dependence relation involved in explanations that unifies various causal and noncausal explanations.
2. Explicate the relationship between explanation and understanding.
3. Provide a satisfactory analysis of the positive and ineliminable contributions that idealizations make to scientific explanations.

Having outlined these three requirements, I now turn to the tasks of presenting the counterfactual account of explanation and arguing that it satisfies this set of requirements better than other prominent accounts of explanation.[4]

4.2 A Counterfactual Account of Explanation

As we saw in chapter 2, by far the most common response to the counterexamples to Hempel's deductive-nomological (DN) model has been to bring in considerations of causes. Although bringing in causation is one way of providing a solution to these problems, it is not the only way. As an alternative, I suggest that causation has traditionally been an attractive response here because considerations of causation reveal the explanatory counterfactual dependence information in many cases of explanation (e.g., the flagpole case, the hexed salt case, etc.). The key to providing an explanation, I argue, is providing a set of information about the important counterfactual relationships in the system, which is often (but not necessarily) done by specifying causal relations in the system. According to the counterfactual account that follows, the explanandum E will counterfactually depend on a feature, property, or event X just in case, had X not been present (and other features been held fixed), then E would not have occurred (or the probability of E's occurrence would have been significantly different). If E still occurs (with roughly the same probability) when X is not present (and other features are held fixed), then I will say that the occurrence of E is counterfactually independent of X.[5] In order to evaluate such claims concerning counterfactual dependence and independence, we first need to specify a set of features of the system that will be held fixed in the background (Woodward 2003).[6] We then allow the features cited in the antecedent of the counterfactual claim to vary and attempt to determine what implications that variation would have for the explanandum.

Rather than offering a more detailed account of the semantics of counterfactuals (e.g., involving some kind of modal realism), following several other philosophers of science—for example, Bokulich (2008, 2011), Khalifa (2017), Lipton (2009), Potochnik (2017), Reutlinger (2016, 2018), and Woodward (2003)—I think we can engage in debates about which types of counterfactual information enable scientists to explain and understand, without having a complete metaphysical account of what makes counterfactual claims true. Indeed, just as a person can reason about what would have happened had she taken a different job without consulting a detailed philosophical account of the semantics of counterfactuals, I think scientists (and philosophers of science) can appeal to answers to "What if things had been different?" questions (Woodward 2003), even if we lack a complete account of how we can be realists about the truth of modal claims. All that is required is that we assume that there are some facts about how the world would (or would not) have been different in various counterfactual situations and that scientific models and theories are often useful in scientists' attempts to explore and evaluate those situations. However, supplementing the views presented here with a more detailed metaphysics of what grounds the truth of counterfactual claims would be one way to fruitfully expand the account in the future.

In emphasizing counterfactual information, my account builds on a key aspect of Woodward's interventionist account of explanation. According to Woodward, "[an] explanation must enable us to see what sort of difference it would have made for the explanandum if the factors cited in the explanans had been different in various possible ways" (Woodward 2003, 11). Indeed, science aims to tell us not only what the world is like, but also about certain relevant ways the world could have been and how things would be different in those counterfactual situations.

I believe Woodward has identified the important kind of information involved in many forms of scientific explanation. However, I disagree with his account's requirement that the counterfactual relations involved in an explanation must always be understood along strictly manipulationist or interventionist lines. The requirement that these counterfactuals must enable one to determine the outcomes of possible interventions restricts Woodward's account to specifically causal explanations. Woodward makes this restriction explicit: "The theory I will be developing is restricted to *causal* explanations. To the extent that there are forms of explanation that are noncausal, . . . they will be outside the scope of my discussion" (Woodward 2003, 187). Yet I think it is a mistake for our account of scientific explanation to be restricted to causal explanations. Indeed, the ever-expanding list of examples of noncausal explanations available in the literature imply that our account of scientific explanation ought to show how noncausal explanations work and what they have in common with the myriad causal (and mechanistic) explanations offered in science.

Consequently, I suggest that we distinguish the question of what caused the explanandum (or what scientists mean when they make causal claims) from the question of what explains the explanandum.[7] In line with this suggestion, Woodward himself admits that perhaps not all scientific explanations are causal, and he suggests a way that his account might be expanded to include noncausal cases:

> One natural way of accommodating these examples is as follows: the common element in many forms of explanation, both causal and noncausal, is that they must answer what-if-things-had-been-different questions. When a theory tells us how Y would change under interventions on X, we have (or have the material for constructing) a *causal* explanation. When a theory or derivation answers a what-if-things-had-been-different question but we cannot interpret this as an answer to a question about what would happen under an intervention, we may have a noncausal explanation of some sort. (Woodward 2003, 221)

In precisely this way, we can retain Woodward's emphasis on providing counterfactual information without requiring that these be *causal* counterfactual relations (i.e., counterfactuals that are grounded in causal dependencies). That is, the counterfactual dependence requirement on explanations is independent of the causal dependence requirement (Bokulich 2008, 2011; Reutlinger 2016, 2018; Rice 2013, 2016, 2017; Saatsi and Pexton 2013). Given its focus on what-if-things-had-been-different questions, the counterfactual account that follows is in many ways an expansion of Woodward's account. However, given that it drops the interventionist requirement that is central to Woodward's view (and adds some additional conditions that are not part of Woodward's view), it also represents a major departure from his account.

In order to see the appeal of this counterfactual approach, consider just what is required in order to deal with the counterexamples raised to Hempel's DN model. First, the flagpole case raises the asymmetry problem. Solving this problem requires that our explanatory dependence relation be asymmetric such that if X explains Y, Y does not thereby also explain X. Counterfactual dependence delivers the required asymmetry. For example, suppose that a car crash would not have occurred if the car had tires with more tread on them. In this case, the car crash counterfactually depends on the tires' treads, and therefore this feature of the tires is a key part of the explanation of the crash. However, it is not true that if the crash had not occurred, then the tires would have had different levels of tread. Therefore, the car crash fails to explain the amount of tread on the tires. The same holds for the flagpole case. If the flagpole had been a different height, then the shadow would have had a different length. But changes in the length of the shadow (due to other factors, such as the sun moving) do not change the height of the pole. So the asymmetry of explanatory dependence can be captured by a counterfactual approach—one just has to be careful to avoid certain "backtracking" counterfactuals.[8]

The second set of counterexamples to Hempel's account involve cases of explanatory irrelevance, such as the hexed salt case. Although it is a true law (or generalization) that all hexed salt dissolves in water, the salt being hexed is completely irrelevant to its dissolving. Once again, the counterfactual account delivers the correct result. Citing hexing as the relevant feature is incorrect because the dissolving does not counterfactually depend on whether the salt is hexed; that is, changing whether the salt is hexed does not change the occurrence of the dissolving. In contrast, changing certain chemical properties of the salt would change the occurrence of the explanandum, and therefore those are the counterfactually relevant features that need to be included in the explanation.

However, it is also extremely important to note that the main problem with the hexed salt explanation is that it cites an irrelevant factor as though it were relevant. This is crucially different from demonstrating that a feature that is irrelevant is in fact irrelevant. Explanations should not provide

A Counterfactual Account of Explanation 97

inaccurate information about which features are relevant or irrelevant to the explanandum. This does not mean, however, that demonstrating that something is irrelevant should never be included in an explanation. In sum, the counterfactual approach delivers all that is required to solve the problems of asymmetry and relevance: counterfactual dependence is asymmetric in the required way, and the counterfactual approach maintains that explanations should not cite irrelevant features as if they were the features on which the explanandum depends.[9]

On the counterfactual account I develop here, counterfactual dependence information can be explanatory without tracking any relationships of causal dependence. For example, counterfactual dependence relations that involve noncausal structural, mathematical, statistical, or topological properties of the system can also be used to provide explanations in science. Consequently, we get the central claim of the counterfactual account: *scientific explanations must provide information about counterfactual dependence relations that hold between features cited in the explanans and the explanandum.* This is a necessary condition for providing a counterfactual explanation, but it is not sufficient on its own.

In addition, I maintain that, in many contexts, scientific explanations must provide two kinds of counterfactual information by showing us both what is and what is *not* important for the occurrence of the explanandum. That is, the explanation should tell us how the explanandum counterfactually depends on certain relevant features, as well as the ways that certain contextually important features can be changed and yet the explanandum will still occur. This information about why certain features are counterfactually irrelevant to the occurrence of the explanandum is essential to many scientific explanations. Indeed, the three types of noncausal explanation surveyed in chapter 3 made essential use of information about which features of the system are counterfactually irrelevant to the phenomenon of interest. As a result, I argue that scientific explanations must enable us to see both the counterfactual dependence and independence relationships that hold between the contextually salient features cited in the explanans and the explanandum.

One objection here might be that requiring all explanations to include information about irrelevant features is implausible because it would rule out most causal explanations in science.[10] Here it is crucial to remember that, as with causal accounts of explanation, the context in which the explanation is sought will make particular features salient or nonsalient to providing an explanation. For example, as I will argue later in this chapter, when our best scientific theories suggest that certain features should be relevant but they are in fact irrelevant, then the explanation will need to account for their irrelevance. This is compatible, of course, with there being explanations that do not mention any irrelevant features because none of those features are deemed contextually salient by our current scientific theories (or other pragmatic considerations). Therefore, my account requires only that the explanation must account for the contextually salient features that are irrelevant to the explanandum. Because in many cases no irrelevant features will be contextually salient, many explanations (e.g., many causal explanations) will be successful even if they do not describe any features that are irrelevant to the explanandum.

Another objection comes from Woodward, who argues that information about irrelevance is implicitly included in his account because showing how the explanandum depends on certain features "at least implicitly conveys that other variables . . . that are not explicitly mentioned . . . are irrelevant" (Woodward 2018, 129). While I grant that this is possible in some contexts, it ignores the fact that in many other explanatory contexts, the explanation will need to *explicitly* include information about irrelevance rather than simply failing to mention irrelevant features. Minimal model explanations and other explanations of universality provide the clearest examples here. Despite granting that irrelevance sometimes plays a role, Woodward goes on to argue that "there is an obvious sense in which the dependence information seems to be doing the explanatory work. . . . One indication of this is that the independence information by itself, apart from dependence information, does not seem explanatory" (Woodward 2018, 130). In response, I suggest that Woodward's test for the necessity of a kind of explanatory information cuts both ways. In many cases, such as in many equilibrium, statistical, and minimal model explanations, information about the features that are relevant (i.e., difference-makers) is unable to provide the desired explanation on its own.[11] In particular, this occurs when we would like to explain why a given pattern (or phenomenon) continues to occur despite various changes to features of the system. This is because the explanandum of interest will often take the following form: "Why does phenomenon P occur across these physical systems despite differences in features $y_1, y_2, y_3, \ldots y_n$?" In these cases of explaining stable patterns, information

A Counterfactual Account of Explanation

about relevant features is often unable to provide the explanation without additional explicit information about why many of the systems' other features are irrelevant. As a result, information about relevance and irrelevance must work together in many scientific explanations.

Unfortunately, most accounts of explanation focus exclusively on the features that make a difference and, therefore, miss the importance of actually showing that certain features are irrelevant to the occurrence of the explanandum. Both kinds of information are often required to provide the desired explanation, and neither should be universally privileged as more important. Some explanations will explicitly represent only those features that are relevant because the irrelevance of other features is easily discovered (or is already known); for instance, many causal models will simply abstract away all causal factors that are known to be irrelevant and focus on specifying the difference-making causal factors (Strevens 2008). In other cases, the features on which the phenomenon depends will be easier to discover, and showing the irrelevance of various other features will be the primary focus of the explanation, such as in cases of explaining universal patterns across extremely heterogeneous physical systems (see Batterman 2002b and Batterman and Rice 2014 for some examples).

The importance of providing a larger set of counterfactual dependence and independence information is also essential for addressing some traditional counterexamples to counterfactual accounts. For example, a thunderstorm is caused by a drop in atmospheric pressure. However, a drop in atmospheric pressure also causes a lower reading on a barometer. This common cause results in a strong correlation between lower barometer readings and thunderstorms. In particular, whether or not the thunderstorm occurs seems to counterfactually depend on a low barometer reading because had the reading been higher, the storm would not have occurred. But surely the lower barometer reading fails to explain why the storm occurs.

While many philosophers have suggested that we must adopt a strictly causal account of explanation in order to resolve this issue, I suggest that what is needed here is additional counterfactual information beyond the dependence between the barometer reading and the occurrence of the storm. Specifically, the complete counterfactual explanation of the occurrence of the thunderstorm ought to show how changes in atmospheric pressure change the occurrence of storms (and perhaps the reading of the barometer as well). In addition, the counterfactual explanation should address how various changes in the barometer (e.g., the barometer being broken) would fail to result in changes in the occurrence of thunderstorms. That is, the explanation should also show why a certain range of changes to the barometer reading are counterfactually irrelevant to the occurrence of the storm. Of course, in this case, these additional counterfactual relationships are made true by causal relationships, but one can imagine other cases where such relationships are grounded in noncausal dependencies (e.g., mathematical relationships). Moreover, I contend that it is the counterfactual relationships that are doing the real explanatory work here, not the additional fact that they are grounded in causal relationships. Only by providing these more complete sets of counterfactual information will a satisfactory explanation be provided. Identifying a single counterfactual dependence relation will almost always be insufficient. This is why my counterfactual account requires a *set* of counterfactual information about the relevance and irrelevance of the features of interest.[12]

Other possible counterexamples involving preemption also illustrate the importance of including a larger set of counterfactual information (i.e., beyond a single counterfactual dependence). For example, suppose that we want to explain why a bottle is broken every day at recess, and we note that every day Suzie throws a rock at the bottles. However, it just so happens that Billy also throws rocks at the same bottles every day (after they've been broken), so even if Suzie hadn't thrown her rocks, the bottles still would have broken. As a result, Suzie's throwing of the rocks is counterfactually irrelevant to the breaking of the bottles, but it seems that we can explain the bottles' breaking by noting that Suzie threw rocks at the bottles each day. First, it is important to note that the counterfactual account that I defend allows information about counterfactually irrelevant features to be essential to providing many kinds of explanations. Thus, the fact that the explanandum is counterfactually independent of Suzie's throwing rocks is not sufficient grounds for excluding it from the explanans—the context might make that counterfactually irrelevant factor particularly salient. My main response, however, is that there is something deficient about the explanation that only cites the fact that Suzie threw rocks at the bottles, and that deficiency has to do with failing to grasp the correct set of counterfactual information. We can see this point by supposing that Suzie's teacher believes that Suzie's throwing of the rocks explains why the bottles are broken, so he keeps Suzie inside during recess. However, Billy still throws a rock and breaks a bottle. The teacher is frustrated and believes he has not yet explained why the bottles are breaking. I think this shows that there is something deficient about the teacher's original belief that the broken bottles are explained by Suzie's throwing the rocks. Instead, what the teacher will now understand is that what explains the breaking of the bottles is that *someone* throws a rock at a bottle each day, and the only way to prevent it from occurring is to prevent all students from throwing rocks at the bottles. Moreover, the bottles' breaking is counterfactually dependent on someone's throwing a rock. Recognizing this fact is what makes the teacher's new explanation complete, whereas the original explanation fails to adequately explain given the context.[13]

What these cases clearly show is that precisely which set of counterfactual information is required to explain will depend on how the context of inquiry specifies the target explanandum, the contrast class, and the set of features that are taken to be contextually important (Woodward 2003, 90). For one thing, many philosophers have noted that how a phenomenon is characterized in a particular context will change the set of information required to explain it (Garfinkel 1981; Potochnik 2017; van Fraassen 1980). For example, we may want to explain why polar bears have fur rather than feathers, or we may want to explain why polar bears have fur of a particular thickness. Although they address the same phenomenon, these two requests for explanation require the consideration of different counterfactual scenarios: the first involves showing why polar bears have fur rather than feathers, and the second involves showing why polar bears have fur of a particular thickness rather than a lighter coat. In this way, the characterization of the explanandum and the contrast class will influence the counterfactual information that ought to be provided by an explanation.

In addition to addressing alternatives to the explanandum, a contrastive why question will often dictate which kind of features the explanans ought to account for (Pincock 2018). For example, suppose that we wanted to explain why all US presidents have been male. One way of making this into a (set of) contrastive why question(s) is to ask, for each presidential election, why the winner was X (a male) rather than Y (a male or female). An explanation that tracks the votes that led to each of these election results could provide an adequate explanation in this case because the following counterfactuals are true: for any one of these elections, had Y been a female and won the election, then not all US presidents would have been male.

However, if we are instead interested in explaining why all US presidents have been male rather than more closely reflecting the distribution of the overall population of the United States, then a very different kind of explanation is required (Pincock 2018, 44). Here, the counterfactual dependencies required to provide the desired explanation must appeal to the sexism inherent in our current societal structures. In particular, we need to consider the following counterfactual situation: had our society been less sexist than it is, a more representative distribution of male and female presidents would have occurred (Pincock 2018, 45). What this shows is that the types of contrastive questions that are of interest to scientific researchers and how those questions are asked will influence which counterfactual situations are salient to providing the desired explanation.[14]

More generally, the explanatory context will help determine "what counterfactual alternatives the explanation's audience takes to be salient" (Potochnik 2017, 146). As Woodward explains:

> When we ask for an explanation of some outcome, we often (perhaps even typically) have in mind accounting for the contrast between that outcome and a specific alternative or set of alternatives to that outcome, rather than all possible alternatives. Often, we are not interested in, and our request for explanation should not be taken as a request to account for, the contrast between the explanandum phenomenon and other alternatives outside this set of interest. In other words, there is a specific what-if-things-had-been-different question or range of such questions that we want answered. Broadly pragmatic considerations play a role in specifying which of these are of interest. (Woodward 2003, 227)

In other words, the pragmatic features of the explanatory context will determine the set of features whose counterfactual relevance and irrelevance need to be accounted for in providing the explanation. This is what enables scientists to cite subsets of the features on which the explanandum counterfactually depends and still successfully explain. If we required the inclusion of all the features necessary for the occurrence of the explanandum, explaining phenomena would be impossible. For example, in most contexts, when explaining why polar bears have thick fur, we need not include that this event counterfactually depends on the big bang. In sum, the context of inquiry will not only specify the phenomenon we want to explain, but also will determine the set of features whose counterfactual relationships to the explanandum need to be accounted for in order to provide a complete explanation. Moreover, because the set of contextually salient features can be chosen in a variety of ways given our interests, background theories, and goals, the counterfactual account allows multiple explanations of the same phenomenon.

A key feature of my account is that, in many cases, the context of inquiry makes it such that various relationships of counterfactual independence are essential to providing a complete explanation. Put differently, *the explanatory context can make features that are not difference-makers explanatorily relevant.* While some philosophers have discussed how idealizing or abstracting away irrelevant features can improve explanations by making them more general (e.g., Potochnik 2007; Strevens 2008), my account is unique in requiring information about the irrelevance of various features to the explanandum be *explicitly* included within the explanans in order to provide a complete explanation in many cases (Rice manuscript). That is, this information about irrelevant features is often not an optional supplement that merely facilitates our focusing on the relevant features that do the real explanatory work. Instead, in many explanatory contexts, a complete explanation of the explanandum can be provided only by including detailed information showing that (and perhaps why) various features of the systems of interest are irrelevant to the occurrence of the explanandum.[15]

To see the difference here in another way, it is worth noting that most philosophers who have discussed irrelevance have argued that showing that a feature is irrelevant is what justifies omitting that feature from the explanation. For example, Strevens argues that causes that do not make a difference ought to be removed from the explanation and claims that "a causal factor cannot be explanatorily relevant unless it is a difference-maker" (Strevens 2008, 146). In addition, Woodward has discussed cases in which there is a set of variables X_i (at some upper level) that are causally relevant to the explanandum E and which are such that, given the values of those X_i, further variations in some other set of variables Y_k (at some lower level) are irrelevant to E (Woodward 2018, 18). However, Woodward's point here is to show that "if we want to explain E, we can just cite the variables X_i, and ignore the lower level variables Y_k" (Woodward 2018, 18). Consequently, like Strevens, Woodward suggests that demonstrating that some features are (conditionally) irrelevant is what justifies leaving them out of the explanation. Several other philosophers who focus on the explanations of patterns, such as Potochnik (2017), have likewise suggested that when explaining a pattern, we can omit or idealize away those features of the system that have little impact on the occurrence of the pattern of interest.

In contrast with these views, my account explicitly requires the inclusion of information about which of the contextually salient features of the system are irrelevant to the occurrence of the explanandum. That is, this information about which features are irrelevant—and perhaps some kind of demonstration of their irrelevance—often plays an essential role within the explanation itself. For one thing, as we saw in chapter 3, without this information about irrelevant features, the stability of many of the patterns we observe is left unexplained. More specifically, I contend that a scientific explanation ought to say explicitly which of the contextually salient features of the system could have been different (and in what ways) and yet the explanandum would still have occurred.

Rather than being a trivial (or minor) revision to existing accounts of explanation, this suggestion cuts at the core of the widespread use of difference-making as the primary criterion for determining explanatory relevance. While other authors have argued that being a difference-maker is not sufficient for being explanatorily relevant (Potochnik 2017; Strevens 2008), I argue further that being a difference-maker is not *necessary* for being explanatorily relevant. While explanation is certainly about establishing which features make a difference to the explanandum (in terms of counterfactual dependence) and which features do not (in terms of counterfactual independence), the mistake of many accounts has been assuming that the context of explanation always entails that features that are irrelevant to the occurrence of the explanandum must be explanatorily irrelevant.

Of course, given the seemingly unbounded set of features that are irrelevant to the occurrence of the explanandum, just which features' irrelevance ought to be included will depend heavily on the explanatory context. While sorting out precisely how contextual features influence the set of features that ought to be included is extremely complicated, I see no reason to think that the pragmatic features that help specify the set of relevant features that ought to be included could not also help specify the set of irrelevant features that need to be accounted for.

To see how context influences which information about irrelevance ought to be included, we can consider two useful examples discussed by Woodward (2018). He first notes that, when explaining the trajectories of the planets, we need not mention the irrelevance of the color of the planets because "a property like color plays no interesting role anywhere in the mechanics or in most of the rest of physics and *no will be surprised by the observation that the influence of gravity on the planets is unaffected by their color"* (Woodward 2018, 130; emphasis added). Indeed, the background assumptions of the explanatory context are doing much of the work here in establishing that the irrelevance of the color of planets to their trajectories need not be accounted for. Because this is precisely what our background theories tell us to expect, the context of these explanations implies that the irrelevance of these features can be left out (or at least left implicit).

This contrasts with many of the cases discussed earlier, where our theories and observations suggest that precisely these kinds of features typically would be assumed to be relevant to the phenomenon of interest. As Woodward notes, in the case of the ideal gas law:

> On the other hand, facts about the detailed trajectories of individual molecules are among the sorts of facts that physics pays attention to: they are relevant to what happens in many contexts and are explananda for many physical explanations. There thus seems to be *a live question about why, to a very large extent, details about individual molecular trajectories don't matter for the purposes of predicting or explaining thermodynamic variables.* (Woodward 2018, 131; emphasis added)

In other words, whether we feel the need to explicitly incorporate information about the irrelevance of various features depends heavily on whether the current state of science tells us that those features are typically relevant to the kind of phenomenon we would like to explain.

As another example, the influence of which features we expect to be relevant is extremely prominent in instances of explaining universality. Indeed, as the physicist Leo Kadanoff (2013) defines it, "Whenever two systems show an *unexpected* or deeply rooted identity of behavior they are said to be in the same universality class" (Kadanoff 2013, 178; emphasis added). That is, it is precisely when physicists believe that the differences between the systems typically *would* make a difference to the behavior of the system that we require that information about the irrelevance of those features be explicitly included in the explanation of universality. These examples suggest that whether or not information about irrelevance ought to be explicitly included in the explanation depends on what our current scientific knowledge tells us about which features we should expect to be relevant (or irrelevant) to the explanandum.

However, once the context of inquiry has specified the target explanandum, the contrast class, and the set of contextually salient features, the ontic features of the target systems will objectively determine the set of counterfactual information necessary to explain. First, if the target explanandum counterfactually depends on certain contextually important features, the explanation will need to show how it depends on those features (i.e., the explanation ought to show how changes in those features result in changes in the explanandum). Moreover, the features cited in the explanans must be sufficient to account for the occurrence of the phenomenon. This is why citing only irrelevant features will always be insufficient to explain. Roughly, the included relevant features on which the explanandum counterfactually depends ought to entail that the occurrence of the explanandum was expected, or at least predictable. Because we can explain highly improbable events, rather than specifying a particular probability that the explanans must confer on the explanandum, I will simply say that the counterfactually relevant features included in the explanans must be adequate to account for why the explanandum occurred. How this criterion gets applied in practice will depend on the nature of the explanandum and other features of the context of inquiry; for example, different disciplines will require different degrees of predictive accuracy or precision (Weisberg 2007b). Second, if certain features are contextually important but the occurrence of the explanandum is counterfactually independent of those features, then the explanation will need to show that the explanandum is independent of changes in those features. Therefore, although context certainly plays a crucial role, pragmatic features or individuals' subjective interests cannot override the fact that certain features will be deemed important by our best scientific theories and only certain sets of counterfactually relevant features will be able to satisfactorily account for the occurrence of the explanandum. In this way, the context of inquiry combines with objective facts in the world to dictate the set of counterfactual information required to explain.

Consequently, unlike Strevens (2008) or Potochnik (2017, 2018), I don't think that an account of explanation is required to choose between being ontological (and focused on objective facts out in the world) or pragmatic (and focused on communication). Strevens is clear that, following Hempel and others, he privileges the ontological sense of explanation (Strevens 2008, 6). In contrast, Potochnik (2017) follows van Fraassen (1980), Achinstein (1983), and others in arguing that communication ought to be the privileged sense of explanation. I see no reason to choose either of these options exclusively—a complete account of explanation ought to address both of them. Given that the goal of this book is to provide normative guidance for practicing scientists, I think our account of explanation ought to focus on how scientists themselves provide explanations and what representations they use to do so (Bokulich 2016). I have little interest in determining the metaphysical nature of a world filled with explanations that human scientists have no interest in or are unable to access with their methods for explaining. This is why my discussion throughout this book is focused on the explanations provided by idealized scientific models. However, just because we ought to focus our attention on the explanations that are actually provided by science doesn't mean that explanation is purely about communicative acts between human beings and their interests. The goal of scientific explanations is to represent or communicate facts about the world that explain why the explanandum occurred. In this sense, at least, it seems that scientific explanation will undeniably involve both aspects of human beings, and their methods of representation, along with some understanding of how the explanations scientists provide "latch on to" objective explanatory facts in the world that are sufficient to account for why the explanandum occured.

Despite this factive requirement concerning the counterfactual information provided by an explanation, as we saw in the previous chapter, it is also important to remember that many scientific explanations also employ essential idealizations. This is because in order to represent, derive, extract, or communicate the explanatory counterfactual relationships, scientists often require models and theories that drastically distort the features of real systems (more on this in later chapters). However, even though idealizations routinely make essential contributions to scientific explanations, in order to count as genuine, the explanation must provide a true set of information about a set of counterfactual dependence and independence relations that holds between the features cited in the explanans and the explanandum.

Another way in which my account preserves the objective sense of explanation is that it only requires a scientific representation that explains to make it possible for an agent who grasps (or understands) that representation to acquire the set of true modal information necessary to explain. Thus, if an agent fails to grasp how the model or theory works, or somehow fails to acquire the modal information necessary to understand why the explanandum occured, that is in a sense *their* fault, not the fault of the explanation provided via the scientific representation. Therefore, although the pragmatics of explanation will influence the specification of the target explanandum and the contextually salient features (Achinstein 1983; van Fraassen 1980), on my account, whether a set of facts explains does not depend on whether an agent actually grasps the explanation or acquires understanding from it.[16]

This means that the influence of pragmatic factors is limited by ontic constraints on explanations. This distinguishes my view from Bas van Fraassen's (1980) pragmatic account of explanation. On van Fraassen's view, the audience determines which type of dependency relations count as explanatory. In contrast, my view of explanation specifies the explanatory dependence relations as relationships of counterfactual dependence and independence. These explanatory relations remain constant across explanations in different disciplines and across changes to the audience. Instead of determining the explanatory dependence relation, I argue that the primary role of the pragmatics of explanation is in determining *which* counterfactual dependence and independence relationships are of interest in a particular context. Holding the explanandum, contrast class, and other pragmatic features of the explanatory context fixed results in a single objective set of counterfactual dependencies and independencies that are required to provide a complete explanation—although different supplements to that set may produce better or worse explanations.

The counterfactual account of explanation combines these ontological and pragmatic aspects of explanation together within a single account. Let me now state the account more explicitly:

> *The Counterfactual Account of Explanation*: In order to explain phenomenon P, an explanans E must include a set of true modal information about P regarding both how the contextually salient features on which P counterfactually depends account for the occurrence of P and show that the contextually salient features on which P does not counterfactually depend are irrelevant to P.

The next task is to argue for the superiority of this counterfactual account over other prominent accounts of explanation. In order to do so, it will be useful to explicitly lay out some of the crucial differences between my counterfactual account and other prominent accounts of explanation provided by Potochnik (2017), Strevens (2008), and Woodward (2003).

Most important, contrary to these accounts, on my account, the explanatory dependence relation is counterfactual dependence rather than causal dependence. This is in direct response to the various kinds of noncausal explanation discussed in the previous chapter. Departing from the standard view's exclusive reliance on causal relations is crucial for capturing the

A Counterfactual Account of Explanation 109

variety of ways that scientists explain (Rice and Rohwer 2020). Relatedly, this account includes no requirement that the counterfactual dependencies and independencies that constitute the explanation be provided by a model that veridically describes difference-making causes (contra Strevens 2008), accurately describes the causal relations in the system (contra Woodward 2003), or accurately represents the causes of interest to the research program (contra Potochnik 2017). More generally, independent of its ability to accommodate noncausal explanations, my account is also crucially different in that it explicitly does not build in that the models (or theories) appealed to within an explanation must accurately represent the explanatorily relevant features. As we will see in later chapters, the features on which the explanandum counterfactually depends may be drastically distorted by models that are used to extract the set of modal information required to explain. This is crucial for accounting for the various ways that pervasively inaccurate models are able to explain and produce understanding. In addition, contrary to these causal accounts, this account allows the patterns of counterfactual dependence cited in the explanation to range over systems with very different causal patterns or mechanisms. This feature is essential for capturing instances of explaining universal patterns that range over real and possible systems whose relevant causes are different (see Rice manuscript). Finally, this account makes explicit the need to account for contextually salient features that are irrelevant to the occurrence of the explanandum. This is an important departure from the exclusive focus on difference-making by causal accounts.

At this point, a few objections are likely to arise. First, one might object that these requirements are too weak to be able to rule out cases that are clearly nonexplanatory. Pincock (2018) raises this sort of objection, arguing that merely appealing to modal information is "too flexible" because it would allow cases like the barometer case to count as genuine explanations (Pincock 2018, 47). I certainly agree that this single counterfactual dependence fails to provide an explanation. This is precisely why it is crucial to my account that a satisfactory explanation must provide a *set* of counterfactual dependence information that can account for the occurrence of the phenomenon and takes into account the counterfactual irrelevance of various features that are made salient by the context in which the why question is asked.

Because of these further requirements, merely citing the barometer reading falls short of providing a sufficient explanation of the thunderstorm, on numerous fronts. First, this single counterfactual does little to account for the occurrence of the thunderstorm. In order to adequately account for the occurrence of the storm, other features on which the storm counterfactually depends (as does the barometer) need to be taken into account, such as the low pressure in the environment. Second, even if the barometer reading were sufficient to account for the occurrence of the storm, the context of inquiry will almost always entail that we are interested in additional features of the system on which the phenomenon counterfactually depends. In particular, the explanation leaves out many of the features that our background knowledge would deem of interest (e.g., the pressure of the surrounding environment that is measured by the barometer). Third, my counterfactual account also requires that the explanation take into account certain counterfactually irrelevant changes to the system that would not result in changes to the explanandum. Here, I suggest, the barometer explanation will fall short by not taking into account the modal information that changing the barometer reading in various ways (e.g., by breaking the barometer or putting it in a pressure chamber) will not change the occurrence of the storm. Consequently, rather than easily granting explanatory status to the barometer case because it provides some modal information, my counterfactual account shows why the explanation is unsatisfactory—it fails to account for the occurrence of the storm, it fails to account for all the counterfactual dependencies that we are interested in, and it fails to provide essential information about the irrelevance of various changes to the system that are of interest to us.

Pincock also objects to modal accounts on the grounds that any account of explanatory pluralism that recognizes distinct types of explanation "owes us a discussion of what all explanations have in common and what nevertheless divides these explanations with this common feature into distinct types" (Pincock 2018, 42). I contend that counterfactual dependence is the dependence relation that unites almost all explanations and distinguishes them from nonexplanations (see Rice and Rohwer 2020 for some potential exceptions). In addition to this common feature, what distinguishes different types of counterfactual explanation from one another are the types of ontological dependencies that underlie, or are responsible for, these counterfactual dependencies. For example, in causal explanations, the counterfactual dependencies occur because of causal dependencies that hold between the explanatorily relevant features and the explanandum.

A Counterfactual Account of Explanation 111

However, in various other kinds of noncausal explanations, the explanatory counterfactual dependencies are underwritten by other kinds of ontological dependence (e.g., statistical dependence or mathematical dependence). In other words, different kinds of explanation can be distinguished by looking at the "metaphysical origin" of the counterfactual dependencies involved in the explanation (Bokulich 2011; Kim 1994). To show how my account can unify these various kinds of explanation under a single account, I now turn to the task of showing how the account applies to various kinds of causal and noncausal explanations.

4.3 Unifying Causal and Noncausal Explanations

A key advantage of the counterfactual account is that it allows us to unify a wide variety of causal and noncausal explanations found across the sciences. We have already seen how the counterfactual account can subsume various instances of causal explanation: causal explanations succeed because they provide the necessary counterfactual information when the counterfactual dependencies involved in the explanation are supported by, or grounded in, causal dependencies. For example, by showing that properties x, y, and z were difference-making causes of the explanandum, we can answer a range of "What if things had been different?" questions concerning how changes in those features would change the explanandum (Woodward 2003). In a similar way, mechanistic explanations provide a plethora of counterfactual information because "when one possesses explanations of this sort, one knows how to intervene into the mechanism in order to produce regular changes in the phenomenon" (Craver, 2007, 160). Indeed, causal relationships underlie many counterfactual dependence relationships. Therefore, discovering the causes or mechanisms that produced (or constituted) the explanandum will often yield the counterfactual information required to explain. Given that many explanations in science are causal, it is important that the counterfactual account can show why citing causal relationships is often explanatory.

Traditionally, the main alternatives to causal accounts have been unificationist theories of explanation. The central idea behind the unification approach is that explanation is a matter of providing a unified account of a wide range of phenomena (Friedman 1974; Kitcher 1981, 1989). Indeed, philosophers have long recognized the value of explanations that are general and unifying (Garfinkel 1981; Jackson and Pettit 1992; Putnam 1975; Sober 1999; Strevens 2008). I argue that the key to such unifying explanations, when they are successful, is that they are able to provide extensive counterfactual information about when certain patterns of counterfactual dependence will continue to hold *and* where they will fail to hold. For example, many models in evolutionary biology provide unifying explanations by showing that various evolutionary patterns counterfactually depend on certain key features of the population (e.g., variations in fitness) and that other contextually salient features (e.g., the different features of various species) are counterfactually irrelevant to the occurrence of those patterns (Ariew, Rice, and Rohwer 2015; Ariew, Rohwer, and Rice 2017; Kitcher 1981, 1993).[17] However, unifying a range of phenomena like this is only *one* way that counterfactual information can be used to explain.

Next, according to most covering-law accounts, the explanans must contain a set of true laws and initial conditions from which the explanandum can be derived or deduced (Hempel 1965; Hempel and Oppenheim 1948). We can understand these cases as aiming for the discovery of patterns of counterfactual dependence (or generalizations) that are *maximally stable* (Mitchell 2009; Woodward 2003). Indeed, the stability afforded by laws often allows scientists to provide a wide range of counterfactual information that can be used to provide sufficient explanations in many cases. As Woodward notes, "The ideal gas law allows us to answer a certain range of what-if-things-had-been-different questions, and in doing so explains. Statistical mechanics does not explain in virtue of doing something different in kind from this, but instead simply provides information that allows us to answer what, in some respects, is a wider, more detailed range of [What if things had been different?] questions" (Woodward 2003, 223). Indeed, in both instances, these laws (and the models that incorporate them) explain in virtue of providing the required set of information about the counterfactual dependencies and independencies in the target systems. However, while deduction from universal laws can sometimes provide the counterfactual information required to explain, neither deduction nor the invocation of laws is necessary to provide an explanation.

In addition, many authors have argued that there are mathematical explanations that cannot be captured by causal accounts (Lange 2013b; Pincock 2012; Woodward 2003). Pincock provides the example of how mathematical features can be used to explain the impossibility of traversing

A Counterfactual Account of Explanation

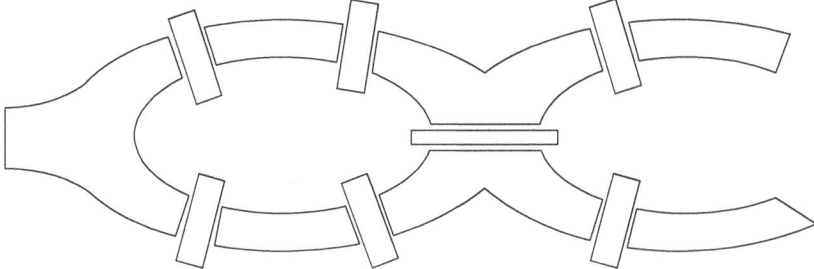

Figure 4.1
The bridges of Königsberg (Pincock 2012, 52).

each of the bridges of Königsberg exactly once, in such a way that the traveler returns to the region they came from (figure 4.1).

We can explain this fact by noting that the bridges form a non-Eulerian graph (figure 4.2). However, as Pincock notes, "the mathematics here is not tracking genuine causal relations, but is only reflecting a certain kind of formal structure whose features in the physical system have some scientific significance" (Pincock 2012, 53). More precisely, we see that if the layout of the structure of the bridges were to form an Eulerian graph, then it would be possible to traverse each of the bridges in this way. This shows that the explanandum counterfactually depends on this mathematical property of the bridges. Moreover, we also get information about the counterfactual irrelevance of other features of the system that might otherwise have been thought to be relevant, such as where one starts, the materials of the bridges, the causes that led to the construction of the bridge, and the city that the bridges are in (Woodward 2018). Consequently, the counterfactual account, unlike an exclusively causal account, can easily accommodate such instances of mathematical explanation.

A possible objection here concerns the ability of an account based on counterfactuals to handle mathematical explanations that involve necessary truths—that is, truths that could not have been otherwise.[18] These cases appear tricky, given that there is no possible counterfactual situation in which a necessary mathematical fact is false, so there is no way to clearly establish counterfactual dependence. There are several possible responses that one might give here. First, one might simply grant that certain kinds of mathematical explanations are outside the scope of a counterfactual account of scientific explanation. Indeed, as I noted at the beginning of the

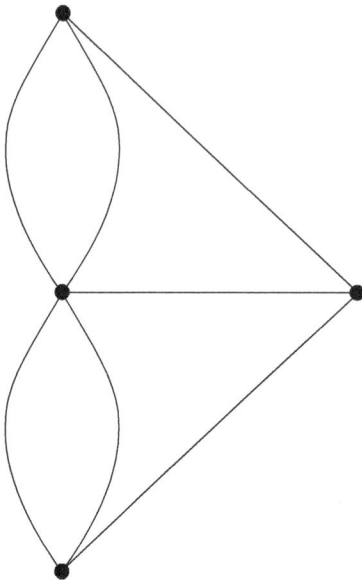

Figure 4.2
A non-Eulerian graph of the structure of the bridges of Königsberg (Pincock 2012, 52).

chapter, my aim isn't to provide a unified account of absolutely all explanations in terms of necessary and sufficient conditions, so there will likely be a few cases outside the scope of the account. However, a downside of choosing this option is that it would entail that there is something importantly different about how science typically explains and how explanations are provided in pure mathematics. Another option would be to follow some philosophers in rejecting many of these mathematical cases as genuine explanations (Lange 2017). While this might rule out some of the problematic cases, it is unlikely that no explanations within science appeal to necessary truths. Consequently, my preferred response here is to note that a counterfactual account can capture many forms of mathematical explanation because we can sometimes think counterfactually about impossible situations involving necessary truths (Baron, Colyvan, and Ripley 2017).

For example, Alan Baker (2005, 229) discusses an explanation of three species of cicadas that synchronize when their adults emerge from the soil to either 13 or 17 years, depending on the geographical area. The explanation of this phenomenon appears to depend on the necessary mathematical fact that 13 and 17 are both prime numbers, which minimizes the intersection of the cicadas' life cycles with those of their predators. While evaluating counterfactual situations in which this necessary truth does not hold is difficult (and of course is practically impossible to actualize), I think we can agree that the counterfactual statement, "If 13 and 17 were not prime, then the cicadas' life cycles would have been different" is true. That is, in many cases, we can entertain impossible antecedents that would involve changing necessarily true mathematical facts. This is enough to illustrate counterfactual dependence on this mathematical fact even if we are unsure of exactly what *would* be true in such a counterfactual situation. More generally, Baron, Colyvan, and Ripley (2017) show how counterfactual reasoning can be extended to many instances of mathematical explanation that involve impossible changes to the system. As a result, I think that many forms of mathematical explanation can be handled by a counterfactual account (but see Colyvan, Cusbert, and McQueen 2018 for some difficult cases involving proofs). Determining precisely which mathematical explanations can be brought under this counterfactual account is one way that the view might be usefully developed in the future.

Next, we can consider cases of equilibrium explanation, such as the optimality explanations discussed in chapter 3 (Potochnik 2007, 2009b; Rice 2012, 2013; Sober 1983). In these cases, the model explains by showing how the equilibrium point counterfactually depends on various noncausal constraints and trade-offs that hold within the biological system. Another key feature of these explanations is that they show why the explanandum's occurrence is counterfactually independent of various features of the systems that might otherwise have been thought to be relevant (e.g., the initial state, trajectory, or particular causal mechanisms of the system). While most accounts focus exclusively on features that make a difference, the counterfactual account also emphasizes the importance of demonstrating that certain features are counterfactually irrelevant to the explanandum. Consequently, the counterfactual account better captures this key aspect of equilibrium explanations.

Another kind of noncausal explanation discussed in chapter 3 appeals to statistical properties of a system (Ariew, Rice, and Rohwer 2015; Ariew, Rohwer, and Rice 2017; Lange 2013a; Matthen and Ariew 2009; Walsh 2007, 2010; Walsh, Lewens, and Ariew 2002). As I argued there, such statistical derivations are sometimes sufficient to explain because a statistical model can show us how a large-scale statistical regularity counterfactually depends on the statistical properties of the population distribution, such as the mean and variance of the population's fitness distribution (Ariew, Rice, and Rohwer 2015; Ariew, Rohwer, and Rice 2017; Matthen and Ariew 2009; Walsh 2007, 2010, 2015). Furthermore, these statistical models typically provide extensive counterfactual information about why the explanandum would have occurred, regardless of various changes in the underlying causal histories of individuals within the population. This counterfactual information about the lack of dependence on these features is an essential part of how these statistical models explain the stability of these large-scale patterns across causally heterogeneous systems. In contrast with prominent causal accounts of explanation, these explanations that focus on statistical dependence and the irrelevance of the systems' causes are easily subsumed under the counterfactual account of explanation.

Finally, as we also saw in chapter 3, in many cases, a minimal model is appealed to in an explanation of why a universal pattern holds across extremely diverse physical systems (Batterman 2002a, 2002b; Batterman and Rice 2014). In order to explain the universality of these patterns across various real, possible, and model systems, we require the inclusion of information about why most of the features of the heterogeneous systems that display the explanandum are irrelevant to their universal behaviors. Although these minimal model explanations are difficult to square with standard accounts of explanation that focus exclusively on accurately representing difference-making features (Batterman and Rice 2014), such explanations that focus on demonstrating irrelevance are easily accommodated by the counterfactual account.

These examples illustrate how the counterfactual account of explanation can unify a wide variety of kinds of explanation found across the sciences, including both causal and noncausal explanations. The account exploits an asymmetry between providing information about causal dependencies and providing information about counterfactual dependence. While causal relationships will almost always entail (or support) counterfactual relationships, not all counterfactual relationships will entail (or require) causal relationships. By focusing on counterfactual dependence and independence directly, we see that in each of the kinds of explanation discussed here, the explanation is provided by showing how the set of features cited in the explanans are counterfactually related to the explanandum.

I conclude that the counterfactual account satisfies the first requirement for an account of scientific explanation: it specifies the relations involved

A Counterfactual Account of Explanation   117

in explanation as relationships of counterfactual dependence and independence between the explanans and the explanandum. What's more, the counterfactual account satisfies this requirement in a way that other prominent accounts (e.g., causal accounts) cannot—namely, by unifying a wide variety of kinds of causal and noncausal explanations. This provides strong motivation for adopting the counterfactual account, but there are additional reasons that provide further support for this view.

4.4 Connecting Explanation and Understanding

As I have noted, a widely accepted feature of scientific explanations is that they produce scientific understanding in agents who grasp them (Achinstien 1983; Friedman 1974; Grimm 2006; Kitcher 1981, 1989; Lewis 1986; Lipton 1991; Potochnik 2017; Salmon 1984, 1997; Strevens 2008, 2013; Woody 2015). This requirement is made all the more important given the recent resurgence of interest in the nature of understanding within the epistemology literature (e.g., de Regt, Leonelli, and Eigner 2009; Elgin 2007, 2017; Grimm 2006, 2012, Khalifa 2017; Kvanvig 2003; Potochnik 2017; Strevens 2013; Zagzebski 2001). Understanding is a cognitive achievement (or success). A scientific model that provides an explanation can allow scientists (and the rest of us) to attain this epistemic achievement. As a result, a satisfactory account of scientific explanation ought to show how the information provided by scientific representations that explain is related to the cognitive achievement of understanding. Unfortunately, as Jaegwon Kim notes, accounts of explanation are "seldom evaluated by explicitly considering whether they give a satisfactory account of understanding" (Kim 1994, 53). In this section I argue that, unlike many accounts of explanation, the counterfactual account can provide—in a very straightforward way—a clear link between explanation and understanding.

In order to see the nature of this relationship, we need to get a bit clearer about what kind of cognitive achievement understanding is.[19] I provide a more detailed account of understanding in chapter 8. For now, though, a few widely accepted features of understanding will enable us to see how counterfactual explanations produce understanding. The first thing to note is that the object or unit of understanding is typically taken to be a "body of information" rather than individual propositions (Elgin 2007; Grimm 2006; Khalifa 2017). For example, Elgin claims that "understanding is primarily a cognitive relation to a fairly comprehensive, coherent body of information" (Elgin 2007, 35).[20] A second important feature is that understanding involves grasping something further, beyond merely justifiably believing a set of true propositions (Elgin 2007, 2017; Grimm 2006, 2012; Kvanvig 2003). As Stephen Grimm puts it, the epistemic value provided by understanding involves the ability to grasp "how the various parts of the world [are] systematically related" (Grimm 2012, 103). Consequently, understanding requires grasping a body of information involving various systematic relations among parts of a system (or among beliefs about the system). Finally, it is widely accepted that objective understanding is factive in some sense (DePaul and Grimm 2007; Grimm 2012; Kvanvig 2003; Mizrahi 2012). To say that understanding is factive does not require that all the beliefs within one's understanding must be true (this point will be discussed in more detail in chapter 8). The important point for my purposes here is that it is widely agreed that "our understanding of natural phenomena seems conspicuously *factive*—what we are trying to understand is how things actually stand in the world" (Grimm 2006, 518).

So we have three main features of scientific understanding that need to be accounted for: (1) it applies to bodies of information, (2) it requires the grasping of various systematic relationships among the parts of a system, and (3) it is factive in the sense that one only understands when one grasps how things really are.

The counterfactual account captures these features of understanding by proposing that the understanding produced by explanations derives from the agent's ability to grasp a set of true counterfactual dependence and independence information about the phenomenon of interest. In other words, the set of counterfactual information that is required to explain produces understanding by providing a systematic body of information about the dependencies and independencies that hold between features of the system and the explanandum. An explanation that provides this body of counterfactual information allows an agent to grasp how the various parts of the world are systematically related. In addition, the objective nature of the counterfactual account requires that the counterfactual information provided by an explanation is factive, in the sense that if the explanation claims that certain features are counterfactually relevant to the explanandum and others are counterfactually irrelevant, then these statements must be true of the phenomenon that we want to explain and understand. This is how the explanation enables agents to grasp how things really are in a way that illustrates why the explanandum occurred. In sum, grasping the set of true counterfactual information provided by an explanation is what enables us to understand why the phenomenon occurred. In contrast, most other accounts of explanation tell us very little about how explanations actually produce understanding in human agents.

The counterfactual account also coheres with a prominent view in the epistemology literature on understanding (Grimm 2006, 2012; Kim 1994). According to this view, understanding consists of grasping the dependency relations within a complex system that has "parts or elements that depend on, and relate to one another, and that the mind grasps or apprehends when it understands" (Grimm 2012, 105).[21] Moreover, Grimm explicates the kind of cognitive achievement needed for an agent to understand in terms of having the ability to answer Woodward's "What if things had been different?" questions because "if you lack the ability to see how things are connected in this way . . . there remains an important sense in which you will not have grasped how they depend on one another" (Grimm 2006, 533). As Grimm puts it, in order to understand, one must "grasp how the different aspects of the system depend upon one another," such that one can "anticipate how changes in one part of the system will lead (or fail to lead) to changes in another part" (Grimm 2011, 89). If this account of understanding is on the right track, then it is fairly straightforward to see how grasping counterfactual explanations yields scientific understanding.[22]

It is important to note, however, that some causal theories of explanation could plausibly satisfy this requirement in a similar way. For example, a defender of a causal theory of explanation might suggest that explanations produce understanding because of their illumination of counterfactual dependence relations, but that this is just the kind of information provided by specifying causal relations in the system (e.g., Potochink 2017 directly links causal patterns with understanding). While this approach can perhaps meet the requirement of explicating the relationship between explanation and understanding, it does so in a much less straightforward way because the modal information that provides understanding is merely a by-product of specifying the causes that are necessary to explain. In contrast, the counterfactual account directly links the kind of information required to explain to the kind of information required to understand.

The second problem is that appealing to causal dependence fails to capture the ways in which noncausal explanations can produce understanding.

Given the myriad kinds of noncausal explanations (see chapter 3), our account of how explanations produce understanding ought to draw on features that both causal and noncausal features have in common. That is, our account ought to show how both causal and noncausal explanations can produce understanding. Therefore, even if some causal accounts are able to connect causal explanation with understanding why the phenomenon occurred, when we consider the three requirements for an account of explanation *collectively*, there are still strong reasons to favor the counterfactual account because it shows how both causal and noncausal explanations contribute to science's understanding of natural phenomena.

4.5   Accounting for the Positive and Ineliminable Contributions of Idealizations

Finally, a satisfactory account of explanation must show how many idealized models are able to provide explanations even when it is impossible to remove the idealizations from the model (Batterman 2002b, 2009; Bokulich 2012; Rice 2012, 2013; Wayne 2011). Furthermore, the account must provide a satisfactory analysis of the positive contributions made by these ineliminable idealizations to many model explanations. More precisely, the account ought to show why many idealizations cannot (or at least should not) be replaced by abstractions or more realistic assumptions without eliminating the explanation provided via the model. Given that models are essentially idealized and widely used to explain in science, this is one of the most important requirements for an account of scientific explanation (Bokulich 2011; Elgin 2007, 2017).

As we saw in chapter 2, most extant accounts claim that idealized models can still explain when they accurately represent a set of relevant causes and use idealizations to indicate those causal factors that are known to be irrelevant, negligible, or not of interest (Craver 2006; Elgin and Sober 2002; McMullin 1985; Potochnik 2007, 2017; Strevens 2004, 2008; Weisberg 2007a, 2013). This kind of account has been attractive to many philosophers because we need a story about how idealized models can explain and, in some cases, scientists use idealizations simply to indicate what is irrelevant to the explanandum. However, while these accounts allow certain idealized models to explain, their general goal seems to be to show that the idealizations are harmless (Elgin and Sober 2002) because they do not get in the way of the accurate representation of the salient causal difference-makers. Therefore, in some sense, the idealizations themselves do not play any positive role in the explanation provided by the accurate parts of the model—that is, their contribution is strictly passive, in that their role is to focus our attention on the accurate representation of relevant causes that does the real explanatory work.

I think this kind of view fails to adequately capture many of *positive* and *ineliminable* contributions that idealizations make within model explanations. Indeed, I contend that this approach fails to show why we need to use idealizations within model explanations at all, given that simply abstracting away (i.e., leaving out) those features would presumably convey the same information about irrelevance (or lack of interest).[23] As a result, these accounts fail to show what is distinctive about the contributions made by idealizations over and above the contributions made by merely abstracting away irrelevant features. To see the problem in a different way, if the only role of idealizations in scientific explanations were to indicate which causal factors don't make a difference (or are not of interest), then a similar model that simply abstracted away those causal factors (or just explicitly assumed that they were irrelevant) would presumably provide the same information about their irrelevance (or nonsalience), and thus should be capable of providing the same explanation. However, I think it is extremely unlikely that it would be possible to remove or replace the myriad idealizations present in our best scientific models without thereby destroying the explanations they provide—and likely dismantling the model as well. This suggests that the standard approach is missing something really important about the positive contributions idealizations make to scientific explanations.

What we really need, then, is an account of how idealizations make positive contributions within explanations beyond the contributions of abstractions. In addition, our account ought to show why idealizations are often ineliminable, in the sense that removing (or replacing) the idealizations would eliminate (or substantially weaken) the ability of the model to explain the target phenomenon. In what follows, I argue that many ineliminable idealizations are introduced as a means of discovering which features are counterfactually relevant and irrelevant to the explanandum. In other words, I suggest that the standard view misses the positive contributions of ineliminable idealizations that are introduced before we know which features of the target systems are relevant and which are irrelevant. As Elgin puts it, idealizations are often "felicitous falsehoods" because they "afford epistemic access to matters of fact that are otherwise difficult or impossible to discern" (Elgin 2007, 39). More specifically, I contend that idealizations often play ineliminable roles in the explanations provided by scientific models by allowing for the application of mathematical techniques that reveal the counterfactual information required to explain.

To see how ineliminable idealizations can contribute to explanations in this way, it will be useful to revisit the example of minimal model explanations. As we saw in chapter 3, an essential feature of applying the renormalization group in the LGA model case is the invocation of the thermodynamic limit. Without this idealization, we would not be able to use the renormalization strategy to discover the modal information required to provide an explanation of the universality of these patterns of fluid flow. As a result, the thermodynamic limit plays an ineliminable role in how physicists are able to explain and understand the universal behaviors of real fluids. In this case, however, rather than distorting factors that are already known to be irrelevant, the thermodynamic limit plays an ineliminable role by allowing for the application of a mathematical technique that demonstrates (or reveals) the counterfactual irrelevance of a wide range of features.

In addition, idealizations can sometimes be essential for getting certain variables related in the right way within a mathematical model or allowing for the use of a mathematical framework. For example, Pincock discusses a model of deep-water wave dispersion in which the assumption that the ocean is infinitely deep is essential to the mathematics of the model because it "allows us to represent the relationship between the phase velocity of a wave and its wavelength" (Pincock 2012, 96). Other examples include the use of idealizations in physics and biology to allow for the application of statistics, or the use of idealizations in game-theoretic modeling to allow for calculations of equilibrium (Morrison 1996, 2009; Rice 2013). Indeed, as Morrison correctly points out, when considering the use of limits in physics and biology, "we do not introduce [these assumptions] simply as a way of ignoring what is irrelevant to the problem or as a method for calculational expediency. Instead, the mathematical description functions as *a necessary condition* for explaining and hence understanding the phenomenon in question" (Morrison 2009, 130; emphasis added). More generally, in many cases, idealizations are essential for the application of the mathematical frameworks used to represent and relate the features that the explanandum (counterfactually) depends on.

A Counterfactual Account of Explanation 123

Without these idealizations, the models and mathematical techniques that reveal these relationships of counterfactual dependence and independence would be inapplicable. That is, the idealizations in these cases are *constitutive* and *deeply intertwined* within the explanations provided by these models—they are not peripheral bystanders that merely help to isolate the accurate representation of relevant causes. Therefore, rather than simply emphasizing the irrelevance of the distorted features, many idealizations are explanatorily essential mathematical operations that are constitutive of the processes that enable scientists to demonstrate (or discover) which features are counterfactually relevant and irrelevant to the explanandum.

In sum, there are at least three ways that idealizations contribute to counterfactual explanations:

1. Emphasizing the counterfactual irrelevance of certain factors.
2. Being constitutive of a mathematical framework that allows for the representation, or extraction of, the explanatory counterfactual dependencies.
3. Being constitutively involved in (mathematical) modeling techniques that demonstrate the counterfactual irrelevance of certain features to the explanandum.

There are certainly several other ways that idealizations make ineliminable contributions to explanations, but these will have to be investigated a bit later on in the book. For my purposes here, it is sufficient to show that the counterfactual account captures several cases from science that are missed (or misunderstood) by other accounts of how idealized models can explain. While prominent causal accounts (e.g., Strevens 2008; Weisberg 2013) have accounted for (1), they miss the positive and ineliminable contributions of idealizations of (2) and (3). Moreover, by capturing (2) and (3), the counterfactual account shows how idealizations are able to make positive contributions to scientific explanations beyond the contributions of abstractions. The idealizations in these cases are essential to the core mathematical techniques or frameworks used within the explanation. Replacing these idealizations with abstractions or more realistic assumptions would destroy our ability to use those mathematical techniques and thereby destroy the model's ability to reveal the counterfactual information required to explain. It is for precisely this reason that these idealizations, and their positive contributions, are ineliminable.

Yet in order to see these positive contributions, we need to look beyond the stringent accuracy, isomorphic, or mapping requirements imposed by most accounts of how scientific models explain. Put differently, in order to capture these kinds of contributions, we need to separate our account of explanation from the kinds of accuracy conditions imposed by the standard approach. By allowing for the distortion of difference-making, relevant, or contextually salient causal factors (or mechanisms) and focusing on the counterfactual information revealed by various modeling techniques used in scientific practice, the counterfactual account allows us to better capture the positive and ineliminable contributions that idealizations make within the explanations provided by scientific models.

I conclude that the arguments given here provide strong reasons for adopting the counterfactual account of explanation. The account identifies the kind of dependence (and independence) information provided by explanations that unifies a wide variety of causal and noncausal explanations. It also provides a direct link between the information provided by explanations and the cognitive achievement of understanding. Moreover, it shows how idealizations make positive and ineliminable contributions to explanations beyond the contributions of abstractions. Because the counterfactual account satisfies this set of requirements better than other prominent accounts of explanation, we should adopt and continue to develop the counterfactual account of explanation.

4.6 Comparison with Other Counterfactual Accounts of Explanation

In this final section of this chapter I briefly highlight two of the key features that distinguish my account from some of the other counterfactual proposals in the literature (Bokulich 2008, 2011, 2012; Reutlinger 2016, 2018). Both of these features will play a crucial role throughout the remaining chapters.

4.6.1 The Relevance of Irrelevance

Like my account, Reutlinger (2016, 2018) argues that focusing on counterfactual information provides the most promising way of unifying causal and noncausal explanations. Reutlinger proposes the following criteria for providing a satisfactory explanation, where E is the target explanandum and the explanans consists of generalizations $G_1 \ldots G_m$ and auxiliary statements $S_1 \ldots S_n$ (Reutlinger 2016, 6):

A Counterfactual Account of Explanation

1. *Veridicality condition:* $G_1, \ldots, G_m, S_1, \ldots, S_n$, and E are (approximately) true or, at least, well confirmed.
2. *Implication condition:* $G_1, \ldots, G_m$ and $S_1, \ldots, S_n$ logically entail E or a conditional probability $P(E|S_1, \ldots, S_n)$, where the conditional probability need not be "high" in contrast to Hempel's covering-law account.
3. *Dependency condition:* $G_1, \ldots, G_m$ support at least one counterfactual of the form: had $S_1, \ldots, S_n$ been different than they actually are (in at least one way deemed possible in the light of the generalizations), then E or the conditional probability of E would have been different as well.

Note that all that condition 3 requires is that the explanation support at least one counterfactual statement. While I agree that the counterfactual approach is the most promising strategy for unifying causal and noncausal explanations, if we purchase that unification at the cost of such minimal requirements for being an explanation, then not only will we end up with an account that is too permissive, but also our account of explanation will fail to tell us much about how counterfactual information is able to provide a satisfactory explanation. I will illustrate these objections by looking at Reutlinger's discussion of how renormalization is used to explain universality.

There are two main problems with how Reutlinger's account handles the renormalization case. First, as I said above, finding only one feature of the system that the phenomenon counterfactually depends on is rarely sufficient for explanation—a larger set of counterfactual information typically is required to explain. A useful example is offered by van Fraassen (1980). Suppose that Laura's alarm clock goes off and wakes her up. While it is certainly true that "if Laura had not gone to sleep, she would not have woken up," citing the fact that Laura fell asleep is insufficient to explain why she woke up. In general, citing only one counterfactual dependence between a single feature (or auxiliary assumption) and the explanandum will almost always be insufficient to provide an explanation.[24]

In the renormalization case, Reutlinger argues that his counterfactual dependency condition is satisfied because "there is a physically possible Hamiltonian H* such that if a physical system S had the original Hamiltonian H* (instead of its actual original Hamiltonian H), then S with original H* would be in a different universality class than a system with original Hamiltonian H" (Reutlinger 2014, 12; see also Reutlinger 2018). While this counterfactual is certainly true, it does almost nothing in terms of explaining why all the systems in the universality class display the same macroscale behaviors. Simply noting that there is a physically possible Hamiltonian that would not be in the universality class gives us no information about why the Hamiltonians that are in the universality class are in that class, nor does it tell us why other possible Hamiltonians fall outside that universality class. The explanation of this universal pattern is provided only by digging into the details of the renormalization group to identify precisely the set of features (i.e., degrees of freedom) that are relevant and irrelevant to the occurrence of the universal macroscale behaviors. In sum, merely noting that there is a possible Hamiltonian that would fall outside the universality class fails to provide a complete explanation of why the diverse systems within the class display the universal pattern. This is why my account requires an explanation to provide a set of counterfactual dependence information that is sufficient to account for why the explanandum occurs.

What's more, simply identifying features on which these behaviors counterfactually depend fails to explain why the pattern is universal across those systems despite rather drastic differences in their physical features (Batterman 2002b; Batterman and Rice 2014).[25] This highlights the second issue with how Reutlinger's account handles renormalization group explanations: it completely misses the counterfactual information about irrelevance that is most important in explanations of universality. Moreover, Reutlinger argues for what he calls a "minimality condition," which serves to "guard against including irrelevant factors into the explanans" (Reutlinger 2018, 79). However, the explanation of universality given by applying the renormalization group is provided largely by demonstrating that most features of the systems within the universality class are irrelevant to their universal behaviors (Batterman 2002b; Batterman and Rice 2014). By ignoring the role of counterfactual irrelevance, Reutlinger's account completely misses the most essential explanatory information provided by the renormalization group and other explanations of universality.

In contrast, my counterfactual account maintains an essential role for counterfactual information about changes to irrelevant features. As I noted above, this is a rarity among accounts of explanation because most accounts focus exclusively on identifying relevant (e.g., difference-making) features. For example, Craver claims that when models explain by describing a mechanism, "the components must be *relevant* to the phenomenon to be explained" (Craver

2006, 371). Woodward also claims: "Good explanations should both *include* information about all the factors which are such that changes in them are associated with some change in the explanandum . . . and *not include* factors such that no changes in them are associated with changes in the *explanandum*" (Woodward 2010, 291). Like these causal accounts, several noncausal accounts have also focused exclusively on identifying "change-relating" relationships of counterfactual dependence (Bokulich 2011; Reutlinger 2016; Saatsi 2019; Saatsi and Pexton 2013). In contrast, I have argued that many explanations of stable patterns must explicitly include information about features that are counterfactually irrelevant to the occurrence of the explanandum.

One possible response here is to appeal to Woodward's use of the term *invariance* (Woodward 2003, 250). For Woodward, a relationship is causal (and explanatory) if it is stable across some interventions on other features that are irrelevant to the explanandum. However, on Woodward's account, explicit information about the irrelevance of these features has no clear place in the actual explanation; rather, it only plays a role in establishing the explanatoriness of the causal dependencies that do the real explanatory work (Gross 2015, 181). As evidence of this, Woodward suggests in many places that "an explanation of an outcome must cite factors on which that outcome depends. . . . The explanation should not cite factors that are irrelevant to the outcome" (Woodward 2003, 372). The problem is that in many cases—such as in the renormalization group explanations of universality—information about the features that are relevant is unable to provide the desired explanation on its own. In general, this occurs when we would like to explain why a given pattern (or phenomenon) is stable despite various changes to features of the systems.

Another seeming exception to ignoring the role of irrelevance information is Strevens's idea that idealizations that indicate which causal factors are irrelevant (or are not difference-makers) can make idealized models *better* explanations than their nonidealized counterparts (Strevens 2008). While I will have more to say in chapter 5 about why I think this account of idealizations is inadequate, for now the important point is that while Strevens's account makes it possible for information about irrelevance to improve the quality of an explanation, irrelevance information is not required to provide a complete explanation. Instead, according to Strevens, a causal model that included only difference-making causes would still provide a satisfactory explanation of the phenomenon. Moreover, the main benefit that Strevens and others have focused on for showing that certain features are irrelevant is that it allows us to better focus on the relevant features. Indeed, Strevens tells us that something "cannot be explanatorily relevant unless it is a difference-maker" (Strevens 2008, 146). This again sidelines the importance of information about irrelevant features as merely an optional means to the end of accurately representing the causally relevant features that do the real explanatory work.

A key feature of renormalization group (and minimal model) explanations is that they demonstrate that most of the features of the systems are counterfactually irrelevant to their universal behaviors. Similarly, in the case of equilibrium explanations, understanding that the initial conditions and trajectory of the system are counterfactually irrelevant to the occurrence of the explanandum is a crucial part of the explanation. In addition, statistical explanations often show that the particular causes and interactions within the system are irrelevant to the observed statistical results. Therefore, while explanations certainly need to identify the relevant set of features on which the explanandum counterfactually depends, in many explanatory contexts, information about why many features are counterfactually irrelevant to the explanandum is absolutely crucial.

4.6.2 Extracting Modal Information from Idealized Models

Like my account of explanation, Bokulich (2008, 2011, 2012) develops an account of explanation that draws on Woodward's use of counterfactual dependence, but that removes the requirement that explanations cite causal dependencies. Bokulich's examples primarily involve appeals to fictional entities. For example, classical orbits are often appealed to in explanations of ionization patterns (Bokulich 2008). Because these orbits do not exist, they cannot be causes of the phenomenon of interest. However, because they can enter into counterfactual dependence relations with the explanandum, they can be cited in explanations of the phenomenon. Consequently, like my account, on Bokulich's account, a model explains "by showing how there is a pattern of counterfactual dependence of the relevant features of the target system on the structures represented in the model." (Bokulich 2008, 146).

While I agree with much of Bokulich's view, there are some crucial differences worth noting. One issue is that, like Reutlinger, Bokulich presents a counterfactual account that ignores the crucial role of information about irrelevance by focusing only on counterfactual dependence. Another issue for

A Counterfactual Account of Explanation 129

Bokulich's view is that it restricts the counterfactual information provided in an explanation to counterfactual relationships represented *within the model itself*. This not only greatly limits the source of counterfactual information that might be used in the explanation, but it would likely rule out many of the most interesting instances of using idealized modeling techniques to extract modal information. As we have seen, in many cases, the modal information involved in providing a complete explanation is discovered only if we expand beyond the relationships represented within the model itself and include the counterfactual information that can be extracted by scientists' use of the model, along with various mathematical modeling techniques.

Consequently, an important feature of my counterfactual account that is highlighted by these contrasts with Bokulich's account is the distinction between the counterfactual relationships represented in a particular idealized model and the counterfactual dependencies and independencies that constitute the explanation offered by appealing to the model (or set of models). In other words, I argue that scientific models ought to be distinguished from the explanations extracted by using those models in various ways. The models are certainly essential parts of the explanation, but the counterfactual relationships appealed to in the explanation are not limited to those that are explicitly represented within a particular idealized model.

This is not to suggest that *none* of the modal information involved in the explanation will be represented within the model itself.[26] To see why, I should note that I take the so-called model system to roughly include the explicit and implicit assumptions of the model, the equations of the mathematical model, and the deductive consequences of those assumptions and equations. For example, Parker's dung fly model, discussed in chapter 3, includes the explicit assumptions of the model (e.g., that more copulation time will increase the amount of eggs fertilized); the implicit assumptions of the model (e.g., that the population is effectively infinite in size); the equations of the model (e.g., the function relating copulation time to fertilization); and the deductive consequences of those features of the model (e.g., that the modeled flies will optimize their fertilization rate by copulating for 41 minutes). While these features of the model are most clearly specified in instances of mathematical modeling, I think that similar ideas can be applied to nonmathematical models as well.

Now, several modal dependencies certainly will be represented within such a model system, and some of those dependence (or independence) relations may feature prominently in the explanation provided by appealing to the model. The point that I want to stress is that the modal structure of the model itself does not exhaust the amount of modal information that can be inferred from the model. In particular, the modeler's background assumptions, interpretation of the results, or additional mathematical modeling techniques (e.g., the renormalization group) may be required to discover the modal information required to provide the explanation. Furthermore, in many instances, "explanatory accounts are so complex that they do not consist in and *cannot be captured by a single representation*" (Brigandt 2013, 490). Therefore, it isn't just that individual models need to be distinguished from the explanatory information that can be extracted from using those models in various ways; very often multiple idealized scientific representations will be appealed to in the course of providing a complete explanation of a phenomenon (Green 2013; Hochstein 2017).

In sum, the main issue with Bokulich's account is that it aligns a bit too closely with the standard approach's focus on accurate representation of relevant features. In earlier work, her account required that "in order for a model M to explain a given phenomenon P, we require that the counterfactual structure of M be *isomorphic* in the relevant respects to the counterfactual structure of P" (Bokulich 2011, 39; emphasis added). In contrast, on my account, models can be used to provide the relevant counterfactual information without meeting this further condition of isomorphism. I think it is clear from the examples presented in this book that strict isomorphism between the relationships in the model and the relationships in the target system is much too strong. More generally, I agree with Weisberg that isomorphism is simply too rigid of a relation to account for the way that models relate to their target systems (Weisberg 2013, 137–142).

In line with this suggestion, in more recent presentations of her view, Bokulich has suggested that "a veridical representation of the causes is not what is most important. Instead, the relevant consideration is whether the representation licenses correct inferences and provides scientists with true modal information" (Bokulich 2016, 276). In order to license these explanatory inferences, she argues that a model must satisfy a "justificatory step" that involves "specifying what the domain of applicability of the model is" in order to show that the model is "able to adequately capture the relevant features of the world" (Bokulich 2011, 39).

A Counterfactual Account of Explanation

I certainly agree that such a justification for explanatory inferences from idealized models is required, and this need not be provided by an appeal to the accurate representation of causes. However, I disagree that the only (or best) way to provide that justification is by appealing to some other kind of accurate representation relations between the model and the relevant features of its target systems. As I will argue in the following chapter, scientists are often not in a position to know which parts of their scientific models accurately represent the features of their target systems. Moreover, in many cases, even the relevant features have to be distorted in order to extract the modal information required to explain. As a result, I think an account of the justification for these explanatory inferences needs to be provided that does not closely parallel the appeal to accurate representation of relevant features advocated by the standard approach. On this point, Bokulich seems to agree: "Certainly having an accurate representation is one way to get such modal information, but the success of idealized and fictional models in science teaches us that it is not the only way" (Bokulich 2016, 271). What is needed, then, is a more detailed account of what some of the other ways of justifying the use of idealized models to reveal explanatory information are. With this account of explanation in hand, I take up this challenge in the next two chapters.

5 Models Don't Decompose That Way: The Holistic Distortion View of Idealized Models

Having moved beyond the requirement that all explanations must describe causes, it is now time to more directly address the standard approach's view of the role of idealizations in science. As we saw in chapter 2, most accounts of scientific modeling—across a range of debates—adopt some version of the following three claims:

1. *Target decomposition assumption*: The real-world system is decomposable, such that the contributions of the features that are relevant to the occurrence of the target phenomenon (e.g., the difference-makers) can be isolated from the contributions of features that are irrelevant (or are largely insignificant) to the target phenomenon.
2. *Model decomposition assumption*: The scientific model is decomposable, such that the contributions of its accurate parts can be isolated from the contributions of its inaccurate (i.e., idealized or abstracted) parts.
3. *Mapping assumption*: When successful, the accurate parts of the model can be mapped onto the relevant, important, or contextually salient parts of the real-world system and the inaccurate parts of the model only distort the irrelevant, negligible, or uninteresting parts of the real-world system.

This decompositional strategy allows one to accommodate idealizations by suggesting that the accurate parts of scientific representations are what do the real epistemic work, while the inaccurate parts of those representations are justified by distorting only what is irrelevant, negligible, or not of interest. As we saw in chapter 2, this decompositional strategy is central to several of the most influential accounts of how to model complex systems, how models explain, how idealizations contribute to model explanations, and how idealized modeling is compatible with scientific realism (Rice 2017).

In this chapter I argue that many of our best scientific models cannot be decomposed in the ways required by the decompositional strategy. To be clear, I am not arguing that decomposition is a completely wrongheaded strategy—such decomposition is epistemically convenient when it obtains. However, I will contend that the decompositional strategy cannot be the whole story because it requires three assumptions that are not typically met by our best scientific models. As a result, decompositional accounts of modeling, explanation, idealization, robustness, and realism need to be supplanted by an alternative approach to idealization and modeling in science.

Therefore, after arguing against the decompositional strategy, I will argue for what I call the *holistic distortion view* of how idealizations contribute to scientific models. According to this view, idealized models in science ought to be characterized as holistically (i.e., pervasively) distorted representations of their target systems, including both relevant and irrelevant features. In other words, idealized models in science are typically greater than the sum of their accurate and inaccurate parts because they result from the complex interaction of various modeling assumptions and idealizations that produce a pervasive misrepresentation of most of the features of their target systems. The explanatory use of these holistically distorted representations is justified, I will argue, because they allow for the application of mathematical modeling techniques that provide epistemic access to the kinds of information that scientists are interested in. Because this holistic distortion view moves us away from focusing on accurate representation relations, I will argue in chapter 6 that a promising alternative is to appeal to *universality classes* in order to link idealized models with their target systems in ways that can produce explanations and understanding (Batterman and Rice 2014; Rice 2017, 2018). Discovering instances of universality can enable scientific modelers to justifiably extract relationships of counterfactual dependence and independence between certain key features and an explanandum, even if those relationships hold for drastically different reasons in the idealized model system than they do in real-world systems. Therefore, even if the model pervasively distorts its target systems—including the explanatorily relevant features—it is still possible to use the model to explain and understand because many of the counterfactual relationships displayed by (or extracted from) the model system will be similar to those of the model's target systems.

In the following section I argue that many of our best scientific models cannot be decomposed in the ways required by the decompositional strategy. In response, in section 5.2 I present my holistic distortion view of idealized models. Section 5.3 concludes the chapter.

5.1 Against the Decompositional Strategy

In order for the decompositional strategy to succeed, the three conditions given above must be met. However, in this section I provide examples of how the model decomposition assumption and the mapping assumption can—and often do—fail to hold. Moreover, I argue that these cases are representative of much larger classes of scientific models that will systematically fail to meet the requirements for applying the decompositional strategy. While I think that concerns can also be raised regarding the target decomposition assumption, I think philosophical accounts of modeling and idealization ought to be kept independent of metaphysical commitments about the nature of real-world systems whenever possible.[1] Consequently, I will simply note that there is no clear argument for adopting such a strong metaphysical assumption regarding the decomposability of real-world systems and instead focus my critique on the model decomposition assumption and the mapping assumption.

5.1.1 Many Scientific Models Don't Decompose That Way

The first problem with the decompositional strategy is that many of our best scientific models cannot be decomposed into the isolable contributions made by their accurate and inaccurate parts. According to the decompositional strategy, we should be able to isolate the contributions of the accurate parts of our models from the contributions made by their idealized parts. This implies that we should, at least in principle, be able to remove or replace the idealizations within our scientific models while leaving the contributions of the isolated accurate components intact. In other words, if scientific models are truly decompositional in this way, then the idealizations within our best scientific models should be eliminable in the sense that they could in principle be removed (or replaced) without affecting the parts of the model that accurately describe the relevant parts (or features) of the model's target systems. However, in this section I argue that for a wide range of scientific models, idealizations cannot be quarantined in this way and their distortions are pervasive because they are essential to the foundational mathematical frameworks of those models. Although a model's having ineliminable idealizations and being unable to decompose the contributions of its accurate and inaccurate parts are conceptually distinct, we can test the model decomposition assumption by considering what occurs to the contributions of the (purportedly) accurate parts of the model when the idealizations are eliminated. In many cases, the idealizations are essential for the mathematical frameworks used in the model, which results in the pervasive distortion of most of the systems' features, which in turn leads to the representation of the relevant (or important) parts of the system being distorted through the lens of the idealized mathematical framework. Consequently, the parts of the model that are supposed to be accurate representations of relevant features can make their contributions only within the context of the idealized mathematical modeling framework that pervasively distorts them (and many other features). When this is the case, the contributions of the idealized parts of the model cannot be isolated from the contributions made by the accurate parts of the model, but are instead intertwined and embedded within the pervasively distorted representation provided by the scientific model.

As a first example, we can consider the ideal gas law, which states that $PV = nRT$, where P is pressure, V is volume, T is temperature, n is the number of moles of gas, and R is the constant. This highly idealized equilibrium model is derived from simpler laws of gases (e.g., Boyle's law and Charles's law) and can be used to calculate macroscopic changes of measurable variables in real gases. The ideal gas law is at the heart of the kinetic theory of gases, which utilizes the Maxwell-Boltzmann distribution for the velocities of molecules. This distribution is derived by imposing a particular probability distribution on the micostates of the system and then averaging over those microstates to discern macroscale properties of gases. More specifically, the Maxwell-Boltzmann distribution requires that one assume that the molecules are in constant random motion, do not interact, and have velocities that are statistically independent of one another. This enables one to model the speed of the molecules using a Gaussian distribution (i.e., a bell curve) by applying the central limit theorem. Importantly, however, the Maxwell-Boltzmann distribution applies only to ideal gases. In real gases, there are several additional factors (e.g., van der Waals interactions, vertical flows, relativistic speed limits, and quantum exchanges) that make the speeds of the gas particles often very different from those specified by the Maxwell-Boltzmann distribution.

In fact, the exact calculations provided by the ideal gas law require a long list of idealizing assumptions, including the following (Moore 2003, 22):

1. The gas consists of a large number of identical molecules in constant random motion.
2. The volume occupied by the gas molecules is infinitesimally small compared to the volume of the container (i.e., the molecules do not take up any space).
3. The velocity (components) of each of the particles are statistically independent of one another.
4. The molecules exert no long-range forces on one another, and there are no intermolecular forces between the molecules.
5. Collisions between the molecules and the walls of the container are perfectly elastic (or are simply assumed not to occur).
6. The gas obeys the processes of classical Newtonian mechanics.

Of course, each of these assumptions will fail to hold in many real-world gases, and no real-world gas will satisfy all six of them. In fact, the above equation "strictly applies only in the zero-density limit" (Moore 2003, 25). Despite these distortions, the ideal gas law can be—and routinely is—used to explain and understand various behaviors of real gases (Woody 2015). Indeed, many real gases behave close to the ways predicted by the ideal gas law, and it has been used to explain fundamental features of gas behavior, such as diffusion and pressure.

We might, then, ask the following question: What would the ideal gas law look like without these idealizations? Put slightly differently, which parts of the ideal gas law are the dissociable accurate parts that are describing the relevant (e.g., difference-making) parts of reality? I contend that we cannot really answer either of these questions because the idealizing assumptions made by the model are introduced to apply foundational mathematical techniques that result in a *pervasively distorted* representation of actual gases. Put in a different way, the contributions made by the purportedly accurate parts of the model only make their contributions within an idealized mathematical modeling framework that pervasively distorts those features. As a result, the contributions of the idealized and (purportedly) accurate parts of the model are intimately intertwined within a pervasive misrepresentation of the model's target systems.

The issue here is that—contra Strevens (2008) and Weisberg (2007a)—the idealizing assumptions within the ideal gas law are not simply noting that certain features of the gas are known to be irrelevant. Instead, they pervasively distort the components and interactions of real gases in order to allow for the application of mathematical tools (e.g., statistical modeling techniques) that enable scientists to extract explanatory information about the system. Without these idealizations, the mathematical foundations of the model would not be applicable. It is precisely because these idealizations are essential to the foundational mathematical techniques involved that they are ineliminable from the explanations provided by the model and their distortions are so pervasive. The important point is that there is not an isolable part of the ideal gas model that accurately represents some dissociable set of difference-makers in real gases and is unaffected by the list of idealizing assumptions given above. Instead, an ideal gas is a fundamentally different kind of system—an idealized model system that allows physicists to investigate, explain, and understand real-world gases but does not accurately represent some dissociable set of difference-makers of real gases. The fact that these idealizations are essential to the mathematical frameworks employed by the model demonstrates their pervasive effect on the overall description provided by the model. It also shows that it will be impossible to isolate the contributions made by some accurate parts of the model from the contributions made by these inaccurate (i.e., idealized) parts. The distortions introduced are simply too pervasive because they are constitutive of the foundational mathematical techniques required for the model to provide the explanation.

Parallel applications of mathematical modeling can be found in biology (Ariew, Rice, and Rohwer 2015; Rice 2012, 2013, 2016; Walsh, Lewens, and Ariew 2002). As we saw in chapter 3, this is largely because the foundational assumptions made by R. A. Fisher (and others) in developing population genetics were inspired by the assumptions underlying the kinetic theory of gases (Morrison 1996, 2004, 2015). Indeed, Fisher himself tells us that "the whole investigation may be compared to the analytical treatment of the Theory of Gases" (Fisher 1922, 321). Like Maxwell and Boltzmann, Fisher's general approach was to make various idealizing assumptions about the nature of the individual components of the system and their interactions in order to develop general statistical models of the large-scale behaviors of populations (Ariew, Rohwer, and Rice 2017; Morrison 2004). Assuming that the individual-level events of the population are random and statistically independent allowed Fisher to apply the central limit theorem, which tells us that such a sample will conform to the normal distribution (i.e., a Gaussian bell curve). Then, by assuming that the population is infinitely large, one can apply various laws of large numbers to eliminate sampling error (in this case, genetic drift).[2] Finally, Fisher averaged over the individual-level events in order to identify statistical features of the overall distribution of genotypic trait types and their fitnesses. In combination, these idealizing assumptions allowed Fisher to assume that biological populations approximated a normal distribution of trait types whose mean value was the expected outcome (presumably due to natural selection). In other words, Fisher's statistical assumptions allowed him to model evolving populations in such a way that no knowledge of the parts or their interactions is required. As Morrison puts it, Fisher saw that "the kinetic theory had shown that knowledge of particular individuals was not required in order to formulate general laws governing the behavior of a population" (Morrison 1996, S319).

In addition, with Fisher's statistical modeling, we again see that when certain idealizing assumptions are made about the components and interactions of the system, scientists can use various mathematical techniques to construct models with which to investigate, explain, and understand real-world systems. In order to do so, however, often requires that they move to a drastically distorted representation of those systems in order to apply those modeling techniques. The success of Fisher's approach was due to his replacing actual populations with highly idealized model populations that relied on the statistical assumptions that he adopted from gas theory. As a result of his idealizing assumptions, Fisher constructed an infinite model population in which evolution did not involve migration, genetic recombination, genetic interaction, or drift. This resulted in a pervasively distorted representation of the evolutionary processes in any real-world population. Indeed, Fisher's work on population genetics was able to provide various explanations "only by invoking a very *unrealistic and abstract* model of a population" (Morrison 2015, 24). These kinds of statistical modeling techniques would later become the foundation of modern population genetics (Ariew, Rohwer, and Rice 2017). The most important parallel, for my purposes, is that once again, the contributions made by the idealizing assumptions cannot be isolated from the contributions made by the accurate parts of these models because the idealizing assumptions are necessary for applying the foundational mathematical frameworks used by these scientific modelers.

While some idealized models can perhaps be decomposed into the contributions made by their accurate and inaccurate parts, I argue that this decomposition is impossible for many of our best scientific models because the idealizations are essential to the foundational mathematical frameworks used within those models. As a result, without the idealizations, the mathematical techniques, derivations, or inferences would not be applicable, and so the explanation and understanding provided by the model would be inaccessible.[3] In addition, the essential role played by these idealizations in the mathematical foundations of many scientific models shows that the contributions of the inaccurate parts cannot be isolated in the way required by the decompositional strategy. Instead, the purportedly accurate parts can make their contributions only within an idealized modeling framework that drastically distorts the model's target systems—including the explanatorily relevant features. Therefore, contrary to the decompositional strategy, in many cases idealizations are not innocent bystanders that can be justified by their distorting only irrelevant or insignificant features; instead, they are deeply invested collaborators that allow the application of various mathematical modeling techniques.

5.1.2 Many Models Distort Difference-Making Features

The second challenge to the decompositional strategy is that even if we assume that the real-world system and the idealized model are decomposable in the required ways, the model's idealizations often directly distort relevant (e.g., difference-making) features of the model's target systems. As a result, we cannot map the accurate parts of the model onto what is relevant and its inaccurate parts onto what is irrelevant.

For example, the Hardy-Weinberg equilibrium model is used to explain and understand various features of heredity and variation. The model establishes a mathematical relation between genotypic frequencies that captures the genetic structure described by Mendelism. The model tells us that if we have a pair of alleles, A_1 and A_2, at a particular locus, and in the initial population, the ratio of A_1 to A_2 is p to q, the distribution for all succeeding generations will be

$$p^2 A_1 A_1 + 2pq A_1 A_2 + q^2 A_2 A_2$$

This is the case regardless of the distribution of genotypes in the initial population (or generation).

This mathematical model describes the frequencies of different genotypes at a single locus in an infinitely large population, in which mutation, selection, sampling error, and migration do not occur; all members of the population breed; all mating is completely random; all organisms have the same number of offspring; there is no intergenerational overlap; and all these conditions are held constant. However, each of these distorted features makes a difference to the evolution of most (if not all) real-world biological populations. For example, in all real finite populations there will be some nonnegligible chance that trait frequencies will diverge from the values predicted by selection; that is, drift makes a difference to every real-world population (Rice 2013). As a result, removing sampling error (i.e., drift) from this mathematical model distorts a difference-making feature of every real-world population. In addition, the Hardy-Weinberg model assumes that there is no intergenerational overlap, but this is untrue of almost every real-world population and is a difference-making feature for many evolutionary outcomes (Hartwell et al. 2000; Levy 2011; Relethford 2012).

As was the case in the examples above, many of these idealizing assumptions are necessary for the mathematical foundations of the model (Morrison 2015). For example, without the assumptions of infinite population size and random mating, the stability of genotypic frequencies across generations will be violated—that is, without these assumptions, the very mathematical framework used by the model that allows it to capture key features of heredity and variation will no longer apply.

Additional examples were already presented in chapter 3. For example, in the case of biological optimality explanations, the idealizations involved in deriving the explanandum include (among others) (Rice 2013):

1. The models' smooth mathematical curves and simplified (e.g., discrete) equations are idealized when compared to the selection processes within the target systems.
2. The strategy sets are intended to capture the relevant set of alternatives rather than the set of strategies actually competing in the population's past.
3. The models' optimization assumptions (e.g., maximization of energy intake) do not accurately represent an actual selection process in the

system, but only capture a general optimizing tendency of the system in the long run.
4. There are idealizations regarding the processes of inheritance (e.g., assuming that offspring will perfectly resemble their parents and reproduction is asexual).
5. It is assumed that selection pressures in the population do not change over time.
6. The models assume that there is no intergenerational overlap.
7. The individual-level events that determine fitness are assumed to be random and statistically independent in order to allow the application of the central limit theorem.
8. Infinite population size is assumed to allow for the use of various laws of large numbers that eliminate statistical error (i.e., drift).

This collection of idealizations entails that optimality models fail to accurately represent the evolutionary processes that occur in any real-world biological system (Rice 2012, 2013, 2018). Furthermore, these idealizations not only distort nonselective features such as drift, sexual recombination, and migration, but also drastically distort the difference-making selection processes that produced the adaptive trait. This is because optimality models typically invoke a selection process that does not—and could not—occur in any real-world biological population. Consequently, optimization models provide pervasively distorted representations of their target systems, including features that are assumed to be the difference-makers for the target explanadum. What's more, the features distorted by these models are precisely the features of selection that are of interest to the adaptationist research program in which these models are formulated.

More generally, biological modelers frequently make strategic use of several idealizations that distort the actual processes involved in mating, genetic recombination, inheritance, and selection in order to apply various mathematical modeling techniques. In doing so, they often distort features of real-world systems that are known to make a difference to their evolutionary outcomes (Potochnik 2017; Rice 2012, 2013, 2016). Indeed, using idealizations that distort difference-making factors is pervasive in biological modeling, even within mechanistic modeling (Love and Nathan 2015).

Furthermore, the idealization of difference-makers is widespread in several other sciences as well (Batterman 2002b; Batterman and Rice 2014;

Bokulich 2008, 2011, 2012; Morrison 2015). For example, as we saw in chapter 3, minimal model explanations in physics distort or ignore almost all of the causal (or causal-mechanical) features of their target systems and use essential idealizations that distort difference-making causal processes in order to apply mathematical modeling techniques (e.g., renormalization). As a result of their distortion of difference-makers, the distortions introduced by the inaccurate parts of these models cannot be isolated to the distortion of the irrelevant parts (or features) of their target systems (Rice 2017, 2018). Therefore, once again, for a wide range of cases across numerous sciences, one of the three core assumptions required to apply the decompositional strategy fails to hold.

A defender of the standard view might try to account for these kinds of cases by arguing that the model need not accurately represent difference-making causes; instead, it only needs to identify a set of relevant conditions that are more or less satisfied in the target system.[4] For example, both Strevens (2008) and Weisberg (2013) suggest that sometimes accuracy regarding some causal difference-makers can be sacrificed in order to accomplish other modeling goals (e.g., providing additional generality). Potochnik (2017) also argues that many difference-making causes can be set aside if they are not of interest to the current research program. In contrast with these views, my claim is that these models can reveal a lot of information about the relevant features of the system (i.e., those on which the explanandum counterfactually depends) *even if the model drastically misrepresents or completely ignores the causal features of interest to the research program and distorts the way in which the features cited in the explanans actually produce the explanandum.* All that is required is that the scientists who appeal to the model in their explanations are able to identify the required relationships of counterfactual dependence. For example, a biological optimality model can explain despite the fact that the conditions of the model are not satisfied by any real-world target system, the model ignores causal factors that do make a difference to the explanandum, and the model distorts the causal factors involved in selection that are of interest to the adaptationist research program in which the model is formulated.

Another possible response is for decompositional views to claim that the fact that these models reproduce the explanandum shows that the idealizations and the factors that they distort do not make a big difference to the values actually observed. This is precisely what Elgin and Sober (2002) mean by claiming that the idealization are harmless. This response misinterprets my view in two ways. The first issue is that my claim here is that the idealizations distort precisely those features that *do* make a difference to observed behaviors in the model's target systems, in that without those features, the real system (or a veridical causal model) would not produce the observed effect. That is, the features of the real-world system that are distorted by the model make a difference to the outcome in actual cases. More important, however, I argue that the idealizations play an essential role in the behavior of the model in precisely the way that Elgin and Sober would deem problematic. This is because I argue that those idealizations *do* make a big difference to the values predicted by the model, in that without those idealizations, the model would no longer predict the explanandum. This, I argue, shows that the idealizations play a fundamental role within the explanation provided by the model rather than their being harmless in the way suggested by the standard view. In short, the features distorted make a difference in the real system, and the idealizations are essential to the model that is used to provide the explanation.

Consequently, I argue that the explanations provide by many of these models violate all three tenets of the standard approach presented in chapter 2. First, as I argued in chapter 3, the explanatorily relevant features in many of these cases are not causes of the explanandum. Second, these models do not provide an accurate representation of the explanatorily relevant features, but instead distort the very features that are of interest and are known to make a difference to the explanandum. Third, there is no attempt to restrict the use of idealizing assumptions to the distortion of irrelevant, insignificant, or nonsalient features.

As a result, we cannot isolate—or quarantine—the distortions introduced by these idealizations to some set of features that are irrelevant or otherwise not of interest. Instead of attempting to accurately represent an isolable set of relevant features, these scientific modelers use a variety of idealizations to apply mathematical modeling techniques that allow for the discovery of explanatory information. Therefore, the contributions made by the idealizations cannot be isolated from the contributions made by the accurate parts of the model. Rather than restricting idealizations to the distortion of irrelevant or uninteresting features, the representations these models provide of the explanatorily relevant features are deeply intertwined with the distortions introduced by the collection of idealizations necessary for applying the mathematical modeling techniques used within the explanation.

To simply reject these highly idealized models as nonexplanatory would be to render incomprehensible much of what contemporary science has purported to explain. Furthermore, attempting to force these cases into some version of the standard view would mischaracterize the distinctive roles that idealizations play within the explanations provided by these models. Therefore, in order to account for these cases, we require an alternative account of the role of idealizations within scientific models that are used to explain and understand real phenomena.

5.2 An Alternative Approach: The Holistic Distortion View

In the rest of this chapter I will argue for what I call the *holistic distortion view* of idealized models (Rice 2017, 2018). In its most general form, holism is just the thesis that the whole is more than the sum of its parts. I contend that many idealized models in science ought to be characterized as holistic distortions that cannot be decomposed into the contributions made by their accurate and inaccurate parts. As the examples surveyed throughout this book show, in many cases there is no principled way of separating out the accurate and inaccurate parts of models. Accounts that attempt to restrict idealization to the distortion of irrelevant features require that we are able to decompose models in ways that are often epistemically inaccessible to scientists (or anyone else)—especially at the early stages of investigation, when essential and pervasive idealizations are typically introduced. Because they are essential to various modeling frameworks, these idealizations often become deeply entrenched in the modeling techniques repeatedly used to explain and understand complex phenomena. Characterizing idealized models as holistic distortions explicitly recognizes these epistemic limitations of scientific practice and forces our philosophical accounts of how idealized models contribute to the epistemic aims of science to find alternative ways to justify their use. The point isn't that idealized models always distort all the features of the target systems, but rather that we are rarely in a position to demarcate the accurate parts of a model from its idealized parts. Consequently, philosophical accounts of idealized modeling must find alternative ways to characterize how scientists can justifiably use these models to explain and understand complex phenomena.

It is important to note, however, that my holistic distortion view does not directly entail other more traditional forms of holism concerning semantics, metaphysics, or confirmation. Instead, I only advocate for a more holistic approach concerning philosophical attempts to justify the use of idealized models in science. Specifically, I contend that many (if not most) idealized models should be characterized as holistically distorted representations of their target systems that are greater than the sum of their purportedly accurate and inaccurate parts. The decompositional strategy mistakenly ignores the myriad ways in which the explanations and understanding provided by scientific models are typically the result of a rich and complicated mixture of various modeling assumptions whose contributions cannot be studied in isolation. Therefore, my holistic distortion view is a *methodological* prescription for philosophers' attempts to understand how to model complex systems, how models explain, how idealizations contribute to model explanations, robustness analysis, and how idealized modeling is compatible with scientific realism. In developing these various philosophical views, we ought to characterize idealized models as pervasively inaccurate representations.

The second part of my view is that idealizing assumptions that result in holistic distortions are often essential to the explanations provided by scientific models because they allow for the application of various mathematical modeling techniques used to extract modal information (Cartwright 1983, chap. 7). Once again, a particularly instructive example is physicists' use of the thermodynamic limit in which (roughly speaking) the number of particles of the system approaches infinity (Batterman 2010, 7). For a wide range of modeling techniques in physics, the thermodynamic limit is an essential mathematical operation (Batterman 2002b; Sklar 1993). For example, as Morrison explains in the case of modeling phase transitions:

> The occurrence of phase transitions requires a mathematical technique known as taking the "thermodynamic limit," $N \to \infty$; in other words we need to assume that a system contains an infinite number of particles in order to understand the behavior of a real, finite system . . . [because] the assumption that the system is infinite is *necessary* for the symmetry breaking associated with phase transitions to occur. In other words, we have a description of a physically unrealizable situation (an infinite system) that is *required* to explain a physically realizable phenomenon (the occurrence of phase transitions). (Morrison 2009, 128)

Or, as the physicist Leo Kadanoff puts it, "The existence of a phase transition requires an infinite system. No phase transitions occur in systems with a finite number of degrees of freedom" (Kadanoff 2000, 238). Like the cases discussed in this chapter, the thermodynamic limit is not introduced simply as a way of ignoring what is irrelevant to the explanandum or as a method for calculational expediency. Instead, these idealized mathematical descriptions function as a necessary condition for using the mathematical techniques required to explain and understand the phenomenon of interest (Batterman 2002b; Kadanoff 2000; Morrison 2009; Sklar 1993).

As I have noted in this chapter, additional instances of the use of idealizations to allow for the application of mathematical techniques can be found across biological modeling, such as the use of optimization models to explain phenotypic traits (Rice 2012, 2013, 2016). Rather than distorting irrelevant factors so that scientists can focus on the accurate representation of difference-making features (e.g., natural selection), these models purposefully move scientists away from even attempting to accurately represent some isolable part of the dynamical processes that led to the explanandum. Instead, in these cases, the idealizing assumptions enable scientists to apply various mathematical modeling techniques that allow them to calculate exactly how changes in the parameters involved in the system's constraints and trade-offs will result in changes in the expected equilibrium point of the system. The explanation is then provided by showing how the optimal strategy counterfactually depends on those constraints and trade-offs—despite the fact that the optimal strategy is the expected outcome within the idealized model for very different reasons than in the model's target systems.

The point is that a model can be used to show that there is a relationship of counterfactual dependence (or independence) between X and Y, even if it drastically distorts many aspects of X and Y, it pervasively misrepresents the processes or mechanisms that connect X with Y in the real system, and it tells us little about how it would be possible to surgically intervene on X in order to bring about changes in Y. Whether the model enables scientists to accomplish other goals (e.g., manipulation or control) may require additional features to be accurately represented, but all that is required for the model to be (or contribute to) an explanation is that it enable scientists to discover how the explanandum counterfactually depends on (or is counterfactually independent of) a set of features within the model's target systems.

More generally, as Pexton notes, in many scientific models, there are "ineliminable misrepresentations of the true ontology of a system. . . . [They] are necessary in some models because we require them to frame a system in a certain way in order to extract modal information" (Pexton 2014, 2344). It is this last part, about the way in which idealizations allow scientists to frame the system as a whole to extract modal information about the system, that I think has been largely missed by the philosophical literature, although Cartwright (1983, chap. 7) and Morrison (2015) are important exceptions.[5] The source of this failure is that most philosophers (and many scientists) have sought to quarantine idealization in order to show that the falsehoods used in science are harmless because they distort only irrelevant or unimportant features and do not get in the way of the accurate parts of scientific models that do the real work.

As an alternative project, I suggest that philosophers of science ought to provide accounts of the *pervasive* and *unique* contributions that idealizations that distort relevant (e.g., difference-making) features can make to models conceived of as holistic distortions (e.g., by enabling scientific modelers to use certain mathematical modeling tools that extract the explanations and understanding they are interested in).[6] Rather than attempting to isolate individual idealizations and show that their distortions are irrelevant (one at a time), we need to analyze how the practice of constructing holistically distorted models is able to consistently make positive contributions to achieving the epistemic goals of scientific inquiry (e.g., explanation and understanding) without having to provide an accurate representation of an isolated set of difference-making features.[7]

Furthermore, in contrast with several other accounts of idealization that focus on scientists' cognitive limitations, my holistic distortion view focuses our attention on the *representational* limitations of science that necessitate the use of idealization and abstraction. Idealization and abstraction can certainly help limited human agents understand in a variety of ways, such as by reducing complexity or focusing our attention on the most important features of a phenomenon (e.g., Mitchell 2009; Potochnik 2017; Wimsatt 2007). I will explore these contributions a bit more in later chapters when I discuss the nature of scientific understanding. However, I argue that much of the distortion that is introduced in science isn't motivated so much by our cognitive limitations as it is by the limitations of the mathematical, theoretical, and representational tools that scientists have at their disposal. That is, often the choice to use a highly distorted representation of the target systems is not driven by our choosing the representation that is easiest for our

Models Don't Decompose That Way 149

minds to comprehend or understand (one need only read a few articles in advanced mathematical physics to start to doubt this idea). Instead, I argue that it is at least as common that a highly distorted representation is chosen because it is amenable to the scientific modeling tools that scientists have available. Moreover, in many cases this choice is made for scientists because they have no other way of representing, explaining, or understanding the phenomenon.[8] In short, the strategic use of holistic distortions rarely occurs in a context in which our cognitive limitations are used to choose from a large set of models that range from highly inaccurate to highly accurate. Instead, scientists typically are confronted with complex phenomena and a limited set of highly distorted modeling frameworks with which to investigate that phenomenon. The challenge, then, is not determining *whether* scientists ought to use a holistically distorted model to explain or understand the phenomenon, but instead *which* holistically distorted representation will be amenable to the modeling tools they have available that will enable them to extract the information required to explain and understand the phenomenon.

Here, then, is the core of my holistic distortion view. First, in a wide range of cases, idealized models pervasively distort the fundamental nature of the entities, processes, and features of their target systems—including both relevant and irrelevant features. That is, their distortions are holistic rather than piecemeal. Second, these idealizing assumptions often move scientists to an entirely different representational framework, in which the mathematical tools necessary to explain and understand the phenomenon of interest are applicable. These different mathematical modeling frameworks represent different features and patterns of the system (e.g., statistical patterns at the population level) in different ways (e.g., as continuous processes or isolable factors) and allow for the use of different techniques for deriving the behavior of the system (e.g., using statistical limit theorems).

As a result, these idealizations are often necessary for the models to provide a particular explanation (or understanding) of the target phenomenon—one that is sometimes the only explanation available. A model without these idealizing assumptions would be unable to use the mathematical modeling techniques that are required for extracting the explanatory information of interest. The goal, therefore, should be to justify scientists' use of these holistically distorted representations in terms of the access that they provide to explanations and understanding that otherwise would be unobtainable. Analyzing these unique contributions of idealizations that distort relevant features sits in stark contrast to trying to show that the inaccurate parts of scientific models distort only what is irrelevant, insignificant, or not of interest.

In sum, the four main claims of my holistic distortion view of idealized models are the following (Rice 2017, 2018):

1. Many (if not most) idealized models in science ought to be characterized as holistically distorted representations of their target systems.
2. Idealizing assumptions that result in holistically distorted representations typically make essential contributions by allowing for the application of various mathematical modeling techniques.
3. The use of such holistically distorted models ought to be justified by their ability to provide epistemic access to explanations and understanding that otherwise would be inaccessible.[9]
4. Given the diversity of modeling techniques, goals, and explanations found in scientific practice, we require a more pluralistic approach to investigating the relationships between models and their target systems that justify their use in providing explanations and understanding.

These four claims provide the foundation for a fundamentally different way of thinking about idealized modeling in science that avoids the mistakes of the decompositional strategy. By making idealizations essential, central, and pervasive contributors to scientific representations, the holistic distortion view makes clear the challenge of showing how idealizations that distort relevant features positively contribute to the heart of scientific theorizing.

5.3 Conclusion

The decompositional strategy is central to the standard view and is pervasive across a wide range of debate within the philosophy of science. I have argued that the assumptions underlying these decompositional approaches will fail to hold for a wide range of cases of scientific modeling. In response, I have proposed the holistic distortion view, which offers an alternative approach to characterizing and justifying the use of idealized models. Going forward, I suggest that philosophers of science take up the challenge of showing how scientific modelers are justified in using idealized models that holistically distort their target systems by analyzing the ways in which various modeling techniques can produce explanations and understanding that otherwise would be inaccessible. In other words, philosophers of science need to investigate the ubiquitous and unique epistemic contributions that various holistically distorted scientific representations make to scientific inquiry, rather than continuing their attempts to ignore, remove, isolate, or quarantine the roles that idealizations play within scientific theorizing. I take up this challenge in the next few chapters.

6 Using Universality to Justify the Use of Holistic Distortions to Explain

As we saw in chapter 2, most accounts of explanation, idealization, and modeling have sought to justify the use of idealized scientific representations by showing that they only distort features that are irrelevant, insignificant, or not of interest. However, as I argued in chapter 5, the distortions introduced by many (if not most) idealizations in scientific theorizing are far more pervasive than the standard view assumes. As a result, trying to quarantine idealization to the distortion of irrelevant features will fail to justify many of the idealized modeling techniques that are used to explain and understand in actual scientific practice. In response, I argued that idealized models ought to be characterized as holistically distorted representations. That is, models are themselves idealizations (as a whole), rather than having idealizations as isolable parts.

The next challenge is to show how such holistically distorted representations can justifiably provide explanatory information, given that most accounts of explanation require accurate representation of difference-makers in order for models to explain. This chapter focuses on how to justify the use of holistically distorted models to explain (and understand) without appealing to accurate representation relations between the model and a set of relevant features. I propose that an alternative method for justifying scientists' use of idealized models to explain and understand is to appeal to universality classes.

In support of this view, I provide several examples in which scientific modelers explicitly appeal to universality classes in order to justify their appeals to idealized models in their attempts to explain and understand real-world phenomena. In other words, I will argue that the justifications offered by practicing scientific modelers are often in line with those proposed by the universality account. These cases also reveal a variety of ways that idealizations make positive and ineliminable contributions to the explanations and understanding provided by holistically distorted models. In particular, we will see that scientific modelers are extremely adept at *leveraging* holistically distorted models by discovering universal behaviors that are stable across heterogeneous real, possible, and model systems and then strategically introducing idealizations in order to bring existing modeling frameworks to bear on new problems.

The following section outlines how to justify the use of holistically distorted models to explain by appealing to universality classes. Next, in section 6.2, I demonstrate the generality of the universality account by discussing a variety of parameters of universality classes. Then, in section 6.3, I survey several cases in which scientific modelers explicitly appeal to universality classes in order to justify the use of their highly idealized models to explain complex phenomena. Finally, in section 6.4, I compare the concept of universality with several other notions of stability that have been evoked in the philosophical literature. Section 6.5 summarizes the conclusions of the chapter.

6.1 Appealing to Universality Classes to Justify the Use of Idealized Models to Explain

Rather than thinking in terms of models that contain isolable idealizing assumptions, my holistic distortion view argues that we should think in terms of the models themselves as being holistically distorted representations (i.e., they are idealized model systems that pervasively distort the features of their target systems). Given that I have argued that these models distort both the relevant and irrelevant features of their target systems, it is a bit unclear how such holistically distorted models can be used to explain and understand. In response, I suggest that idealized modeling techniques that involve holistic distortion of real-world systems can be justifiably used to extract explanatory information because many idealized model systems are known to approximate the patterns of counterfactual dependence and independence present in real-world systems—but perhaps they do so only in their idealized limits, and perhaps for very different reasons than their target systems.[1] Justifying these explanatory inferences, however, does require that we can establish some kind of connection between the behavior of the ideal case (i.e., the system described by the highly idealized model) and the behavior of the real cases (i.e., the real systems in which the phenomenon occurred).

While I advocate a more pluralistic approach to understanding the relationships between idealized models and real-world systems, one particularly promising strategy is to appeal to an extremely convenient feature of our universe: many patterns of behavior are universal in the sense that systems with (perhaps very) different physical features will display similar behaviors (Rice 2017, 2018). As we've seen in earlier chapters, universality is a concept used in mathematical physics to describe the stability of certain features (e.g., critical exponents) across very different physical systems (e.g., magnets and fluids) (Batterman 2002b; Morrison 2015). The view that I defend here requires us to generalize the concept of universality (and universality classes) beyond its use in characterizing systems that share critical exponents. In what follows, I use the term *universality* to simply mean the stability of certain patterns or behaviors across systems that are heterogeneous in their features. Universality classes are, then, just the group of systems that will display those universal patterns or behaviors. For example, universality can connect the behavior of systems as diverse as magnets and fluids (Batterman 2002b). It can also show why a large class of biological systems will all reach the same equilibrium point (e.g., a 1:1 sex ratio) despite differences in the kinds of species, methods of inheritance, size of the population, etc. (Batterman and Rice 2014). Universality has also been used to connect various patterns found in statistical physics with similar universal behaviors found in biology (Rice 2017). Indeed, some of the most striking observations in science are the universality of various patterns across systems that are extremely different in their physical features (Frank 2009; Goldenfeld and Kadanoff 1999; Tao 2012). This universality guarantees that a class of systems will display the same general patterns of behavior even if the components, interactions, dynamics, and physical features of those systems are very different.

In addition, within these universality classes there are often various model systems—that is, universality classes that include a range of real-world systems will often also include some of the (imaginary or possible) model systems described by scientific models. When this is the case, the model systems and the real-world systems will display similar behaviors (or patterns of behavior), despite having perhaps drastic differences in their fundamental components, interactions, and other features. When this occurs, I argue that we can appeal to universality classes to justify scientists' use of highly idealized models to explain the behavior of real systems—even if the model drastically distorts or leaves out features that make a difference in real-world cases. This process of justifying the use of idealized models by appealing to universality is clearly described by many scientific modelers. For example, Gisiger says:

> Universality has been described as a physicist's dream come true. Indeed, what it tells us is that a system, whether it is a sample in a laboratory or a mathematical model, is very insensitive to details of its dynamics. . . . From a theoretical point of view, to study a given physical system, one only has to consider the simplest mathematical model possibly conceivable in the same universality class. (Gisiger 2001, 173)

After reviewing how this works in physics, Gisiger shows how the same approach can be used to justify the use of extremely idealized models to explain the behavior of complex systems in biology. The key result is recognizing that "one only has to choose a simple, or simplistic, model in the same universality class as the system under study" (Gisiger 2001, 175). As a result of being in the same universality class as the systems of interest, these idealized models will display many of the patterns of behavior of interests to scientists despite their pervasive distortion of the real features of those systems. This connection, I contend, is often what enables scientists to justifiably use holistically distorted models to explain various behaviors of their target systems. In other words, universality can show us that there is a class of systems, which includes the idealized model system and its target systems, that will exhibit similar patterns of behavior, and this link is often what allows scientists to justifiably use holistically distorted models to explain and understand the behavior of real systems (Batterman and Rice 2014; Rice 2017, 2018).

It is important to distinguish this appeal to the empirical fact of universality to link scientific models with their target systems from the use of mathematical techniques to explain universality itself. For example, as we saw in chapter 3, the renormalization group has been used to explain why a wide range of fluids and magnets all display universal behaviors near their critical points (Batterman 2002b). While providing additional demonstrations that explain universality itself will typically provide greater degrees of understanding and justification (and may be required to provide a complete explanation for some explananda), providing a complete explanation of why an instance of universality occurs is not a requirement for appealing to the empirical fact of universality to connect the behaviors of various real, possible, and model systems in ways that allow idealized models to be justifiably used to discover explanatory information.[2] That is, scientists can justifiably use idealized models within a universality class to explain the behaviors of real-world systems in that class, even when they do not have a complete explanation of why that universality class occurs. Indeed, many universality classes exist that contain various real, possible, and model systems, but whose universality has yet to be explicitly delimited (i.e., specifying precisely the range of systems in the class) or completely explained (i.e., explaining why those patterns are stable across that class of systems).

At this point one might object that claiming that a model is within the same universality class as real systems just amounts to the claim that the model shares some relevant features or conditions that are responsible for the phenomenon with the real systems (Elgin and Sober 2002). Lange (2014) raises this kind of objection by suggesting that this view just collapses into what Batterman and Rice (2014, 351) refer to as "common feature accounts." There are several things to say in response here. Most important, the claim defended here, and in Batterman and Rice (2014), isn't that idealized models will *never* have any features in common with their target systems—they likely will. Instead the claim is that idealized models *need not* accurately represent a set of relevant features in order for the model to be justifiably used to explain. For one thing, many models that contribute to scientific explanations simply ignore key difference makers for the explanandum. Furthermore, even if we grant that the model includes the relevant features of the system, this is compatible with providing a drastically distorted representation of those features and the ways in which they contribute to the occurrence of the explanandum. In short, having features "in common" with a target system is crucially different from providing an accurate representation of those features and their contributions to the explanandum. Moreover, even in cases where the model does accurately represent some relevant features, an accurate representation of those features often fails to play any significant role in justifying the explanatory use of the idealized model. In short, the question of what justifies scientists in appealing to an idealized model in order to explain is importantly distinct from the question of which features the model accurately represents (Batterman and Rice 2014; McKenna forthcoming). Consequently, just because there are common features between a model and its target systems does not entail that accurate representation of those features is what justifies using the model for explanatory purposes.[3] Common feature accounts don't just claim that the model has features in common with its target systems. Rather, they argue that the model's having those relevant features in common with its target systems is what justifies using the model to provide an explanation. In contrast, I contend that scientific models that drastically distort (or perhaps even ignore) explanatorily relevant features of their target systems can still be justifiably used to explain when they are in the same universality class as their target systems. All that matters is whether analyzing the model enables scientists to answer the relevant set of questions about what would occur in various counterfactual situations (i.e., whether the model enables scientists to discover the salient patterns of counterfactual dependence and independence).

The crucial point is that this extraction of modal information can be accomplished whether or not the model accurately represents a set of difference-making features within its target systems. Indeed, one of the main advantages of an account based in universality classes is the ability to account for similar patterns of behavior that arise out of different features in different systems. As the mathematician Terence Tao explains concerning various instances of universality:

> These macroscopic laws for the overall system are often largely *independent* of their microscopic counterparts that govern the individual components of that system. One could replace the microscopic components by *completely different types of objects and obtain the same governing law at the macroscopic level*. When this occurs, we say that the macroscopic law is *universal*. The universality phenomenon has been observed both empirically and mathematically in many different contexts. (Tao 2012, 24)

When different features give rise to the same pattern in different cases, citing the features that those systems have in common (if any) would completely miss the point because we often want to know why very different real, possible, and model systems nonetheless display the same behaviors! As I argued in chapter 3, our conception of stable patterns should not assume that every instance of a pattern must be produced by the same difference-makers, causes, mechanisms, or relevant features. What's more, not only do the idealizations in most of the cases discussed in this book distort features that make a difference to the explanandum in their target systems, but those idealizations also make essential contributions to the ways in which the models produce the behaviors of interest. As a result, we need to capture stable patterns that occur for different reasons in the model than they do in the model's target systems.

Given the account of explanation defended in chapter 4, the universality class of interest will consist of the set of real, possible and model systems that display the patterns of counterfactual dependence and independence of interest (i.e., those counterfactual relationships that are made salient by the explanandum, contrast class, and modeling context). In some cases, the idealized models that are within the same universality class as their target systems will just be those that accurately represent the difference-making features of those systems. For example, this will occur when the same set of difference-makers is necessary and sufficient for the occurrence of the target explanandum. This enables the universality account to subsume the cases used to motivate the standard view in which models that accurately represent difference-making features are used to explain and understand—although I suspect that in many of these cases, the universality of those patterns of dependence rather than accurate representation of difference-making factors is doing the real justificatory work. However, in many other cases, similar patterns of dependence will be produced by different sets of difference-makers and extremely heterogeneous components, causal interactions, mechanisms, and processes across a range of possible systems. In these cases, scientists often construct a pervasively distorted model that includes only a few minimal features in order to apply the mathematical modeling techniques and theories they have on hand. As the physicists Goldenfeld and Kadanoff put it, in many cases "we can exploit this kind of 'universality' by designing the most convenient 'minimal model'" (Goldenfeld and Kadanoff 1999, 87). The resulting idealized model is then used to investigate how various (hopefully measurable) features are counterfactually related to the phenomenon of interest. When the holistically distorted model is—or perhaps can be shown to be—in the same universality class as the systems of interest, scientists can justifiably use the holistically distorted model to discover information about which features are counterfactually relevant and irrelevant to the explanandum.

For example, the ideal gas law need not accurately represent the ontology of real gases or the actual processes that relate pressure and temperature in order to show us how pressure counterfactually depends on temperature (and vice versa). All that is required is that the model enable us to see how changes in one feature will result in changes in the other, and universality can guarantee that those macroscale patterns of counterfactual dependence will be similar even though the model is a holistically distorted representation of the components, processes, and interactions that make a difference to these patterns in real gases. This is because the macroscale patterns of counterfactual dependence between measurable features of real gases will be realized (or at least approximated) by *any* system in the universality class—including the pervasively distorted model system. This enables the holistically distorted model to be used to extract modal information about the phenomenon despite misrepresenting the difference-making (e.g., causal) features that produce those patterns of behavior in real gases.

As a second example, a biological optimization model can show how changes in the constraints and trade-offs among various features of the system will change the equilibrium point of the evolving population, even if the model describes a selection process that is nothing like the one that produced the explanandum in its real-world target systems (Rice 2012, 2013, 2016, 2017). In these cases, the models show how the target explanandum is counterfactually dependent on certain constraints (e.g., the set of strategies available) and trade-offs (e.g., between average energy intake and average predation risk), even though the model represents these explanatorily relevant features in a highly distorted way. Moreover, the optimization model describes a selection process (connecting these features to the target explanandum) that drastically distorts the selection process that occurs in real biological populations. Despite distorting these explanatorily relevant features and the processes that mediate their contributions to the explanandum, biologists have discovered that (and in some cases demonstrated why) these kinds of counterfactual dependencies are stable across a range of real, possible, and model systems. That is, although it describes a system that is drastically different from its target systems, the optimization model—by virtue of being in the same universality class—still displays certain macroscale patterns of counterfactual dependence between certain constraints and trade-offs and the target explanandum that can be used to provide an explanation. Moreover, by discovering that these patterns are stable across systems that are heterogeneous in many of their details, including having different trajectories, species, interactions, and inheritance processes, the universality classes that contain these models also enable scientists to extract a good deal of information about the counterfactual irrelevance of many features to the occurrence of the explanandum.

In sum, highly idealized models can allow scientists to discover counterfactual relationships that hold between certain features and the target explanandum without having to accurately represent the entities, processes, ontology, or difference-making causes of their target systems. Importantly, as we saw in chapter 4, information about counterfactual dependence and independence is essential to a wide variety of causal and noncausal explanations in science. The key thing to notice is that holistically distorted models can provide accurate modal information because universality guarantees that the model system's patterns of behavior will be similar to those of the target systems, even if the actual entities, causal interactions, and processes of those systems are extremely different. Therefore, even if the model drastically and pervasively distorts the fundamental nature of the relevant features of real-world systems in order to use various mathematical modeling techniques, it can still be used to explain because many of the patterns of counterfactual dependence and independence that hold in the model system will be similar to those of real-world systems—those counterfactual relations will just hold for perhaps very different reasons in the model system, and perhaps only in limiting cases. Generalizing the concept of universality (beyond its use in describing critical exponents) allows us to capture this stability of various patterns of behavior that are largely independent of the physical features of a heterogeneous class of systems. In this way, universality enables us to see how idealized models can be justifiably used to explain even when they are holistically distorted representations of their target systems.

One of the key projects of the philosophy of science should be to understand how scientists' use of highly idealized models has been so successful at generating explanations and understanding of our world. Unfortunately, the justification offered by the standard view is typically inapplicable to the cases that we find in actual scientific practice because those models pervasively distort difference-making (or otherwise relevant) features. In response, I suggest that there are at least two links between idealized models and their target systems that can be used to justify scientists' use of idealized models to explain and understand real-world phenomena: (1) having an accurate representation of difference-makers and (2) being within the same universality class.

While appealing to universality is a promising approach, I want to emphasize that I think that one of the key lessons of this discussion is that philosophers need to be more pluralistic about the kinds of relationships that can hold between idealized models and real-world systems that are sufficient for explanation and understanding. Some idealized models will accurately represent difference-makers of their target systems, some will be in the same universality class as their target systems, and others will perhaps exploit some alternative connection to their target systems. Uncovering these additional methods for justifying the use of holistically distorted models to explain and understand is one of the ways that the view proposed here can be expanded going forward. Given the diverse array of modeling techniques, explanations, and goals found across various scientific disciplines, we should expect there to be an equally diverse set of ways that scientific models can relate to real-world systems and justifiably provide the explanations and understanding of interest to scientists.

6.2 Parameters of Universality

In order to appreciate how widely this universality account can be applied, it is important to note that instances of universality can differ with respect to a range of parameters. For example, different universality classes can have different scopes (or ranges) because they will include more or less real, possible, or model systems. In some modeling contexts, the universality class may need to include only one target system and one idealized model that is used to explain the behavior of that system. In other cases, the universality class will need to include a wider range of real and possible systems in order for the idealized model to be justifiably used to explain. For example, if the explanandum is a pattern that occurs across many real-world systems, the universality class will need to be general enough to include most (but perhaps not all) of those real systems. In addition, as we saw in chapter 4, every explanandum will involve the specification of a contrast class and a set of relevant counterfactual scenarios (or "What if things had been different?" questions). Consequently, in order for a model to provide the counterfactual information required to explain, the universality class of which it is a member will need to include the possible systems involved in answering the contrastive "why" question of interest. That is, the universality class will need to be general in the right way (and to the right degree) in order to include the possible systems required to evaluate the counterfactual dependencies and independencies involved in providing the explanation.[4]

It is worth emphasizing here how this notion of universality classes capturing the relevant set of possible systems is importantly different from what other authors have said about generality's role in explanation. In previous chapters I have argued that universality classes' role is most salient (or important) when the pattern of interest is stable across systems that are very different with respect to their difference-making causes and mechanisms. This is why the universality view is importantly different from Craver and Kaplan's (2020) appeal to causally relevant details of the mechanism. On my view, the stable, explanatorily relevant features will often be noncausal features that are stable across systems with very different causal mechanisms. That is, the universal features will often not be part of any causal mechanism that produces or constitutes the phenomenon. Similarly, in contrast with Strevens (2008) and Potochnik (2017), the universal patterns of interest need not be produced by the same causal patterns across the systems within a universality class. That is, universality classes enable us to capture instances in which the same patterns of behavior are produced by radically different causes. It is in precisely these contexts that universality classes help us think about the importance of stable patterns across drastic wholescale changes to the systems' causes and mechanisms. What the cases discussed in earlier chapters of this book show is that often the facts that do the explaining are, as Lange (2013b, 491) puts it, "modally stronger" than causal dependence. This is why appeals to universality classes are crucially different from appeals to causal patterns (Rice manuscript). Contrary to Newton's second rule of scientific reasoning, our accounts of explanation and idealized modeling should not assume that similar behaviors are always the result of similar causes or mechanisms.

In addition, not all instances of universality will involve distinguishing patterns of behavior at a macroscale from various details at smaller scales. Instead, as we will see in the next chapter, many instances of universality take place *across* scales or at some intermediary mesoscale (Batterman 2019). Furthermore, while many of the cases analyzed in this book focus on macroscale explananda, the conclusions drawn from those cases do not depend on assuming that there are distinct levels of science. Instead, I agree with many philosophers that this "layer cake" view of reality is deeply mistaken (Batterman 2015, 2017; Potochnik 2009a, 2017). Moreover, I think that discussions of "levels of reality" or "levels of explanation" problematically intertwine debates about the nature of explanation and modeling with debates about reductionism. While universality does have important implications for the reductionism debate (more on this in chapter 7), the use of universality classes to justify the use of idealized models to explain is importantly independent of taking a stand in that debate. In order to avoid these kinds of confusions, it is important to recognize that a more generalized conception of universality requires only the stability of certain features with respect to perturbations of various other features—there is no need to assume that the stable features will always be at the macroscale and the perturbations will always be at some distinct microscale. Universality is about stability of behaviors across perturbations, not about levels, part-whole relations, or the multiple realizability of properties.

Finally, it is crucial that we recognize that the independence of universal behaviors from the features of a class of systems comes in degrees. Appeals to universality have been most prominent in cases where most of the microscale features are largely irrelevant to the system's macroscale behaviors. However, universality classes can connect idealized models and real-world systems even when the phenomenon of interest depends on particular physical parameters of the system (at various scales) taking on a more specific range of values. The important point is that universality does not require complete independence (or autonomy) of the universal behaviors from other nonuniversal features of the system. Just how much independence is required in a particular instance of idealized modeling will depend on the explanandum, contrast class, and other features of the modeling context. For example, in some cases, we will want to justify the use of a model that appeals only to features at some macroscale and ignores any details at smaller scales. Here, the universality class will need to show a rather high degree of autonomy of the macroscale patterns from the smaller-scale features of the system. However, in other cases, we may be interested in justifying a model that includes only a small set of interactions between the microscale features of the system and the more macroscale behaviors. In this kind of case, the universality class will need to show that the macroscale behaviors depend only on the few smaller-scale features included in the model. The general point is that universal patterns need only be independent of *some* changes to the features of the system to *some* degree. Patterns can be universal even if they are not completely autonomous of all the smaller-scale physical features of the systems in the universality class.

6.3 Explicit Appeals to Universality to Justify the Use of Idealized Models to Explain

In order to further motivate the universality account, I present in this section a variety of examples in which scientists *explicitly* appeal to universality

Using Universality to Justify the Use of Holistic Distortions to Explain

classes to justify the use of highly idealized models to explain and understand. This demonstrates that the justifications given by actual scientific modelers are often in line with those suggested by the universality account. In contrast, the justifications offered by proponents of the standard view are often derived by appealing to general philosophical accounts of explanation and reverse-engineering an account of how to justify the use of idealized models to produce those explanations (e.g., Strevens 2008). While some of the accounts discussed in chapter 2—in particular mechanistic accounts (e.g., Bechtel and Abrahamesen 2005)—have been explicitly developed by appealing to scientific practice, I argue that these accounts fail to capture cases in which scientific modelers invoke universality as a means of ignoring the particular causes or mechanisms that produced the phenomenon. I suggest that the justifications proposed by philosophers ought to align closely with the justifications actually given by practicing scientific modelers—and what we find there is that the idealizing assumptions are often introduced because they are necessary for applying the modeling techniques that have been successfully used to provide explanations of similar kinds of universal phenomena in the past. That is, we will see that the appeal to universality classes is used to justify the introduction of idealizations that enable scientists to leverage the modeling techniques that have been developed for universal problems that reoccur across very heterogeneous systems.

6.3.1 Modeling Melting Ice Ponds

As a first example, scientific modelers have appealed to universality classes in order to justify the use of idealized modeling techniques developed for the study of composite materials to try and explain the development of melt ponds in climate science (Golden 2014; Hohenegger et al. 2012; Rice 2017). Melt ponds that form on top of the sea ice are crucial to modeling climate change because ice reflects most incident sunlight, while water absorbs most of that sunlight. As a result, the rate and shape at which these ponds of melted water form on the surface of sheets of ice are crucial to modeling the impacts of climate change. Unfortunately, the development of these melt ponds is extremely difficult to model because they form complex geometries that sprawl out across the ice in ways that are difficult to predict (figure 6.1).

In order to tackle this problem, these modelers explicitly state that their goal is to discover if "melt pond geometry exhibits universal characteristics which do not depend on the details of the driving mechanisms" (Hohenegger et al. 2012, 1157). Instead of aiming to accurately represent the physical

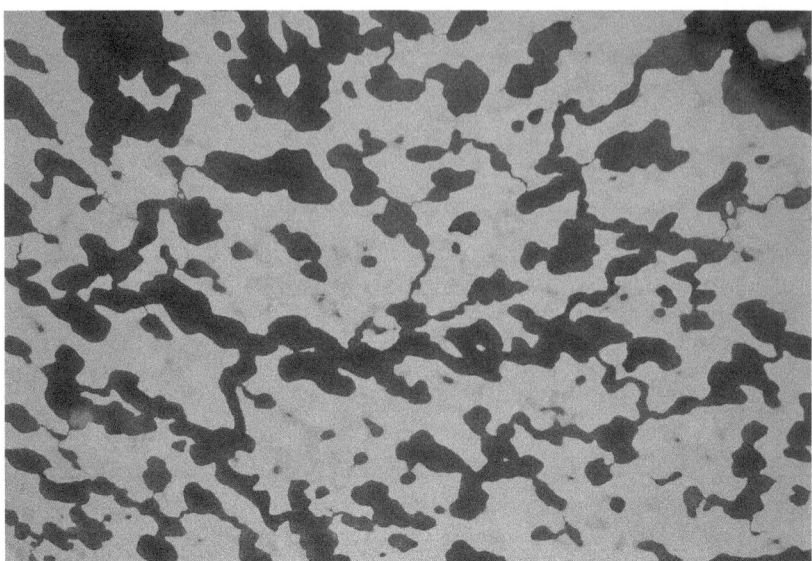

Figure 6.1
The complex geometry of well-developed Arctic melt ponds (Hohenegger et al. 2012, 1158).

mechanisms or processes of the system, by looking at the macroscale geometrical features of these melt ponds, these modelers are able to identify a universality class to which these melt ponds belong. Discovering this universality class enables them to justifiably apply idealized mathematical modeling techniques used in other areas of physics to explain similar phenomena.

In particular, these modelers show that there is a critical length scale at which the melt ponds quickly transition from one phase, having a fractal dimension of 1, to another phase, having a fractal dimension of approximately 2 (figure 6.2). The fact that this universal phase transition occurs in a variety of real and possible systems was derived from huge data sets of observed and simulated melt ponds.

Discovering this phase transition and critical point is crucial because it shows that the development of these melt ponds involves a separation of length scales. This is an important discovery because "a separation of scales in the microstructure of a composite medium is a necessary condition for the implementation of numerous homogenization schemes to calculate its effect properties" (Hohenegger et al. 2012, 1160). In other words,

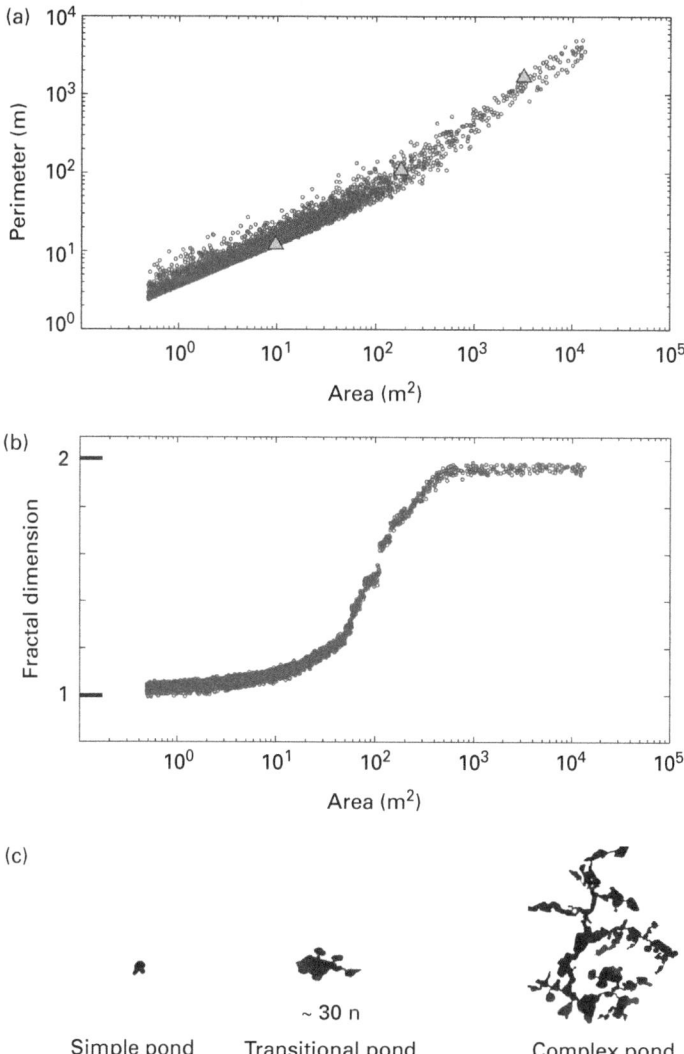

Figure 6.2
(a) Area versus perimeter data for observed melt ponds. (b) Graph of the fractal dimension D as a function of area A. (c) Typical ponds in each regime are shown: a simple pond with $D=1$, a transitional pond with a horizontal length scale of about 30 m, and a large, complex pond with $D \approx 2$ (Hohenegger et al. 2012, 1159).

discovering this universal feature of scale separation is crucial because it is a necessary condition for applying a particular kind of idealized modeling technique. Homogenization is a mathematical modeling technique used to find a homogenous medium whose macroscale behavior is the same as, or highly similar to, a given heterogeneous medium (I discuss homogenization techniques in more detail in the next chapter). Similar to applications of the renormalization group, applying these homogenization techniques results in a pervasively distorted representation of the nature of the real systems by representing a very heterogeneous system as a homogeneous system—but one that preserves the macroscale behaviors that are of interest to scientific modelers. In other words, these modelers discover that Arctic melt ponds display the universal features of a well-known universality class of systems: scale separation and phase transitions. Discovering these facts enables them to justifiably use idealized mathematical modeling techniques that had been used to study similar kinds of phase transition phenomena in other systems within that universality class. As Kenneth M. Golden explains in a later paper, the behavior of these melt ponds

> is similar to critical phenomena in statistical physics and composite materials. It is natural, therefore, to ask if the evolution of melt pond geometry exhibits *universal characteristics* that do not necessarily depend on the details of the driving mechanisms. . . . Fundamentally, the melting of Arctic sea ice is a phase transition phenomenon. . . . We thus look for features of melt pond evolution that are mathematically analogous to related phenomena in the theories of phase transitions and composite materials. (Golden 2014, 13)

As a result of discovering that these melt ponds are in this universality class, these modelers are able to justifiably use various idealized models within that universality class to extract information about the macroscale behaviors of these real-world systems that are independent of their mechanisms, components, and causes. Using this technique enabled them to discover that the change in phase counterfactually depends on the melt pond being self-similar in terms of the perimeter-area ratio. That is, "the distinction between transitional ponds and complex ponds ($D \approx 2$) is achieved by looking at the self-similarity of melt ponds" (Hohenegger et al. 2012, 1160). A melt pond is self-similar with respect to its area-perimeter ratio if there exists a subpond such that the perimeter-to-area ratio of the entire melt pond is approximately the same as that of the subpond. Consequently, changes in this noncausal geometrical feature of the system can be used to explain changes in the different phases of melt pond development. Moreover, these homogenization modeling techniques showed that the details of the particular mechanisms of these systems are counterfactually irrelevant to these universal behaviors. Thus, by discovering that these melt ponds are within a previously discovered universality class, these modelers are able to justifiably apply an idealized mathematical model to extract explanatory information that enables them to understand how (and why) these melt ponds develop. Moreover, these scientific modelers suggest that their results "will help explain the complex transmitted radiation fields observed under melting sea ice" and aid in "understanding . . . communities [of phytoplankton], their evolution, and their interaction with solar radiation" (Hohenegger et al. 2012, 1161). In other words, these mathematical modeling techniques enable access to explanations and understanding that otherwise would have been inaccessible. Furthermore, these modelers justify their use of explanatory information derived from their pervasively distorted model by demonstrating that the idealized mathematical model and the real-world target systems are in the same universality class.

6.3.2 Modeling Bacterial Growth

As a second example, several models from statistical physics that exhibit universal behaviors have recently been applied to problems in biology (Rice 2020). For example, many kinds of biological growth can be modeled by an extremely minimal computational model known as the Eden Growth Model (Eden 1961).[5] The Eden Growth Model is a lattice model with occupied and unoccupied sites. The simplest version of the model involves two basic growth rules:

1. Start from a single, occupied lattice site—called the "seed" site.
2. In each growth step, randomly occupy an empty site that is the nearest neighbor to an occupied site.[6]

Running this algorithm N times produces clusters containing N cells whose resulting topological features and limiting behaviors can be used to explain various features of growing biological tissues (figure 6.3).

Yet, as Eden himself explains, the model involves a myriad of idealizing assumptions:

> It has been necessary to make a large number of simplifying and special assumptions so that *the resemblance between the model and the growth of any complicated*

Figure 6.3
A typical Eden Growth Model cluster simulated on a two-dimensional lattice (Agyingi et al. 2018, 7). The darker cluster represents bacterial growth. The lighter cluster represents a wound on the skin.

metazoan or specialized organ or tissue is slight. We shall assume that each cell is identical with every other cell, that each cell is connected to at least one other cell, that the location of each cell is specified by a node in some regular lattice. For purposes of simplicity the model will be restricted to two dimensions. Any migration of cells, differentiations into specialized cell types, variations in cell size, and simpler properties of organisms will be neglected. Indeed, cells will be assumed immortal and a very special time-to-division distribution function is used. (Eden 1961, 224–225; emphasis added)

That is, the Eden Growth Model distorts most of the features of any real biological growth process (e.g., the cells, their interactions, and the structural features of the population), and instead focuses on the influence of the overall topological structure on the features of the cluster in the limit. What is striking is that, despite this pervasive distortion, across a range of biological and nonbiological cases, many features of the growing population can be explained by appealing to the universal limiting behaviors of various Eden Growth Models. For example, similar growth models have been used to investigate epidemics (Alexandrowicz 1980), the spread of forest fires, combustion fronts of burning paper (Zhang et al. 1992), and urban growth (Benguigui 1995).

This wide range of applications is possible because the Eden model has been shown to be in the Kardar-Parisi-Zhang (KPZ) universality class (Kardar, Parisi, and Zhang 1986). This class of systems is governed by the KPZ growth equation:

$$\frac{\partial h}{\partial t} = v\nabla^2 h + \frac{\lambda}{2}(\nabla h)^2 + F + \eta(x, t)$$

where v is the damping coefficient, λ is the growth parameter, F is constant drift, and η is an uncorrelated white noise. As Kardar, Parisi, and Zhang explain, their proposal of this modeling equation is "guided by the idea of universality" and seeks to "write down the simplest nonlinear local differential equation governing the growth of the profile to such processes as vapor deposition or the Eden model" (Kardar, Parisi, and Zhang 1986, 889).

While scientists and mathematicians are still analyzing this universality class, the following features have been identified as important for membership (Corwin 2016, 233):

1. *Locality*: Changes depend only on neighboring sites.
2. *Smoothing*: Large valleys are quickly filled in.
3. *Nonlinear slope dependence*: Vertical effective growth rate depends nonlinearly on local slope.
4. *Space-time independent noise*: Growth is driven by noise, which quickly decorrelates in space and time and does not display heavy tails.

Identifying this set of necessary features for displaying these universal behaviors is key to delimiting the class of systems that will (or will not) display those behaviors. Moreover, analyzing idealized models within this universality class is crucial for understanding how the limiting behaviors of interest counterfactually depend on these minimal features.[7] As a result, finding this universality class and the features important for membership provides an important link between the model systems and many real-world cases of biological growth. As mathematician Ivan Corwin explains, "a variety of physical systems and mathematical models, including randomly growing interfaces, certain stochastic PDEs [partial differential equations], traffic models, paths in random environments, and random matrices all demonstrate the same universal statistical behaviors in their long-time/large-scale limit. These systems are said to lie in the *Kardar-Parisi-Zhang (KPZ) universality class*" (Corwin 2016, 230).

As a specific example, Bonachela et al. argue that "Similar values for the exponents measured in different bacterial strains indicate that, despite the varying microscopic details and interactions specific to each strain, they share the same basic biological and physical ingredients at a larger scale of observation. That is, they show the same *universal behavior*" (Bonachela et al. 2011, 307). A large part of demonstrating that their idealized growth model is in a universality class that includes their target systems involves matching the characteristic exponents of the KPZ universality class. These exponents are a dynamic scaling exponent $z=3/2$, a growth exponent $\beta=1/3$, and a fluctuation (or roughness) exponent $\alpha=1/2$ (Bonachela et al. 2011; Corwin 2016). In discovering the match between these exponents, they determine that their growth model and many actual systems of bacterial growth are likely part of the KPZ universality class (Bonachela et al. 2011, 307).

However, these modelers also find that one of the exponents of their real bacterial colonies does not reliably match the exponents of the KPZ class. As a result, they look for a model in an alternative universality class that could capture all the observed exponents of the real systems of interest. This shift to a new universality class is achieved by constructing an idealized growth model in which the thermal noise is replaced by a quenched noise. This results in a universality class that they refer to as the *quenched* KPZ class (qKPZ) (Bonachela et al. 2011, 311).

By identifying these universality classes, these modelers are able to identify features of the system that the large-scale limiting behaviors of bacterial growth counterfactually depend on (e.g., their exponents, fractal dimension, and type of noise). Changing these noncausal features of the system puts the system in a different universality class that displays different limiting behaviors. This counterfactual dependence information is crucial to explaining why those limiting behaviors occur across the diverse systems included in the universality class.

Discovering these universality classes also helps these modelers determine which features of the biological systems are irrelevant to their universal behaviors. For example, by investigating their idealized models, they find that "irregular cell shape, long-range cell motility, and extracellular compounds are not necessary for our idealized cell groups to exhibit the same universal behavior as real bacterial colonies. This result suggests that such factors may not be responsible for the behavior of the colony patterns observed in experiments" (Bonachela et al. 2011, 313). In other words, investigation of the models within this universality class is also able to reveal explanatory information about the counterfactual irrelevance of many features to the occurrence of the explanandum. This counterfactual information is important to the explanation because some scientists have assumed that those features would be important to the macroscale behaviors of the growing bacterial colony (i.e., those features are contextually salient). In addition, recognizing the irrelevance of these features is an important part of explaining why the universal patterns of behavior are stable across bacterial colonies that are very different in terms of these physical features.

In sum, by discovering that these highly idealized growth models are within various universality classes, these modelers discover counterfactual information that allows them to explain the patterns of behavior observed across a range of bacterial colonies. Once again, according to these scientific modelers, discovering that their idealized models are in the same universality classes as various real-world systems is what justifies their use of these highly idealized models to investigate and explain various limiting patterns of bacterial growth. Moreover, appealing to the behaviors of those idealized models is justified, even if the models and the various real-world systems display those patterns because of very different causes, mechanisms, and interactions. In addition, when the real systems of interest fall outside of the universality class of the idealized model, these modelers try to construct an idealized model in a different universality class that can capture those real-world systems.

6.3.3 Gaussian Universality

A more familiar example of universality is the Gaussian distribution. It is an extremely striking and useful fact that samples taken from a wide range of real, possible, and model systems will all approximate a Gaussian curve whenever there are sufficiently many trials and the outcomes of those trails are largely statistically independent. These universal behaviors are widely observed because the statistical features of Gaussian curves are largely independent of the physical details of the components, causes, or dynamical processes that operate in the system. Indeed, the central limit theorem tells us that the distribution of *any* random and statistically independent sample will approximate the Gaussian distribution. This is a crucial part of mathematically demonstrating the universality of these behaviors. As Corwin explains, "The universality of the Gaussian distribution was not broadly demonstrated until [the] work of Chebyshev, Markov and Lyapunov around 1900. The *central limit theorem* (CLT) showed that the exact nature of [the chance set up] is immaterial—any sum of independent identically distributed random variables with finite mean and variance will demonstrate the same limiting behavior" (Corwin 2016, 231).

Even before the Gaussian universality class was more explicitly delimited and explained, scientific modelers had long recognized the explanatory value of making various idealizing assumptions to allow for the application of statistical modeling techniques (Ariew, Rice, and Rohwer 2015; Ariew, Rohwer, and Rice 2017; Morrison 2015; Stigler 2010). In chapter 3 we saw how such an approach was leveraged by Galton, Boltzman, and Fisher to develop statistical models of inheritance, gas behavior, and biological evolution, respectively. What is important to note is that these statistical modelers attempted to exploit universal mathematical features of complex systems that are independent of the details of their physical causes, mechanisms, and interactions. Doing so enabled them to study highly idealized statistical models that distorted many of the difference-making causes of real-world systems, and yet still allowed them to investigate the universal macrobehaviors approximated by the model systems and various real-world systems in their large-scale limits.

By making certain idealizing assumptions (e.g., about the size and independence of the sample population), scientific modelers can use statistical modeling techniques that would not otherwise be applicable. The application of these modeling techniques, in turn, enables scientists to provide explanations of various real-world phenomena that would otherwise be inaccessible. The universality of the limiting behaviors captured by statistical laws, such as the central limit theorem, are what provide the necessary link between these highly idealized model systems and the various real-world systems whose behaviors they are used to explain. As a result, by discovering that various idealized statistical models and real-world systems are all in the Gaussian universality class, scientists can justifiably appeal to results derived from those idealized statistical models to explain and understand the behaviors that they observe in real-world populations. As Tao explains regarding the Gaussian distribution:

> The law is universal because it holds regardless of exactly how the individual components fluctuate or how many components there are (although the accuracy of

the law improves when the number of components increases)....That the macroscopic behavior of a large, complex system can be almost totally independent of its microscopic structure is the essence of universality. (Tao 2012, 25)

Moreover, in this case, scientists have worked out detailed theorems for determining the degree to which a system will approximate these limiting behaviors (see Ariew, Rohwer, and Rice 2017 for a discussion). In particular, the larger the number of events, and the more statistically independent those events are, the more closely the behavior of the system will approximate the Gaussian distribution. This allows the universality class that justifies the use of highly idealized statistical models to explain to stretch even further, and justify applications to systems where the conditions of the statistical limit theorems are violated to some degree. The justification for applying the idealized statistical models will certainly be weakened when this occurs, but for some applications, it will be sufficient to justify using the statistical model for purposes of explanation and understanding (see chapter 3 for some examples). This illustrates how the justification provided by universality can extend to systems that meet the features required for membership in the universality class only to some degree, but still *approximate* the behaviors of systems within that class.[8]

In sum, delimiting the Gaussian universality class has enabled scientific modelers to justify the use of various idealized statistical models that are in that universality class to explain the behaviors of various real systems in or near that class—even when the models drastically distort features that are difference-makers for the explanandum. This is accomplished by using various modeling techniques to identify the statistical features of the population on which the phenomenon counterfactually depends (e.g., its mean and variance) and showing that the results are stable across a variety of changes to other features of the system, such as the system's components, causes, or mechanisms (Ariew, Rice, and Rohwer 2015). Even though this takes place within an idealized modeling framework that (sometimes drastically) distorts the processes that mediate these dependence relationships, investigating idealized models within the Gaussian universality class has enabled scientists to provide a plethora of explanations across a wide range of sciences.

6.3.4 The Universality of the Tracy-Widom Distribution

As a final example, I want to look at a case where scientists are just beginning to understand the features of a more recently discovered universality class. While some instances of universality are well understood, other instances are the focus of extremely active research in mathematics (Tao 2012, 25). For example, the universality of the Tracy-Widom distribution has been used to justify various kinds of interdisciplinary applications of mathematical modeling techniques. The Tracy-Widom distribution is the probability distribution of the normalized largest eigenvalue of a random matrix (Tracy and Widom 1993, 1994). In more physical terms, in contrast to Gaussian distributions that focus on statistically independent variables, the Tracy-Widom distribution describes a crossover function between a phase of strongly coupled components and a phase of weakly coupled components (see figure 6.4).

What is surprising is how universal the Tracy-Widom distribution is, and mathematical modelers are beginning to understand some of the features

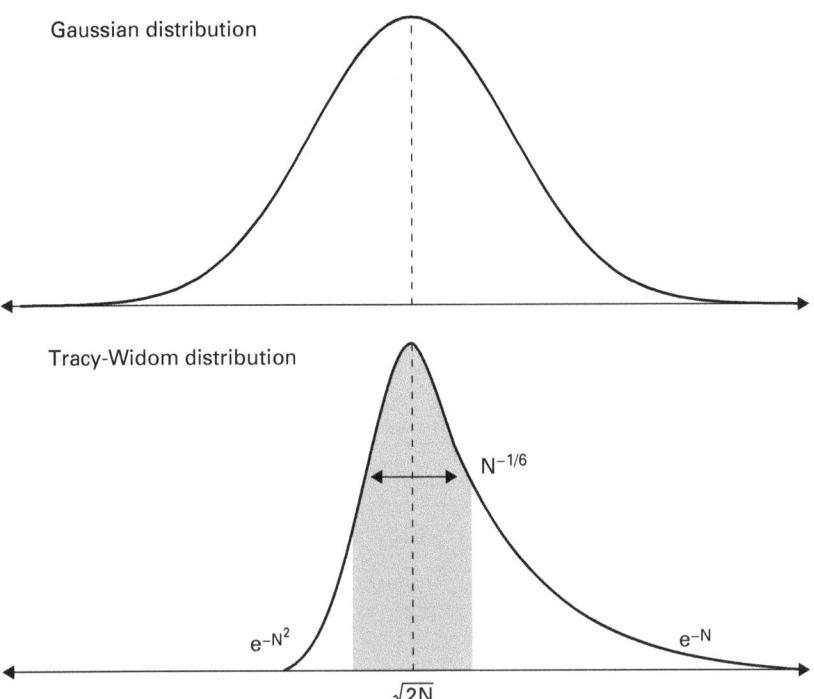

Figure 6.4
The Tracy-Widom distribution in comparison with a Gaussian distribution.

on which this universal limiting behavior depends. For example, Majumadar and Schehr found that Tracy-Widom distributions

> have emerged in a number of *a priori* unrelated problems such as the longest increasing subsequence of random permutations, directed polymers and growth models in the Kardar-Parisi Zhang (KPZ) universality class . . . sequence alignment problems, mesoscopic fluctuations in quantum dots, height fluctuations of nonintersecting Brownian motions over a fixed time interval, . . . finance, . . . experiments on nematic liquid crystals . . . and in experiments involving coupled fiber lasers. (Majumadar and Schehr 2013, 7)

In other words, scientists have determined that this universality class includes a diverse range of real, possible, and model systems that display similar limiting behaviors but are drastically different in terms of their components, causes, and interactions (Majumdar and Nechaev 2004). In addition, "it is now known that a number of discrete models of the KPZ universality class follow the exact Tracy-Widom distribution" (Buchanan 2014, 543). That is, many of the models in the KPZ universality class described here are also in the Tracy-Widom universality class. Consequently, we see that multiple universality classes can overlap (i.e., a single system can be a member of multiple universality classes). As we will see in chapter 7, this allows multiple idealized models in different universality classes to be used to explain and understand the same phenomenon. In addition, because an idealized model can be a member of multiple universality classes, the same model might be used to investigate different universal behaviors in different target systems.

The widespread applicability of models that display the Tracy-Widom distribution occurs because this limiting distribution is universal with respect to changes in the microscopic details of the system (Majumadar and Schehr 2013; Wolchover 2014). As Schehr put it in an interview: "The fact that it pops up everywhere is related to the universal character of phase transitions . . . this phase transition is universal in the sense that it does not depend too much on the microscopic details of your system" (Wolchover 2014, 6). Discovering this enabled these mathematical modelers to identify a set of mathematical modeling techniques that could be used to investigate the universality of various limiting behaviors of systems that display the Tracy-Widom distribution. Indeed, they "realized there was a deep connection between our probability problem and this third-order phase transition that people had found in a completely different context" (Wolchover 2014, 6). In other words, they used the universality of the behaviors of interest to justify the employment of mathematical modeling techniques that had been successfully used to investigate similar phase transitions in other cases.

As Majumadar and Scheher explain, the unique behavior of the systems being modeled "leads, in the limit $N \to \infty$, to a phase transition at the critical point" (Majumdar and Scheher 2013, 9). This third-order phase transition occurs due to a discontinuity at the point where the critical exponent $\alpha = 1/\sqrt{2}$. Despite ignoring any physical details of the causes or mechanisms of real systems, these mathematical modeling results "carry crucial information concerning phase transitions in the system in the form of singularities and hence are useful and important objects to study" (Majumdar and Scherer 2013, 28). Indeed, these idealized mathematical models have enabled scientists to start to explain and understand various reoccurring limiting behaviors of physical systems despite drastically distorting (or ignoring) most of the causes or mechanisms of any real-world system in which those phenomena occur.

The Tracy-Widom distribution highlights another important feature of the universality account by showing that this justification will be stronger or weaker in different cases. For example, in some cases, scientists will have provided (1) demonstrations or evidence showing that there is a universality class and (2) some explanation of why certain behaviors are universal across precisely those systems in that class (e.g., by using the renormalization group). In such a case, the appeal to the empirical fact of universality to justify the use of an idealized model within that class will be particularly strong because we know precisely which systems are included in the class and why. This is illustrated in the case of minimal model explanations discussed in chapter 3, the melting ice ponds case, and the case of Gaussian universality.

In a second kind of case, scientists may have only explicitly shown (1) without being able to produce (2). This is illustrated by the bacterial growth case. While these modelers have acquired strong evidence that there is a universality class that contains the real systems and their idealized model system, they have yet to provide a further explanation of why those patterns are universal across just those systems in the class.

I have argued that both of these situations are sufficient to justify the use of highly idealized models to explain and understand universal behaviors that occur in real systems. That is, I have argued that having a further explanation of why those patterns are universal is not necessary for this justificatory story to be told—although having that further explanation would certainly improve the degree of justification. If scientists know that there is a universality class that contains the real systems of interest as well as their idealized model, then they are justified in using the idealized models to extract the dependence and independence information required to explain and understand the phenomenon.

However, there will be numerous cases in science that fall into a third category (e.g., the Tracy-Widom case), in which scientists have not yet provided (1) or (2). Here, I think appealing to the available evidence that some observed pattern is likely to be universal across a set of real and possible systems can still provide some justification for using the model to extract information—even if additional investigation is required to explicitly demonstrate (1) and (2). The point is that the variety of cases surveyed here show that there will be differences in the ways in which universality classes are employed in the justification of idealized modeling, and those different uses will provide different degrees of justification. In short, the quality and quantity of evidence that scientists acquire to support the claim that there is a universality class that includes the real systems of interest and the idealized model will directly influence the degree to which they are justified in drawing explanatory inferences from the behavior of the idealized model.

There are certainly other examples of scientific modelers appealing to universality that could be included here (e.g., Batterman 2000, 2002b), but the examples discussed in this chapter serve to illustrate that scientific modelers often appeal to universality classes in order to justify their use of idealized modeling techniques to extract explanatory information. The general goal of these idealized modeling techniques is to discover universal behaviors that are independent of most of the physical features (e.g., the causes, mechanisms, and process) of the system. Discovering these universal behaviors (and their stability) then enables scientists to investigate highly idealized models that drastically distort the physical features and dynamics of real-world systems, but preserve the universal patterns of behavior that are of interest. By investigating the behaviors of these idealized models, scientists can extract explanatory information about which features are counterfactually relevant and irrelevant to the occurrence of the explanandum.

More generally, suppose that we could show that there are good reasons to believe that the idealized models used by scientists to discover explanatory information are within the same universality classes as the real physical systems whose behaviors they want to explain. If this were generally the case, we could justify scientists' use of those idealized models to explain and understand the behaviors of real systems, despite their being holistically distorted representations of the difference-making components, interactions, and causes of their target systems (Rice 2017, 2018). In this way, universality can provide an alternative link that can be (and often is) used to justify the use of holistically distorted models to provide scientific explanations.

6.4 Universality and Other Concepts of Stability

Before moving on to some additional appeals to universality in cases of multiscale modeling, it is important to understand how the concept of universality relates to some other important concepts regarding stable patterns that have been appealed to in the philosophical literature. One of the central aims of scientific theorizing is to discover and explain stable patterns across various systems. While this general goal of science has been characterized in several ways by philosophers, I will argue that many of them can (and should) be subsumed under the universality account.

To begin with, the concept of universality can be used to capture philosophers' frequent appeals to multiple realizability (Batterman 2000). In fact, scientists from a wide range of fields use the term *universality* essentially in the way that philosophers use the term *multiple realizability* (e.g., Batterman 2000, 2002b, 2017). This is because we can think of a multiply realizable pattern or property as a universal feature or pattern of a universality class. For example, Batterman suggests that in the case of explaining the multiple realizability of pain:

> Suppose that physics tells us that the physical parameters α and γ are the (only) relevant physical parameters for the pain universality class. That is, that N_h, N_r, and N_m have these features in common when certain generalizations or regularities about pain are the manifest behaviors of interest observed in each of humans, reptiles, and martians. Equivalently, physics has told us that all the other microdetails that legitimately let us think of N_h, N_r, and N_m as heterogeneous are irrelevant. We then have our explanation of pain's realizability by wildly diverse realizers. (Batterman 2000, 133)

In other words, the concept of universality can easily capture the notion of stability across heterogeneous realizers, which is central to the concept of multiple realizability. As Batterman explains, "multiple realizability is the idea that there can be heterogeneous or diverse 'realizers' of 'upper level' properties and generalizations. But this is just to say that those upper level properties and the generalizations that involve them—the 'laws' of the special sciences—are universal" (Batterman 2002b, 4).

The problem with multiple realizability is that it comes with a lot of metaphysical baggage from various debates in philosophy of mind that focus narrowly on concepts such as the *supervenience* of higher-level properties on lower-level properties. This is one way that a pattern might be universal, but it is certainly not the only way. Moreover, I suggest that our characterization of science's quest to identify and explain patterns should not be tied to notoriously problematic metaphysical concepts such as supervenience, levels of nature, or instantiation of properties (Potochnik 2009a). Universality provides a more metaphysically neutral way to characterize the notion of stability of patterns across changes. Most important, however, as Bill Wimsatt (2007) points out, multiple realizability and supervenience are rarely appealed to by practicing scientists. This is because "supervenience could be important for an omniscient LaPlacean demon but not for real, fallible, and limited scientists" (Wimsatt 2007, 66). Indeed, discussions of multiple realizability and supervenience tend to focus on cases where *in principle* we have all the knowledge available about the ways that a pattern could be realized. But *in practice* this is rarely (if ever) the case. Thus, it is unclear what kind of normative guidance concepts like multiple realizability and supervenience can offer to practicing scientific modelers who aim to explain and understand complex systems about which they necessarily have limited knowledge. In contrast, scientific modelers widely use the concept of universality to justify applications of modeling techniques—even when they have limited knowledge about precisely where and why that universality obtains.

In addition, discussions of multiple realizability often seem to assume that the stability of the macroscale properties is completely unexplainable or is just a brute fact about our universe (Batterman 2000). However, scientists are often explicitly interested in explaining why various patterns are stable with respect to particular changes to various features of the system. When this occurs, scientists would like an explanation of the stability itself. As Batterman puts it, the real question concerning multiple realizability is, "How can systems that are heterogeneous at some (typically) micro-scale exhibit the same pattern of behavior at the macro-scale?" (Batterman 2017, 4). The concept of universality (classes) focuses our attention not only on identifying which systems will display a particular pattern, but also on identifying the various features of the system that are counterfactually relevant and irrelevant to displaying those patterns. That is, universality focuses our attention on explaining why heterogeneous systems nonetheless display stable patterns of behavior.

What's more, in chapter 7 I will argue that the notion of higher-level properties being stable across various lower-level realizers fails to capture the variety of kinds of stability across perturbations that can occur between different spatial and temporal scales of the system. In contrast, universality provides the tools to characterize the autonomy of particular features at macroscales, microscales, and mesoscales of the system from particular features at other scales of the system. That is, unlike multiple realizability, universality does not require the assumption that the stable behaviors are always at the macroscale and are realized in heterogeneous ways at smaller scales.

Next, many accounts in the explanation literature have focused on the role of generality in explanations (Garfinkel 1981; Jackson and Pettit 1992; Kitcher 1981; Putnam 1975). For example, Strevens argues that explanations ought to be given using the most general causal model that still entails the occurrence of the explanandum (Strevens 2008). Moreover, many philosophers have argued that the explanations provided by higher-level sciences are superior due to their generality (Putnam 1975). The problem with these blanket appeals to generality is that being more general does not always improve an explanation (Potochnik 2009a). Instead, I argue that the preference for generality or detail in our explanations is largely a matter of our particular explanatory interests (Sober 1999). Accounts of explanation and modeling ought not decide that generality or detail are preferable a priori. Rather, we ought to allow that different preferences and contexts will favor one or the other kind of explanation. Furthermore, as Potochnik (2017) points out, we don't just want models and explanations that are as general as possible, but rather we want explanations that are general in the right way and to the right degree. In addition, we often would like our explanations to specify the scope of those patterns (i.e., tell us the range of systems over which they apply) and tell us something about why the pattern occurs across very different systems (i.e., tell us why the pattern is stable).

The universality account captures these ideas concerning the scope of general patterns by requiring that universality classes capture the relevant set of real or possible systems involved in the actual and counterfactual scenarios salient to providing the desired explanation. In other words, the universality class that is appealed to needs to include the relevant set of counterfactual situations that are made salient by the explanatory context. Moreover, discussions of universality typically focus on providing reasons for the stability of the pattern across heterogeneous systems. Merely appealing to generality or similar causes as a criterion for explanation and modeling fails to capture these important features of how many general phenomena are explained in actual scientific practice. The point is that rather than simply favoring general explanations, we require an account that recognizes the importance of delimiting the group of systems that will display a pattern and showing why the differences among those systems are irrelevant to their universal behaviors. This is not to say that other accounts of explanation that appeal to generality cannot be supplemented in a way that would incorporate these features, but given that universality focuses our attention on these kinds of questions for free, so to speak, I think that we have reason to favor the use of the concept of universality over blanket appeals to the value of generality.

Finally, a common way to conceive of stability across changes is to use Woodward's concept of invariance (Woodward 2003). As I noted in chapter 4, one issue with the concept of invariance is that Woodward's account of explanation restricts the relevant notion of invariance to invariance under interventions: "Invariance under at least one testing intervention (on variables figuring in the generalization) is necessary and sufficient for a generalization to represent a causal relationship or to figure in explanations" (Woodward 2003, 250). That is, for him, a relationship is causal and explanatory just in case it is stable across some interventions. In a similar way, Potochnik (2015, 2017) argues that we ought to capture scientists' interests in stable patterns by focusing on the role of invariant causal patterns in science.

While causal patterns are certainly an important aspect of scientific theorizing, as we have seen, many of the explanations and patterns of interest to scientists are noncausal. What's more, many of the stable patterns of interest to scientists range across systems that are extremely heterogeneous with respect to their causes—even if we abstract to more macroscales. In fact, in many cases, a pattern is particularly interesting to scientists precisely *because* it is stable across systems whose causes, mechanisms, and other physical features are drastically different (Lange 2012). As Bokulich notes, the key here is that many explanations in science account for this universality by decoupling the explanation from the particular types of causal stories that might realize it. It is not because the model explanation is idealized, leaves out many causal details, or because it is formulated in terms of an abstract mathematical model, that makes it non-causal. The [explanation] is non-causal because it is not a representation of the causal processes at all. (Bokulich 2018, 161)

As a result, given that invariance is typically tied to manipulationist accounts of causal explanation, universality provides a broader notion of stability across changes that captures a wider range of causal and noncausal cases—including patterns that range across systems that are very different with respect to their causes.

In addition, while Woodward's notion of invariance helps identify how certain patterns of dependence can be used to explain, he provides little detail about how to explain *why* those patterns are invariant. As Batterman points out: "Woodward is not concerned to answer why-questions about the universality or degree of universality of the regularities that he discusses. That is, he does not, as far as I call tell, ask the question why the regularity has the robustness that it has or has it to the degree that it has" (Batterman 2002b, 59). This is crucial because answering the question, "Why is this pattern universal despite perturbations to these particular features?" is central to scientific theorizing and requires information about the irrelevance of most of the features of the systems in which the pattern occurs. Simply citing that a pattern is stable across a range of cases (i.e., specifying its scope) fails to provide an explanation of why that pattern is stable across those particular perturbations (Batterman 2002b; Batterman and Rice 2014). Consequently, our account of the role of stable patterns in science ought to highlight the crucial role of demonstrating and explaining the stability of patterns across changes to irrelevant features. The concept of universality does precisely that.

6.5 Conclusion

In addition to being able to capture several notions of stability, the main advantage of the concept of universality is that universality classes provide a clear link between the holistically distorted models used by scientists and the complex systems whose behaviors they would like to explain without having to rely on accurate representation relations or mirroring of physical processes. While I have focused on just a few cases here, the use of universality classes to justify the employment of highly idealized models to explain can be (and has been) applied across the sciences.

For example, in introducing the importance of cellular automata models, Wolfram tells us that "Universality implies that many details of the construction of a cellular automaton are irrelevant in determining its qualitative behavior. Thus, complex physical and biological systems may lie in the same universality classes as the idealized mathematical models provided by cellular automata" (Wolfram 1984b, 1). In fact, Wolfram goes on to provide evidence that all one-dimensional cellular automata fall into four basic classes that yield four characteristic limiting behaviors: a homogeneous state, a set of separated simple stable or periodic structures, a chaotic pattern, and complex localized structures (Wolfram 1984a, 1984b). These results are significant because "All cellular automata within each class, regardless of the details of their construction and evolution rules, exhibit qualitatively similar behavior. Such universality should make general results on these classes applicable to a wide variety of systems modelled by cellular automata" (Wolfram 1984a, 419).

In addition, physicists studying self-organized criticality (SOC) have argued that "Universality *justifies* the simplified models in SOC, which ignore all but a few details of what they are modelling, such as ricepiles or evolution, and allows them to display the 'right' behavior" (Pruessner 2012, 19; emphasis added). These researchers also suggest that the main challenge in modeling these critical behaviors is "the identification of universality classes containing models that display solid scaling behavior" (Pruessner 2012, 5). In other words, the fact that these highly idealized models are within the same universality class as their target systems justifies their use, despite the fact that the models ignore or distort most of the features of the phenomenon.

This is particularly useful for scientific modelers, given that the causal dynamics or mechanisms of many complex systems are simply unknown or inaccessible. For example, many practical modeling situations are similar to the one discussed by Barzel and Barabási:

> In many systems of practical importance the analytical form of the dynamics is unknown; hence, we cannot predict the system's behaviour from [the dynamical equations of the system]. Yet, the link we established between the universal exponents . . . and the macroscopically accessible . . . distributions allows us to determine a system's universality class even without knowing the analytical formulation of its dynamics. (Barzel and Barabási 2013, 679)

In other words, identifying a set of universality classes can allow scientific modelers to propose highly idealized models that can be used to model the stable macroscale behaviors of a wide range of heterogeneous systems, even when the dynamical mechanisms or processes of those systems are unknown or unmeasurable.

In conclusion, identifying universality classes and understanding the details of various instances of universality can help justify trusting the results of highly idealized models that are holistically distorted caricatures of any real-world system. In light of these cases, I suggest that philosophers of science continue to explore the ways in which scientific modelers actually justify their uses of idealized models and move away from relying exclusively on the accurate representation requirements derived from the standard view. Doing so will better capture the justifications that should be (and are) offered by scientific modelers and will help develop a more pluralistic account of the multiple model-world relationships that can be used to justify appeals to holistically distorted models to explain and understand the phenomena we observe.

7 Multiscale Modeling and Universality

Instead of asking whether explanations or ontology at one level can be reduced to explanations or ontology at another level, some philosophers of science have started to analyze the unique ways that scientists model phenomena across multiple spatial and temporal scales (Batterman 2013, 2015, 2017; Bursten 2018a; Green and Batterman 2017; Winsberg 2006, 2010). Following this trend, instead of focusing on the ontological, compositional, or part-whole relationships that figure prominently in most philosophical discussions of reduction and emergence, this chapter focuses on questions of how best to model complex systems whose relevant features span a wide range of spatial and temporal scales that interact with one another. Understanding this kind of *multiscale modeling* requires us to pay far more attention to the details of the modeling techniques used to study these systems than is typically found in other philosophical discussions of supervenience, levels of organization, or reduction (Batterman 2017; Bursten 2018a; Jhun 2019; Winsberg 2010). What's more, we will again see that the concept of universality plays a crucial role in justifying the application of these multiscale modeling techniques to discover explanatory information. However, instead of focusing on justifying the use of a single idealized model to explain or understand, here the focus will be on using universality to justify the use of multiple conflicting models across a wide range of scales to study the same phenomenon.

Scientific modeling typically begins by identifying a set of temporal and spatial scales at which the important features of the phenomenon are thought to occur. This is methodologically useful because most mathematical modeling techniques are designed to represent processes only within a narrow range of scales—what scientists often refer to as the *characteristic scale* of the processes being modeled.[1] In addition, the observations used in constructing a scientific model are typically made at particular spatial and temporal scales because making measurements across multiple scales is often difficult or impossible. Mathematical models are then developed to describe the features and processes at the characteristic scales at which observations can be (or have been) collected.

The issue is that many of the complex phenomena that scientists would like to model depend on features and processes occurring across an extremely wide range of scales. For example, in biology, the phenomenon to be modeled can often occur across spatial scales ranging from the molecule (10^{-10} m) to the living organism (1 m), and time scales can span from nanoseconds (10^{-9} s) to years (10^8 s) (Castiglione et al. 2014, 2). In fact, these scales only capture the biological scales involved in modeling processes within an individual organism (figure 7.1). Other complex biological phenomena (e.g., the behavior of an ecosystem) often depend on features

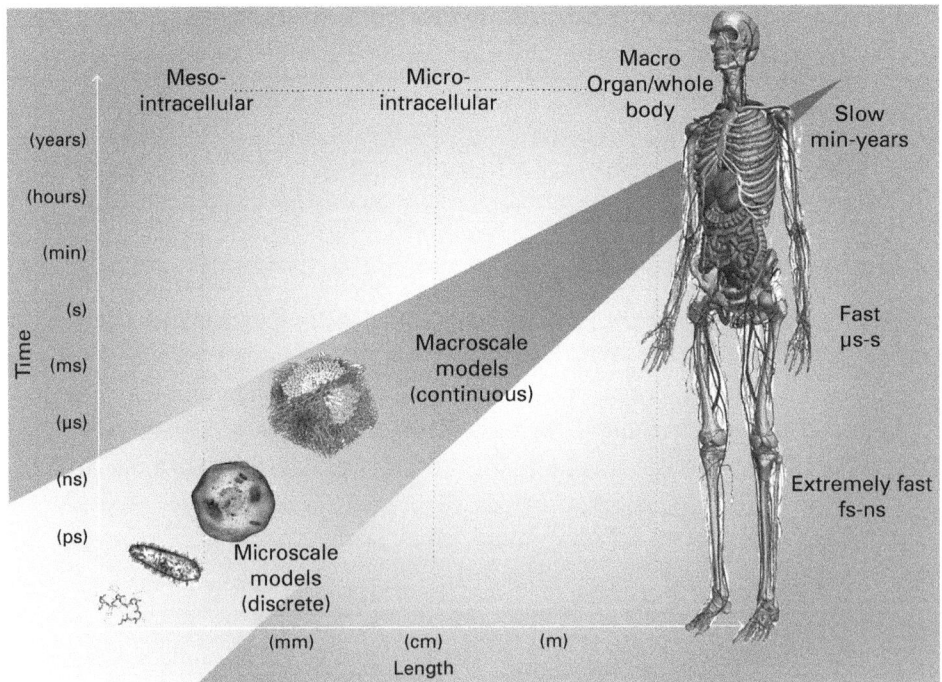

Figure 7.1
Multiscale models of the human body target complex processes that span a wide range of length and time scales (Castiglione et al. 2014, 3).

and processes that range in spatial scale from micromolecules to the whole ecosystem and in temporal scale from nanoseconds through multiple generations (Dada and Mendes 2011, 87).

As a result, "To understand the behaviour of a biological system, whether it is natural or engineered, requires models that integrate the various interactions that occur on these diverse spatial and temporal scales" (Dada and Mendes 2011, 86). For example, the task of just modeling the human heart requires using at least three types of models to capture processes at the scale of ion channels, cells, and tissues (Dada and Mendes 2011). Consequently, "One of the major themes in current biological modelling is the development of theoretical frameworks which allow for the investigation of systems characterised by multiple length scales" (Belmonte-Beitia et al. 2013, 1). Indeed, this kind of multiscale modeling has become widespread across the sciences.

Modeling phenomena that span such a wide range of spatial and temporal scales raises a variety of unique modeling challenges. One issue that often arises is the use of multiple conflicting models to explain and understand the same phenomenon. These models make incompatible claims about the features, causes, ontology, and dynamics of their target systems. When this occurs, the challenge is to show how such a situation can avoid inconsistent metaphysical claims about the models' target systems while still yielding genuine explanations (and understanding). As we saw in chapter 1, this is often called the *problem of inconsistent models* (Chakravartty 2010; Massimi 2018; Morrison 2011, 2015; Rice 2019b).

Another challenge faced by multiscale modelers is called the *tyranny of scales* (Batterman 2013; Green and Batterman 2017; Wilson 2017). This problem arises because, while most approaches to modeling complex systems are designed to handle only a narrow range of characteristic scales, many complex phenomena depend on features and processes of the system that occur across multiple scales of the system (which interact). In other words, while scientific modelers often know that the phenomenon of interest involves relevant features across distant spatial and temporal scales, the modeling techniques that they have available are typically restricted to representing features at only one spatial or temporal scale (or a few).

Therefore, when it comes to multiscale modeling, "despite the attractiveness of this method, it faces many challenges, such as the gaps between models of different scales and inconsistencies between different methodologies"

(Qu et al. 2011, 23). The problem of inconsistent models focuses on the inconsistency of the various modeling methodologies used to capture features at different scales. The tyranny of scales focuses on the challenge of bridging the gaps between different types of models used to represent features across very different scales.

In this chapter I argue that the problem of inconsistent models and the tyranny of scales can both be (at least partially) solved by moving away from the standard view's focus on accurate representation and toward the account based on universality classes defended in the previous chapter. In section 7.1 I show how the focus on accurate representation relations leads to the problem of inconsistent models, and in section 7.2 I argue that my account based on universality classes can avoid those issues. Then, in sections 7.3 and 7.4, I use the concept of universality to tackle the tyranny of scales by analyzing three multiscale modeling techniques that focus on scale-dependent, scale-invariant, and interscale modeling. Section 7.5 extracts some more general lessons for philosophical debates concerning modeling, explanation, and reduction.

7.1 The Problem of Inconsistent Models

As I noted in chapter 1, in many cases scientific modeling involves multiple conflicting idealized models being simultaneously used to explain and understand the same phenomenon (Chakravartty 2010; Longino 2013; Mitchell 2009; Morrison 2011, 2015; Weisberg 2007a, 2013). For example, Wimsatt notes: "In biology and the social sciences, there is an obvious plurality of large-, small-, and middle-range theories and models that overlap in unclear ways and usually partially supplement and partially contradict one another in explanations of interactions of phenomena at a number of levels of description and organization" (Wimsatt 2007, 179–180). In addition, Weisberg describes how the last eighty years have seen several theoretical models that were developed to account for molecular structure that "make different approximations and idealizations," which results in the models saying "different things about the phenomenon of covalent bonding" (Weisberg 2008, 933–934). Wendy Parker also describes how "complex climate models generally are physically incompatible with one another—they represent the physical processes acting in the climate system in mutually incompatible ways and produce different simulations of climate"

(Parker 2006, 350). Wilson (2017, 202–203) has also discussed various cases in physics in which multiscale modeling results in "descriptive conflicts" with respect to the properties of materials. Similarly, Longino shows that studies of human aggression and sexuality resulted in several incompatible methodological approaches that each "parses the causal space" differently by focusing on certain causal interactions and explicitly ignoring or idealizing others (Longino 2013, 126).

As a more specific example, Morrison (2011, 2015) describes the use of more than thirty idealized models to study nuclear phenomena (figure 7.2). The problem is that the assumptions made by any one of these models is in conflict with fundamental claims made by the others. For example, the liquid-drop model and the shell model make contradictory assumptions about the fundamental nature of the elements and interactions involved (Morrison 2011, 347–349). In addition, "some models assume that nucleons move approximately independently in the nucleus . . . while others characterize the nucleons as strongly coupled due to their strong short range interactions" (Morrison 2011, 347). What's more, these conflicting assumptions are typically necessary for the models to produce the behaviors of interest to physicists: "nuclear spin, size, binding energy, fission and several other properties of stable nuclei are all accounted for using models that describe one and the same entity (the nucleus) in different and contradictory ways" (Morrison 2011, 349).

As these examples illustrate, rather than simply modeling different features (or aspects) of the target system in complementary ways, genuinely inconsistent models often make contradictory assumptions about the target system, yield incompatible causal claims, and represent the system's basic ontology in fundamentally inconsistent ways. Furthrmore, in many instances some of these models will be discrete while others will be continuous, some will be agent-based descriptions of smaller scales while others will describe cellular automata at more macroscales, and so on (Bascompte and Solé 1995; Deisboeck et al. 2011). The problem of inconsistent models is seeing how we can interpret these sets of conflicting idealized models constructed within different modeling frameworks as providing genuine explanations and understanding of the same phenomenon.

I argue that this problem arises largely because of the overwhelming focus on accurate representation of relevant (e.g., difference-making) features, which is central to the standard approach outlined in chapter 2.

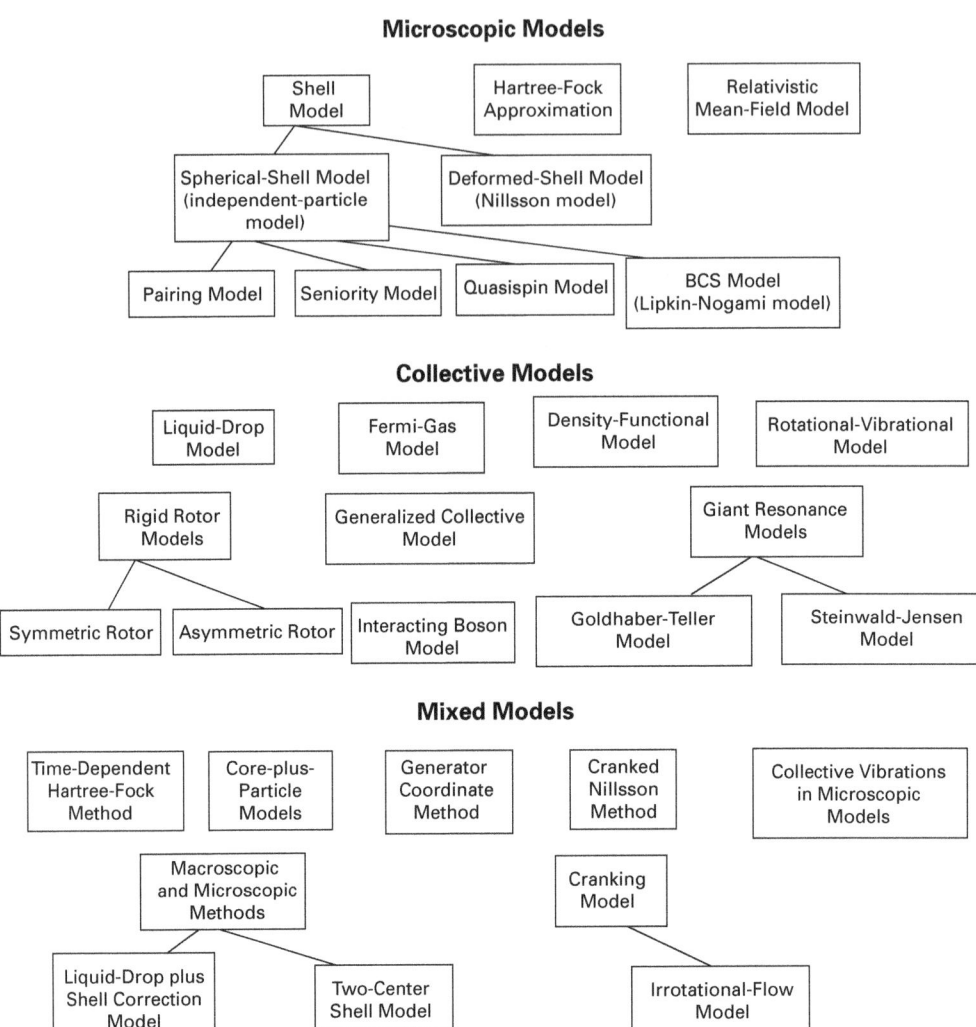

Figure 7.2
The various kinds of conflicting idealized models used to investigate the nucleus (Morrison 2011, 348).

Given these accuracy requirements, many philosophers have noted that the use of multiple inconsistent models for the same phenomenon raises serious challenges for scientific realism (e.g., see Massimi 2018 for an extended discussion). Indeed, if we were to adopt the standard approach, then when multiple conflicting models are used to explain and understand the same phenomenon, they ought to be interpreted as making conflicting claims about the causal interactions among the relevant features of the models' target systems. In other words, multiple conflicting models that are used to explain and understand the same phenomenon ought to be interpreted as aiming to provide an accurate representation of the relevant causes or mechanisms that gave rise to the explanandum. This is problematic, given that the models used in science often make conflicting assumptions about the fundamental ontology of the system (Morrison 2011, 2015), represent incompatible causal structures (Longino 2013), and distort difference-making causes in a variety of ways (Batterman and Rice 2014; Rice 2013, 2017, 2018).

One possible response to this plethora of conflicting models is to appeal to robustness analysis. For example, following Levins (1966) and Wimsatt (2007), Weisberg suggests that the proliferation of multiple conflicting idealized models for the same phenomenon "requires some way of identifying which aspects of models make trustworthy predictions and can reliably be used in explanations. I have argued that one way for theorists to get a handle on this proliferation is through the use of *robustness analysis*" (Weisberg 2013, 174). According to Weisberg, robustness analysis is a two-step procedure. The first step examines a group of models to determine if they all predict a similar result, called a "robust property." Then the models are analyzed to determine if they share a "common structure which generates the robust property" (Weisberg 2013, 158).

While robustness analysis may be helpful for analyzing the relationships between multiple models in certain cases, it will rarely be applicable in cases of using multiple conflicting idealized models from different modeling frameworks. This is because robustness analysis is prefaced on there being a common structure (e.g., common parameters or assumptions) among the models that is responsible for generating the robust results in each of them. However, the problem of inconsistent models typically arises in cases where the various idealized models do not share a common structure that gives rise to their predictions. That is, in most cases in which the problem of inconsistent models arises, it is because the models are explicitly formulated within different modeling frameworks that use different fundamental assumptions, different idealizations, different abstractions, and different dependence relationships. For example, Eric Winsberg describes cases in nanoscience in which the modeling methods used "embody mutually inconsistent frameworks. They each offer fundamentally different descriptions of matter, and they each offer fundamentally different mathematical functions describing the energetic interactions among the entities they describe" (Winsberg 2010, 80). Hence, it is unlikely that the multiple models will generate the same predictions regarding the target phenomenon—let alone as a result of a common structure that they all share. As a result, while robustness analysis may be helpful in certain instances, in most cases, the multiple conflicting models will just be too different for robustness analysis to be helpful.

Parker (2011) suggests a somewhat different use of robustness to test whether the results from a set of models have empirical significance. The important difference from Weisberg's discussion is that Parker explicitly focuses on what she calls "multimodel ensemble studies," in which the models differ in their equations, their parameter values, their spatiotemporal resolution, their solution algorithms, and their computing platforms (Parker 2011, 582). As a result, in these cases, the models will often be inconsistent in just the ways discussed previously. However, while alternative ways of employing robustness analysis might show that the results should improve our confidence in the predictions derived from the models, these applications fail to address the inconsistent representations provided by the set of models. For one thing, as Parker notes, the sets of models subjected to robustness analysis in these cases do not cover a wide-enough range of possibilities to be able to infer anything about the truth of the models or to confirm the truth of the hypotheses derived from them (Parker 2011, 584–586). In particular, Parker argues that when it comes to determining how to interpret what these climate change models get right, "it is not yet possible to make the argument . . . from robustness to likely truth" (Parker 2011, 589). In short, improving our confidence in various predictions still fails to address the inconsistent representations offered by the set of models. More generally, I suggest that robustness analysis is simply not the right tool for solving the problem of inconsistent models. Although alternative versions of robustness analysis may be useful for improving our confidence (or security) in a prediction in certain cases, showing that the predictions of a set of inconsistent models is robust fails to address the question of how a scientific realist ought to interpret the use of multiple inconsistent models to explain and understand the same phenomenon.

Another possible response to the challenge posed by inconsistent models comes from perspectivalism (Chakravartty 2010; Giere 2006; Massimi 2018). The perspectivalism proposed by Giere (2006) suggests that we should interpret models as contextually constructed representations of a system from the perspective of a particular theory. In short, perspectivalism suggests that we need to recast claims about how models represent in the following form: from the perspective of theory T, model M represents system S in a particular way (Giere 2006). However, I agree with Massimi (2018) and Morrison (2011, 2015) that, on its own, perspectivalism does little to solve the problem of inconsistent models for realism. Specifically, although perspectivalism shows us how multiple models might represent the same system in different ways, it does not tell us how sets of inconsistent models from different perspectives can be used to consistently explain and understand the same phenomenon. In particular, perspectivalism does not say how the information acquired from conflicting perspectival representations can be combined or integrated into a unified explanation (or an improved overall understanding) of the phenomenon.

Finally, some philosophers have suggested that we can overcome the challenges raised by multiple conflicting models by noting that the models used to study the same phenomenon often represent different aspects or features of the phenomenon (Elgin 2017; Potochnik 2017). For example, Elgin (2017) discusses the nuclear modeling case described here as follows:

> If what one model highlights is that in some significant respects the nucleus behaves like a liquid drop, and another model highlights that in some other significant respects it behaves as though it has a shell structure, there is in principle no problem. There is no reason why the same thing should not share some significant properties with liquid drops and other significant properties with rigid shells. (Elgin 2017, 270)

According to Elgin's account, the features that these models enable us to understand ought to be exemplified (i.e., instantiated) in some way by the models (i.e., what the models enable us to understand must be grounded in what each model has in common with the actual system). In a similar way, Potochnik (2017) argues that multiple conflicting models can be used to understand a phenomenon, so long as those models target different causal patterns that are embodied in the phenomenon.

This approach works fine when the salient features or aspects involved in the understanding provided by one model are sufficiently different and separable from the features involved in the understanding provided by another model. The issue is that in many instances—including the multiple conflicting models of the nucleus—this assumption cannot be made. As Morrison (2011) notes, cases in which various aspects of the phenomenon can be investigated by idealizing the system in different ways are importantly different from cases of genuinely inconsistent models, such as those in the nuclear modeling case. In this case, each of the models "makes very different assumptions about exactly the same thing" (Morrison 2011, 347). As a result, we cannot interpret these cases as merely exemplifying different aspects of the phenomena of interest. Instead, these models provide contradictory representations of the *same features* of the system. What we need, then, is an account of how multiple kinds of models that represent the same salient features in contradictory ways can each contribute to scientists' explanations and understanding of complex phenomena.

More generally, I suggest that philosophical accounts of modeling (and realism) will continue to be troubled by the problem of inconsistent models, so long as they conflate the following two questions (Batterman and Rice 2014; Rice 2019b):

1. How do idealized models allow scientists to explain and understand a phenomenon?
2. Which of the relevant features (e.g., difference-making causes) of the real system are accurately represented by the idealized models?

In short, the problem lies with the standard view's equating explanation and understanding with the accurate representation of relevant features. So long as these accurate representation requirements are central to how philosophers conceive of explanation and understanding, appealing to only a limited set of difference-makers (Potochnik 2017; Strevens 2008), partially accurate representations (Pincock 2012; Weisberg 2013; Wimsatt 2007; Worrall 1989), or perspectivalism (Giere 2006) will not be able to offer a solution to the problem of inconsistent models.

In addition, Morrison notes that in these cases, "we usually have no way to determine which of the many contradictory models is the more faithful

Multiscale Modeling and Universality 197

representation, especially if each is able to generate accurate predictions for a certain class of phenomena" (Morrison, 2011, 343). That is, in many cases scientists simply cannot answer the question of which of the plethora of idealized models accurately represents which of the relevant features of the models' target systems. More generally, scientists often do not have a choice between accurately representing the target system or idealizing it because we do not know what is true about the system. As Wimsatt suggests, "Isn't it always better to have a true model than a false one? Naturally it is, but this is never a choice that we are given, and it is a choice that only philosophers could delight in imagining" (Wimsatt 2007, 103). While I disagree with Wimsatt that having a true model is always better, the point is that decisions about how to model complex systems are typically not about *whether* to make the model accurate or inaccurate with respect to the relevant features of the system, but about *how* to idealize the model in order to reveal the information that scientists are after. As a result, we ought to consider alternative ways of justifying the use of multiple inconsistent models to explain and understand the same phenomenon. Specifically, we ought to consider alternative ways that do not depend on the accurate representation relations that are central to most philosophical accounts of modeling and explanation.

7.2 Universality and the Problem of Inconsistent Models

As an alternative, I suggest that the universality account presented in chapter 6 suggests a promising solution to the problem of inconsistent models. The solution appeals to multiple overlapping universality classes that include the target systems in which the phenomenon occurs (Rice 2019a, 2019b). When this occurs, sets of universality classes can link sets of inconsistent models with the same real-world phenomenon, even if those models represent the same features (or aspects) of the target system in contradictory ways (figure 7.3).

However, before getting to the details of this solution, some additional remarks about how to individuate universality classes are in order.[2] While our interests will determine which universality class is relevant for our purposes, those interests do not determine which universality classes exist (that is an objective fact about the world). Which universal patterns hold

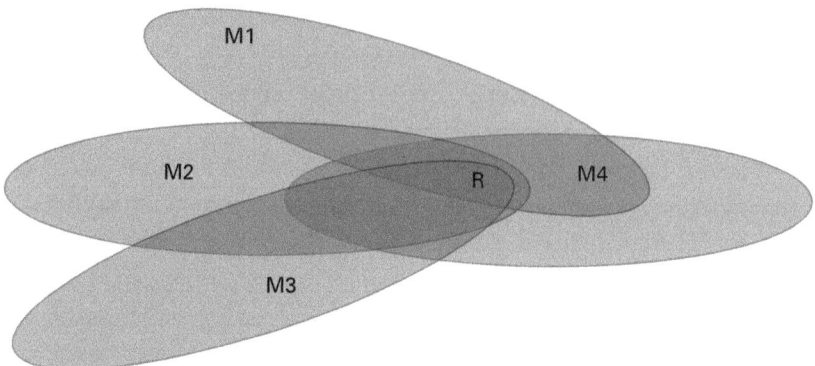

Figure 7.3
Overlapping universality classes that each include a single real-world system, R. M1, M2, M3, and M4 represent multiple conflicting models (or model systems) that might be connected to R via multiple overlapping universality classes (represented by the gray ellipses). M1 and M4 show that multiple conflicting model systems might be within the same universality class as well.

across which systems is determined by the world, not by our interests. The way to differentiate universality classes is to ask three questions:

1. Which features/patterns are stable across the systems in the class?
2. What changes to the systems' relevant features result in the stable features/patterns failing to occur?
3. Which of those systems' features are irrelevant to the occurrence of those stable features/patterns?

In short, what are the universal patterns, and which features (or changes) are relevant or irrelevant to those patterns? Different answers to any one of these questions means that we are considering different universality classes (e.g., if the stable patterns are the same, but they are stable across different changes to the features of the systems, then the two sets of systems will constitute different universality classes). Thus, different models might be used to investigate different universality classes by targeting either different stable patterns or different sets of changes to the features that are relevant to those patterns, or by exploring different ways that various features are irrelevant to the occurrence of the stable patterns. When universality classes overlap, it is because a single (real, possible, or model) system can display multiple patterns that will be dependent on and independent of different sets of features.

Importantly, because universality classes link models with their target systems without requiring accurate representation of, or mapping onto, relevant features, this account need not result in the inconsistent metaphysical claims that plague other accounts of how idealized models connect with their target systems. Specifically, universality classes show that certain counterfactual dependencies and independencies will be stable across a class of real, possible, and model systems that are perhaps drastically heterogeneous in their features (Rice 2017, 2018, 2019b). This allows sets of models to drastically distort the features of the target phenomenon in inconsistent ways and still enable scientists to use those models to extract the modal information required to explain and understand the phenomenon. Moreover, because every target system will involve myriad counterfactual dependencies and independencies that will be stable across different classes of systems, models in different universality classes can capture different sets of modal information about the same phenomenon. In other words, because a single target system may be a member of multiple universality classes, models in different universality classes can connect with the target system in different ways and capture different sets of modal information about the phenomenon of interest.

In addition, different models within the same universality class might be used to explore different counterfactual situations. That is, not every one of the conflicting models will need to be in a distinct universality class that involves different patterns or features of the target system in order for this solution to the problem of inconsistent models to go through. This is because universality classes will typically include many possible, but non-actual systems. As a result, an idealized model within the same universality class as the real system might provide an understanding of the target phenomenon, not by telling us how things occurred in the actual system, but by exploring how (or whether) the phenomenon would have been different in some range of counterfactual situations. For example, one model might investigate what occurs to the universal pattern if feature X, whose real value is known to be 12, takes values from 5 to 10, while another model might be used to investigate what happens to that pattern if X takes values from 20 to 25. As a result, two or more models within the same universality class might provide distinct sets of modal information about the phenomenon of interest by modeling different counterfactual possibilities.

In short, contrary to Elgin's proposal, an idealized model need not exemplify or accurately represent any of the actual features of the system in order to provide understanding of the phenomenon of interest. As I will argue in more detail in chapter 8, scientists often construct highly idealized models for the explicit purpose of exploring a range of nonactual, possible systems with drastically different features. These models may not tell us directly how the real system is, but investigating and altering those model systems in various ways can tell scientists a lot about the range of possible situations and how things would (or would not) be different in those situations, even if multiple conflicting models target the same patterns or features of the system (Rice 2019a).

In some cases a single model will provide enough modal information to explain or understand the phenomenon, but in other cases the modal information provided by multiple models will have to be combined by scientific modelers in order to formulate the desired explanations and understanding. The key point is that claiming that an idealized model can be used to extract some true modal information about the target system—because it is within a universality class that includes the real system—is importantly different from claiming that the model provides an accurate representation of (or exemplifies) any of the real system's relevant features or causal interactions. Consequently, because the universality account can license explanatory inferences without requiring the accurate representation of relevant features, the use of multiple inconsistent models to explain and understand the same phenomenon need not result in the kinds of descriptive inconsistencies that are troubling for scientific realism.

To see how to apply this solution to a particular case, we can consider how biological modelers use multiple conflicting idealized models to describe interactions at various spatial and temporal scales (Davidson, von Dassow, and Zhou 2009; Green and Batterman 2017; Qu et al. 2011; Rice 2019b). For example, when modeling cellular phenomena, "each scale of cell biology not only has its characteristic types of data, but also typical modeling and simulation approaches associated with it" (Meier-Schellersheim, Fraser, and Klaushcen 2009, 4). Consequently, in order to model these behaviors typically requires multiple conflicting modeling approaches (figure 7.4). In addition, because these cellular phenomena take place across multiple scales that interact with one another, biologists often build multiple conflicting models of the interscale interactions as well (Dallon 2010; Green and Batterman 2017; Meier-Schellersheim, Fraser, and Klaushcen 2009).

The resulting sets of models often make conflicting claims about the fundamental ontological components and interactions of the target systems. For

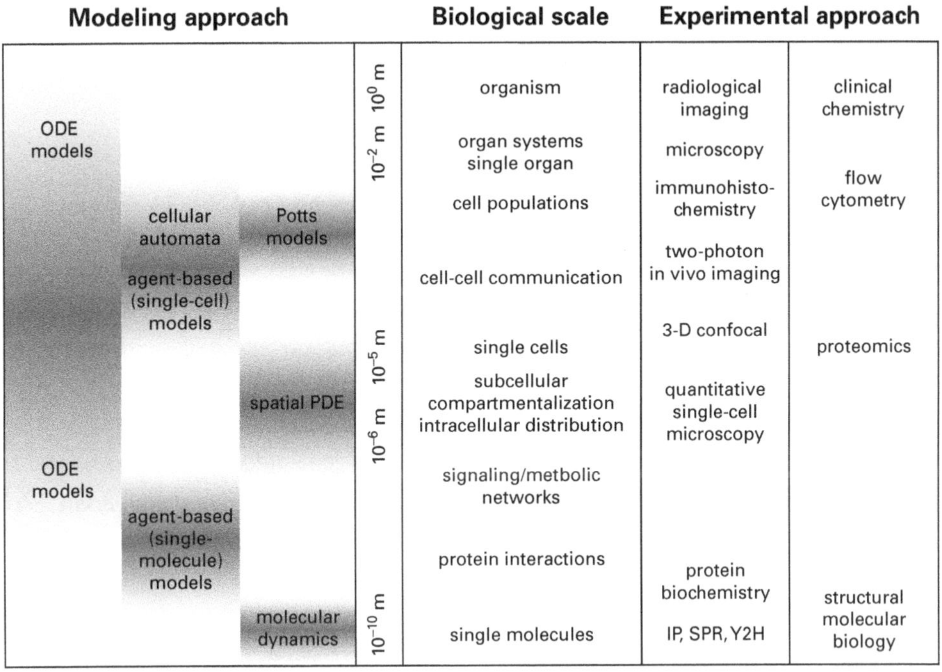

Modeling approach			Biological scale	Experimental approach	
ODE models			10^0 m — organism	radiological imaging	clinical chemistry
			10^{-2} m — organ systems / single organ	microscopy	flow cytometry
			cell populations	immunohisto-chemistry	
	cellular automata	Potts models		two-photon in vivo imaging	
	agent-based (single-cell) models		cell-cell communication		
			10^{-5} m — single cells	3-D confocal	proteomics
		spatial PDE	subcellular compartmentalization intracellular distribution	quantitative single-cell microscopy	
ODE models			10^{-6} m — signaling/metbolic networks		
	agent-based (single-molecule) models		protein interactions	protein biochemistry	structural molecular biology
		molecular dynamics	10^{-10} m — single molecules	IP, SPR, Y2H	

Figure 7.4
Different modeling approaches used for different scales of biological phenomena (modified from Meier-Schellersheim, Fraser, and Klaushcen 2009, 20).

example, "while cellular automata models treat the single 'cells' in their simulations as entities with fixed shape and size, Potts model simulations aim at reproducing the shape changes [that] cells undergo due to mechanical contact with neighbor cells or extracellular matrices" (Meier-Schellersheim, Fraser, and Klaushcen 2009, 6). Furthermore, these sets of models often include both individual-based and continuum models, which make contradictory assumptions about which aspects and scales of the system are relevant (Byrne and Drasdo 2009).[3] The challenge is to figure out how to use the information produced by these conflicting modeling approaches to develop consistent explanations and understanding of the phenomenon of interest.

I argue that a promising response is to consider how these sets of inconsistent models can be related to the same target phenomenon by multiple overlapping universality classes. For example, model M_1 (a Potts model) might be in universality class U_1, which contains the target system, while model

M_2 (a differential equation model) might be in universality class U_2, which also contains the target system (Rice 2019b). Assuming the systems in U_1 display some universal patterns of behavior concerning how the shape of the cell counterfactually depends on changes to neighboring (groups of) cells, then model M_1 can justifiably be used to infer modal information about those dependencies in real cell populations (Meier-Schellersheim, Fraser, and Klaushcen 2009, 6). Similarly, M_2, a differential equation model, can be justifiably used to extract a different set of modal information about the cellular system, given that the systems in U_2 display stable patterns of behavior concerning "how [the number of cells in the population] . . . depends on cell proliferation, death, differentiation and on interactions with other cell types or infectious agents" (Meier-Schellersheim, Fraser, and Klaushcen 2009, 6). Because different universality classes will focus on different universal patterns that are stable across different perturbations, models in different universality classes can be used to extract different sets of modal information about the phenomenon of interest. Importantly, this universality account does not require that the conflicting models be interpreted as accurately representing the relevant features of the target phenomenon. All that is required is that the highly idealized models display certain patterns of counterfactual dependence (and independence) between various features of the system, even if those features and the particular dynamical relationships among them are drastically distorted by the model. Moreover, because the modal information extracted from models within these universality classes need not conflict in the ways that the representations of the models themselves conflict, scientific modelers can use this plethora of modal information to construct consistent explanations and understanding of the phenomenon (Rice 2019b).

More generally, every system will contain several patterns that will be stable across different changes to the features of the system. Therefore, every system will be a member of many overlapping universality classes. Furthermore, many of these universality classes will include the model systems represented by various idealized models. When this is so, even if the model conflicts with other idealized models used to study the same phenomenon, an idealized model can justifiably be used to extract modal information about systems within the same universality class. Importantly, because this link between the model and the real-world systems does not require accurate representation of relevant features, multiple conflicting models need not be interpreted as making conflicting metaphysical claims about the features of their target systems. By linking multiple conflicting models with the same target system via overlapping universality classes we can see how inconsistent models can provide different sets of modal information that can be used to construct various explanations (and understandings) of natural phenomena.

7.3 Multiscale Modeling and the Tyranny of Scales

The problem of inconsistent models is largely theoretical, created by philosophical accounts that have mistakenly required successful explanations (and understanding) to be provided by accurate representations and true descriptions. However, there are several other, more practical, challenges that multiscale modelers must face that are largely independent of such philosophical debates.[4] These practical challenges are generated instead by the practice of multiscale modeling itself and scientific modelers' attempts to cope with the overwhelming complexity of physical systems. In the remainder of this chapter I will shift my attention to some of the more practical methodological challenges that confront multiscale modelers. Interestingly, however, this rather detailed journey through various multiscale modeling practices reveals several important insights for more theoretical philosophical debates concerning modeling, explanation, reduction, and emergence.

Two of the approaches frequently discussed by multiscale modelers are the "top-down" and "bottom-up" approaches (Bechtel and Richardson 1993). Roughly, the bottom-up approach models a system by directly simulating the individual elements and their interactions in order to see how they give rise to behaviors at larger scales. The advantage of the bottom-up approach is that it can sometimes show how various macroscale properties emerge out of smaller-scale components of the system. However, "the disadvantage is that it is computationally intensive, often prohibitively so, and the model itself can become too complicated to be grasped" (Qu et al. 2011, 22).

In contrast, the top-down approach considers the overall system in order to model system dynamics at the most macroscopic scale. The advantage of the top-down approach is that the models of these macroscale behaviors are typically simpler to solve and easier to grasp. The disadvantage is that macroscale models often say little about the smaller-scale properties of the system that are relevant to those macroscale behaviors (Qu et al. 2011, 22). In addition, some modelers describe various "middle-out" modeling techniques that begin with the most important mesoscales and then attempt to incorporate features at both larger *and* smaller scales (see Green and Batterman 2017 for a discussion). The challenge here is identifying which of the various mesoscales is best to start with. As we will see, however, in most cases of multiscale modeling, none of these approaches is sufficient on its own. Instead, successful multiscale modeling typically requires a plurality of modeling techniques to capture different kinds of interscale relationships.

One reason for this is that each of these modeling approaches must confront the problem of how to integrate information across scales. A particularly clear example of this is the recent use of multiscale modeling techniques to model the development of cancer. As Deisboeck et al. (2011) explain, there are several scales involved, each with its own characteristic type of modeling strategies:

Atomic scale—Models at this scale are used to study the structure and dynamic properties of proteins, peptides, and lipids. The most common modeling method used is molecular dynamics (MD) simulation, where atoms and molecules are allowed to interact for a period of time. Atomic-scale models deal with length scales in the order of nm and time scales of ns.

Molecular scale—Models at this scale represent an average of the properties of a population of proteins rather than the dynamics of individual proteins. Cell signaling mechanisms are usually investigated at this scale. Ordinary differential equations (ODEs) are often used to represent biochemical reactions contained in a signaling pathway. Molecular-scale models deal with length scales in the order of nm~μm and time scales of μs~s.

Microscopic scale—This scale is used to represent tissue, cellular, or multicellular properties. Models at this scale must suitably describe the malignant transformation of normal cells, associated alterations of cell–cell and cell–matrix interactions, the heterogeneous tumor environment, and the element of tumor heterogeneity. These models usually use partial differential equations (PDEs) or agent-based modeling (ABM) rather than ODEs to simulate these factors and processes. Tissue-scale models deal with length scales in the order of μ m~mm and time scales of min~hour.

Macroscopic scale—Models at this scale focus on the dynamics of the gross tumor behavior including morphology, shape, extent of vascularization, and invasion, under different environmental conditions. Microscopic details of tissue structure are averaged over short spatial scales to produce a description of the macroscopic-level tissue properties. At this scale, because the number of cells in the model is sufficiently large, it becomes possible and sometimes necessary to treat some or all of the cells as a single continuum. This in turn allows for cell and substrate transport to be modeled with conservation laws for spatiotemporally varying densities (i.e., PDEs), rather than keeping track of individual cell activities. Models generally consider cell responses to gradient fields of diverse origins, such as concentration gradients of diffusible or non-diffusible molecules as well as strain and stress gradients generated by the growing tumor mass. Macroscopic-scale models deal with length scales on the order of mm~cm and time scales of day~year. (Deisboeck et al. 2011, 3–4)

As we can see, complex biological systems often require modelers to use a variety of modeling techniques to capture features across an extremely wide range of spatial and temporal scales. A serious challenge here is that "The use of different modeling approaches introduces gaps among scales. Multiscale modeling, besides modeling the system, needs to address the issue of how to bridge the gaps between different methodologies and between models at different scales" (Castiglione et al. 2014, 7). As I noted previously, accomplishing this task is often very difficult because the modeling approaches used at various scales often conflict with one another such that their data, representations, inputs, and outputs are incompatible.[5] This clearly illustrates the tyranny of scales problem: the available models are each tied to a narrow range of scales, but the phenomena to be modeled is known to depend on features across an extremely wide range of scales (Batterman 2013; Davidson, von Dassow, and Zhou 2009; Green and Batterman 2017; Oden 2006; Wilson 2017).

The tyranny of scales is also prominent in physicists' attempts to model various materials (Batterman 2013). As a report on simulation techniques in engineering by the National Science Foundation states:

Virtually all simulation methods known at the beginning of the twenty-first century were valid only for limited ranges of spatial and temporal scales. Those conventional methods, however, cannot cope with physical phenomena operating across large ranges of scale—12 orders of magnitude in time scales, such as in the modeling of protein folding . . . or 10 orders of magnitude in spatial scales, such as in the design of advanced materials. At those ranges, the power of the tyranny of scales renders useless virtually all conventional methods. . . . Confounding matters further, the principal physics governing events often changes with scale, so that the models themselves must change in structure as the ramifications of events pass from one scale to another. The tyranny of scales dominates simulation efforts not just at the atomistic or molecular levels, but wherever large disparities in spatial and temporal scales are encountered. (Oden 2006, 29–30)

In sum, the modeling techniques developed by scientific modelers are typically tied to rather small ranges of spatial and temporal scales. This, then, results in unique modeling challenges when the phenomena of interest fall into extremely wide ranges of scales that go beyond the representational capacities of any one (or even a small number) of the modeling techniques available. As a result, "examples from multi-scale modeling in both physics and biology show that modelers in both domains must confront the tyranny of scales problem" (Green and Batterman 2017, 32). Indeed, the challenge of confronting the tyranny of scales is widespread across the sciences.

7.4 Tackling the Tyranny of Scales

In this section I investigate a variety of approaches that are commonly used by multiscale modelers in order to deal with the tyranny of scales. Each of these approaches is tied to a particular kind of relationship between scales. Specifically, scale-dependent techniques apply to processes that are largely restricted to particular scales and can be shown to be universal with respect to most changes at other scales. Scale-invariant techniques are applied to relationships that are universal across a wide range of scales. Finally, interscale modeling techniques are used to investigate the relationships between scales whose features are *relatively* autonomous of each other but are still dependent on some dominant features at other scales. I do not claim that this survey of multiscale modeling techniques is exhaustive, but these three kinds of approaches do capture a wide range of the techniques used by multiscale modelers faced with the tyranny of scales. Moreover, what these multiscale modeling techniques have in common is their use of universality to justify their use of highly idealized models to investigate complex phenomena. That is, for each of these multiscale modeling techniques, we will see that universality plays an important role in justifying their application to investigate the various dependencies and independencies that hold among features at different scales.

These cases also illustrate that, instead of having complete dependence or autonomy among the various scales of the system, most complex systems have variable degrees of dependence and independence between particular features at one scale and particular features at another scale. Some features, properties, or processes will be separable with respect to both spatial and temporal scales, others will separate only along spatial scales, others will separate only along temporal scales, and the vast majority will involve partial overlap and interactions across both spatial and temporal scales (Chopard, Borgdorff, and Hoekstra 2014, 4). Consequently, sorting out the details of these multiscale dependencies and independencies is far more complicated than simply specifying the levels, theories, or disciplines involved in studying the system and saying which are completely reducible to or autonomous of each other. Constructing sets of models with which to genuinely explain and understand the behaviors of these systems requires far more nuanced multiscale modeling techniques.

7.4.1 Scale Dependence, Scale Separation, and Universality

A common multiscale modeling technique is to try and identify features of the system that are relatively *scale dependent*. Scale-dependent features or processes are those that occur only at a single scale (or a very small range of scales) and are largely isolated from interactions at other scales. This is certainly the best-case scenario for modelers confronting the tyranny of scales. While many of the processes and interactions may occur across scales, if some of the processes and features of the system can be tied to particular scales, then modelers can use the scale-dependent modeling techniques available for those processes (Bechtel 2015). However, this isn't the end of the story. As we will see, there are still issues concerning how these scale-dependent representations can be combined with models at other scales to provide an overall understanding of the complex system. That is, merely identifying the scale-dependent processes of the system is not enough to tackle the tyranny of scales problem on its own; one still needs to know how to combine the information provided by those scale-dependent models (Batterman 2013; Bursten 2018a, 2018b; Winsberg 2006, 2010).

An important type of scale-dependent modeling appeals to a separation of scales in order to justify using different modeling techniques for processes at different scales. A separation of scales can be defined as a case where the relevant scales differ by an order of magnitude and the features or processes present at those scales are largely autonomous of each other. In other words, sometimes the relevant features of a system appear at very different scales *and* are easily separable from one another. As a result, scientists can effectively model these processes by using modeling techniques designed for those particular scales (and types of processes).[6]

For example, consider a container of fluid where the top is heated to a temperature that is above that of the fluid at the bottom of the container (Bishop 2008). The temperature difference results in heat being transferred from the fluid at the top of the container to the fluid at the bottom by two processes: diffusion and convection. However, as the Reynolds number of the system changes, so does the scale of the dominant features of the system. As Bishop explains, in this case, "For small values [of the Reynolds number], the slower time scale (shorter length scale) diffusive processes are dominant, while at high values the faster time scale (longer length scale) convective processes are dominant" (Bishop 2008, 235). Because the time scales of the processes of diffusion and convection separate in this case, scientists can justifiably use different models for these processes without having to worry too much about the interactions between them. In this way, identifying separations of scales can greatly reduce the degrees of freedom and computational resources required in order to model the system.

In addition, when certain features at a particular scale are largely autonomous of features at other scales, multiscale modelers routinely appeal to universality in order to justify their employment of scale-dependent modeling techniques that ignore features at other scales. For example, in defense of constructing highly idealized agent-based models (ABMs) in economics that target only the scale of agents, Parunak, Brueckner, and Savit argue:

> To an unbiased observer, [these models'] success seems almost magical ... leading some users to hesitate in trusting the results. Universality helps explain this unreasonable success. In spite of all the detail that a simple ABM omits, it captures important qualitative features of the interactions among the entities.... The existence and widespread manifestation of universality can help build confidence in ABM's, as well as guide in their refinement as users gain experience in how universality manifests itself in specific configurations. (Parunak, Brueckner, and Savit 2004, 936)

In short, identifying universality classes and understanding the details of various instances of universality can help justify trusting the results of idealized models that focus only on a particular scale. Using models that incorporate features from only one scale is often justified when many features at that scale are universal with respect to various changes in features at other scales. This is precisely the kind of justification that physicists use when they appeal to models that only describe continuum behaviors and ignore any of the smaller-scale features of real fluids (Batterman 2013). It is also the justification provided in minimal model explanations for using extremely minimal models that caricature the dominant features of the system at more macroscales (Batterman and Rice 2014). The justification for appealing to these scale-dependent models is that the models are within the same

Multiscale Modeling and Universality

universality class as real systems whose universal behaviors at a particular scale are largely independent of changes at other scales of the system. In other words, the justification for using such scale-dependent models often depends on showing that the scale-dependent model is within the same universality class as the target systems in which certain scale-dependent universal behaviors occur.[7]

However, even if we assume that all the features of the real system are dependent on particular—and easily separable—scales, multiscale modelers often must confront the further challenge of using models whose representations of the system overlap and conflict with one another at the same scales. That is, the separation of scales within the real system is not always mirrored by a separation of the scales among the modeling techniques available. Moreover, even though scale-dependent modeling techniques can sometimes be fruitfully applied in cases where the processes in the real system do not separate into characteristic scales (Bechtel 2015), there is still the problem of accounting for how the often contradictory, scale-dependent modeling techniques overlap and interact with one another. For example, Chopard, Borgdorff, and Hoekstra (2014) describe a number of multiscale modeling problems that involve several overlaps among the spatial and temporal scales of the submodels used for various processes (figure 7.5).

The general issue is that, as with the tyranny of scales more generally, the interscale relationships of the real systems typically cannot be mirrored by interscale relationships among the modeling techniques available. In the case of scale-dependent modeling, this is because the various modeling techniques used to model scale-dependent processes will often result in gaps and interactions between models of those processes. As Qu et al. explain, "Multi-scale modeling, besides modeling the system, needs to address the issue of how to bridge the gaps between different methodologies and between models at different scales. There is no straightforward way to go from one scale to another, but there are useful methods that may be used to bridge the gaps" (Qu et al. 2011, 27).

Tracking how these overlaps and gaps between submodels arise is one of the most difficult challenges that multiscale modelers face. Moreover, it is particularly case specific because the details of the processes of the system and the particular types of models available to represent them will determine which further modeling techniques are needed to address these gaps and overlaps.

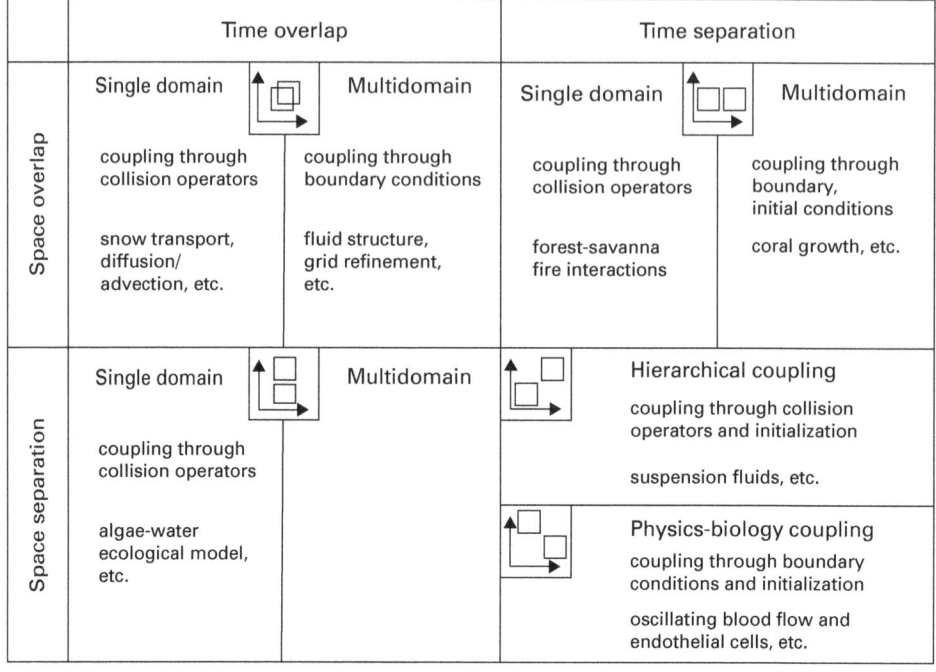

Figure 7.5
A classification of various multiscale problems based on the separation of the submodels used to model features at different scales (or domains). There are a variety of ways that the scales of different submodels might overlap (Chopard, Borgdorff, and Hoekstra 2014, 6).

Ultimately, the success of scale-dependent modeling techniques will depend on how tied to specific scales the processes of interest are and how separable the modeling techniques used for those processes are (Bursten 2018b, 12). In many cases, the interactions across scales will be relatively weak compared to the interactions within particular scales. When this occurs, if the processes involved in the target phenomenon have distinct characteristic scales at which they operate, then the use of multiple scale-dependent modeling techniques has a greater chance of being successful. However, even when it does succeed, scale-dependent modeling still requires that modelers address how the various scale-dependent submodels of the system work together to produce an overall understanding of the behavior of the system. This is why scale-dependent modeling, while useful, is rarely sufficient to solve the tyranny of scales problem on its own.

7.4.2 Scale Invariance, Power Laws, and Universality

The next challenge, of course, is tackling the features of the system that occur across multiple scales of the system (i.e., the features that are not scale dependent). At this point, multiscale modelers often look for features or relationships that are stable, regardless of which scale of the system is under consideration (Henriksen 2015).[8] Identifying such scale-invariant features enables scientists to use a single mathematical modeling equation (which can be rescaled) to capture the same kinds of relationships across multiple scales of the system (Khaluf et al. 2017, 2). As Solé and Bascompte explain, "A striking, widespread feature of many complex systems is that some of their properties are reproduced at different scales in such a way that we perceive the same pattern when looking at different subparts of the same system. This property, known as scale invariance, is widespread in many systems" (Solé and Bascompte 2006, 127).

However, as with scale dependence, it is important to remember that features will only be *relatively* scale invariant across a particular range of scales. That is, there will always be limits to the range of scales that scale-invariant features are stable across (Khaluf et al. 2017). Still, the stability of many features across a relatively wide range of scales allows multiscale modelers to construct models that focus on these scale-invariant relationships, even in cases where relatively little is known about the other features of the system.

Scale invariance comes in many forms. For example, Khaluf et al. (2017) distinguish three main types of scale-invariant systems: scale-invariant spatial structures, scale-invariant topologies, and scale-invariant dynamics (figure 7.6).

One prominent example of scale-invariance is identifying features of complex systems that are self-similar across multiple scales (Barenblatt 1996; Ostling and Harte 2003). Just as different systems can be similar to each other in various ways, different states of the same system can be similar to each other across changes in spatial and temporal scales. For example, Henriksen explains an instance of self-similarity across temporal scales as follows: "In dynamically evolving macroscopic systems, the change to the system in time can be such that each current system is simply a rescaled version of the original system. In such a case, the system is said to remain 'Self-Similar.' These ideas have powerful simplifying implications for our descriptions of the macroscopic world" (Henriksen 2015, 5).

Self-similarity implies that if we are looking at a quantitative property $L(r)$ of a system at scale r, then if we look at the same quantity measured at

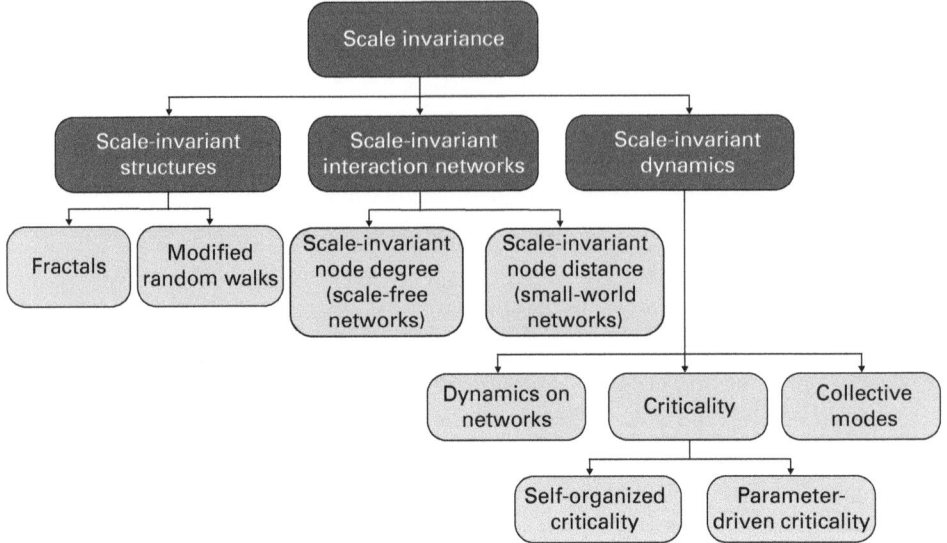

Figure 7.6
Several kinds of scale invariance that can be used by multiscale modelers to simplify their modeling tasks. Fractals and modified random walks are examples of spatial and temporal structure that are scale-invariant. Scale-invariant topologies can be scale-invariant either in their node degree or in their intra-node distance. Scale-invariant dynamics can be induced either by a scale-invariant topology, various kinds of criticality, or via collective modes (Khaluf et al. 2017, 2).

a different scale $r' = \alpha r$, then $L(\alpha r)$ will be proportional to $L(r)$, which can be written as follows:

$$L(\alpha r) = kL(r)$$

where k is a constant. Identifying these kinds of self-similar features allows scientific molders to employ particularly powerful modeling tools to capture the same properties and relations across a wide range of scales. Indeed, according to Solé and Bascompte, "The self-similar character of these systems, sometimes spanning many orders of magnitude, can be exploited in order to extrapolate between different scales" (Solé and Bascompte 2006, 168).

We have already encountered an example of spatial self-similarity in chapter 6 in the case of modeling Arctic melt ponds. In that case, the melt ponds are self-similar when the perimeter-to-area ratio of the overall melt pond is similar to the perimeter-to-area ratio of its subponds (figure 7.7).

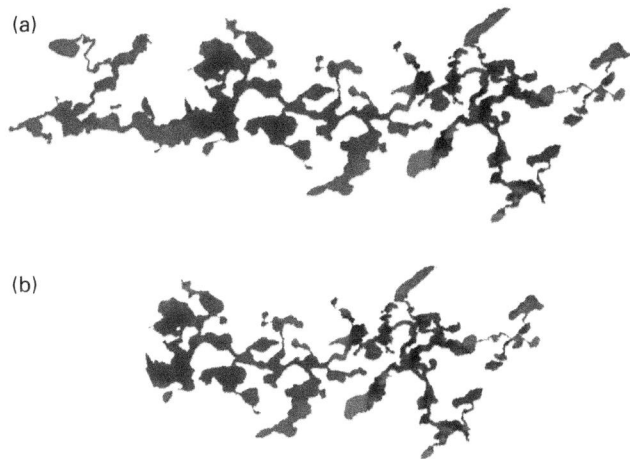

Figure 7.7
A self-similar model pond. (a) is the overall pond and (b) is one of its subponds. They are self-similar with respect to their perimeter-to-area ratio (Hohenegger et al. 2012, 1161).

When this occurs, these modelers are justified in using a number of mathematical modeling techniques from physics that require the system to exhibit some kind of self-similarity. In addition, this self-similarity turns out to be the crucial feature governing the transition of these melt ponds from the phase of having a fractal dimension of 1 to a phase of having a fractal dimension of about 2.

A second kind of scale invariance frequently exploited by multiscale modelers is to identify power laws that are stable across multiple scales of the system (Batterman 2002b; Gisiger 2001; Solé and Bascompte 2006; Stanley et al. 2002; Tao 2012). In fact, some scientific modelers define scale invariance simply as "a hierarchical organization that results in power-law behavior over a wide range of values of some control parameter" (Stanley et al. 2000, 60).[9] Indeed, power laws are scale-invariant relationships that have been found across physics (Batterman 2002b; Tao 2012), economics (Stanley et al. 2000, 2002), biology (Gisiger 2001; Ostling and Harte 2003), and many other sciences.

Power laws take the following general form:

$g(x) = Ax^\alpha$

where A and α are constants. Taking the log of both sides results in

$\log g(x) = \log A + \alpha \log x$

When this is plotted on a log-log scale, the function produces a straight line of slope α. We can prove that power laws of this form are scale invariant by considering a new variable $x = ax'$, where a is a constant. Substituting this into the previous equation gives us:

$$g(ax') = A(ax')^\alpha$$
$$g(ax') = (Aa^\alpha)\, x'^\alpha$$

The overall form of this function is exactly the same as before (i.e., it is a power law with α as its exponent). Therefore, we can "zoom in and out" by changing the value of the scaling constant a without changing the general form of the power law function. This is precisely why power law relationships look the same, no matter what scale of the system is chosen (i.e., they are scale invariant).

In many cases, power laws are of interest because they involve critical exponents that are stable across very heterogeneous systems at different scales. For example, various experiments with physical systems have shown that the critical exponents of the power laws found for various materials only come with certain special values. As a result, all of these systems fall into one of a limited set of universality classes that share these critical exponents. Modelers in physics have long known about the universality of these power-law exponents and have developed various methods (e.g., the renormalization group) in order to explain their universality. More recently, multiscale modelers in several other fields have started to identify power laws with critical exponents, along with mathematical models that are within the same universality class as those real systems.[10]

For example, Gisiger (2001) describes numerous instances of self-similarity and power laws that have been found in biological systems. One particularly striking example is the fossil record, which reveals a few periods of history where a large percentage of species (and families) have become extinct (e.g., the annihilation of the dinosaurs at the end of the Cretaceous period). One might expect, therefore, that the fossil record would show several sharp spikes representing mass extinction events against a fairly constant background of low extinctions. However, this is not what we find. Instead, the extinctions observed in the record do *not* clearly separate into large extinctions and small ones—that is, there is not a clear separation of scales. Instead, "extinctions of many sizes are present: a few large ones, several medium-sized ones and lots of small ones" (Gisiger 2001, 186). Raup

Multiscale Modeling and Universality 215

(1986) sorted these data and computed the frequency of each size of extinction event. Raup's results show that extinction events over time appear to exhibit a power law with a critical exponent of about −2.7 (figure 7.8).

Similar power laws have been found in ecology regarding the species abundance distribution (SAD) (Harte, Kinzig, and Green 1999; Harte, Smith, and Storch 2009; Ostling and Harte 2003; Rosindell and Cornell 2013). This power law describes the relationship between the mean number of species in a patch, S, and the area of the patch, A, as $S = cA^z$ (Ostling and Harte 2003, 219). Such ecological power laws are due to various fractals. As Ostling and Harte explain, "Fractal objects are described as 'scale-invariant' in the sense that they look the same when one 'zooms in' or 'zooms out' on them; i.e. when one looks at them on a smaller or larger scale" (Ostling and Harte 2003, 218). The value of discovering these scale-invariant power laws is that modelers now know that "only one parameter controls the shape of the SAD at different scales" (Rosindell and Cornell 2013, 1102). Moreover, the same species-area curve is universal across a wide range of different species and habitats. This greatly simplifies the task of modeling these systems across such a wide range of scales. More generally, as Ostling and Harte say, "Power-law relationships appear to characterize many patterns observed in ecology, from the level of individuals to the level of ecosystems" (Ostling and Harte 2003, 218).

In addition, the concept of universality is particularly useful for justifying the use of extremely minimal models that focus only on the critical exponents of these scale-invariant power laws. As Gisiger explains:

> After all, what good is a model if one has to include in it an infinite amount of detail to make it reproduce the data. However, here the notion of universality proves helpful. The abundance of self-similarity and power laws . . . is suggestive that ecosystems operate near a critical point. . . . The exponents of the distribution are therefore analogous to the critical exponents defining the dynamics of magnetic systems, for instance. However, . . . these exponents cannot take arbitrary values because of the notion of universality: they are constrained by the specific universality class the system is in. [Consequently], if the model reproduces the critical exponents correctly, then it might be expected that some important features of evolution and extinction have been taken into account. (Gisiger 2001, 190–191)

In other words, as was the case with scale-dependent modeling, the concept of universality is absolutely crucial for justifying the use of these scale-invariant modeling techniques. Indeed, as Gisiger suggests, *"One just has to consider the simplest model conceivable in the same universality class as the ecosystems"* (Gisiger 2001, 191).[11] In other words, the universality of certain

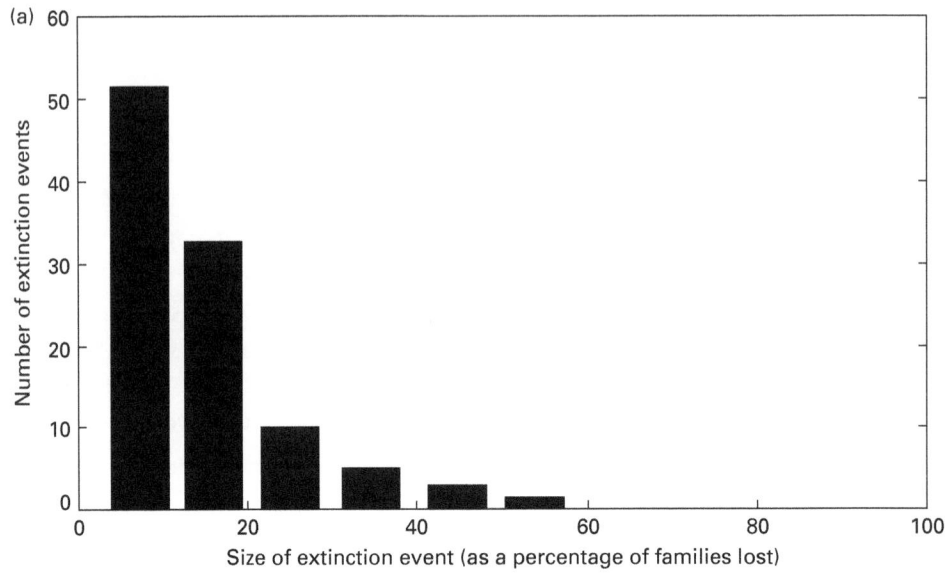
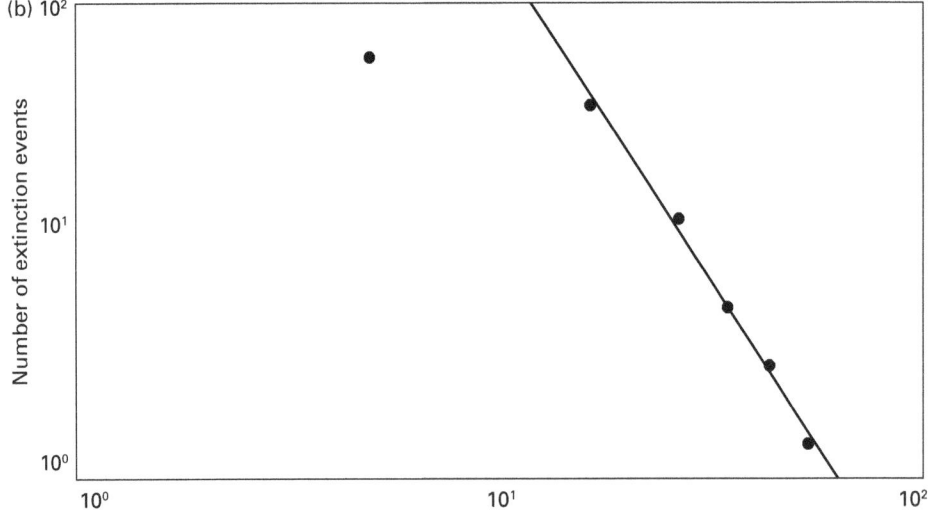

Figure 7.8
(a) represents the distribution of extinction evens as a function of their size for 2,316 marine animal families of the Phanerozoic. (b) is the same distribution plotted on a log-log scale. The straight line has a slope of –2.7, which is the critical exponent of the power law (based on Gisiger 2001, 187).

scale-invariant patterns is a key part of the justification offered for using these highly simplified mathematical models to investigate multiscale phenomena. In contrast to using universality classes to separate features at one scale from those at other scales, universality classes are used in these cases to justify the use of idealized models that capture stable features of the system across multiple scales.

7.4.3 Modeling across Scales: Renormalization, Homogenization, and Universality

Finally, one of the biggest challenges in multiscale modeling is that the relevant scales (and features) of the system can change with changes in space and time. For example, as a tumor grows, different features at different scales become relevant to the behavior of the system and new couplings between scales may arise. The reason these behaviors change is because most complex systems display interactions across scales, such that the macroscale parameters of the system are partially determined by various features of the system across a range of smaller scales (i.e., most macroscales are neither completely autonomous nor completely reducible to microscales). In addition, most complex systems display what Wimsatt calls "interactional complexity" across theoretical frameworks (Wimsatt 2007, 184). What this means is that the interactions across scales also routinely crisscross the representational frameworks available from our current theoretical modeling techniques.[12]

Multiscale modelers would like to understand these kinds of interactions between different scales and across different modeling perspectives. In other words, in addition to discovering separations between scales and various scale-invariant patterns, multiscale modelers often want to understand precisely how the microscale features of the system give rise to stable behaviors at macroscales. Because this can occur in a variety of ways, multiscale modelers use a plethora of tools to try and model these interscale interactions (Qu et al. 2011).[13] Moreover, these "multiscalar approaches follow neither purist top-down nor purist bottom-up tactics in their approaches to modeling, but qualify as intriguing hybrids" (Wilson 2017, 228). An exhaustive discussion of the plurality of interscale modeling strategies used by scientists is beyond the scope of my discussion here. Instead, I will focus on two of the most prominently used techniques that have already shown up in previous chapters: renormalization and homogenization. Both of these multiscale modeling techniques are used to extract macroscale descriptions of the system that preserve (or directly respond to) important information about features that are relevant at smaller scales.

As a first example, we can consider the use of renormalization techniques to understand the relationships between various scales of the system near critical points. As the system becomes critical, it does not exhibit any characteristic length scale because all the scales of the system become relevant. This is typically represented by noting that the correlation length of the system, ξ, diverges to infinity. The correlation length measures the degree to which changes in one area of the system are correlated with changes in other areas of the system. When this correlation length is relatively small, scale-dependent modeling techniques are likely to be successful. However, when the correlation length diverges to infinity near criticality, changes at every scale of the system become correlated with changes across every other scale (and part) of the system. As a result, we require modeling techniques that can capture relevant features of the system (and their interactions) across an extremely wide range of scales.

As we saw in chapter 3, the most prominent technique used to cope with the divergence of the correlation length near criticality is to use renormalization techniques in order to determine which degrees of freedom of the system across multiple scales are relevant and irrelevant to their macroscale phase transition behavior. Renormalization can also be used outside of physics. For example, Solé and Bascompte note that in ecological modeling, "a fundamental result of percolation theory is the *universal* character of the scaling exponents β and ν" (Solé and Bascompte 2006, 142). They then explain that "Percolation thresholds can be estimated by using a powerful technique of statistical physics: the renormalization group" (Solé and Bascompte 2006, 319). In particular, renormalization is an extremely useful tool for finding minimal models within the same universality class as diverse real systems that include only the dominant parameters required to capture the essential features of the system near their critical points. As Solé and Bascompte reiterate: "[Universality] also has implications for the strategy of model building. As noted, despite their simplicity, models such as [a model of chaotic bifurcations] have the key properties (nonlinearity, density dependence) that characterize real populations. Thus, despite the omission of many details from these simplistic models, we may expect that the relevant features are maintained" (Solé and Bascompte 2006, 49).

Multiscale Modeling and Universality

In other words, renormalization is used to extract a set of key minimal features that determine critical behaviors that are universal with respect to changes in the other features of the complex system.[14] The important point here is that renormalization is a technique for identifying the relevant and irrelevant features across *multiple* scales of the system. As Morrison explains in the case of using a block-spin renormalization technique, this process "provides the bridge from the micro to the macro levels and each state in between. Moving from small to larger block lattices gradually excludes the small-scale degree of freedom such that for each new block lattice one constructs effective interactions and finds their connection with the interactions of the previous lattice" (Morrison 2014, 1148). What this means is that as we apply the renormalization transformation at progressively larger scales, the resulting representations of the system enable scientists to distinguish the relevant and irrelevant degrees of freedom across multiple microscales of the system—this is importantly different from averaging over the microstates of system or simply describing the system at some macroscale.

While renormalization is a very powerful multiscale modeling tool, determining how macroscale behaviors are coupled and decoupled with various features at microscales is a far more general challenge beyond just modeling critical points and phase transitions. Indeed, many cases require scaling from microscales to macroscales in ways that cannot be handled by averaging or renormalization methods. When this occurs, multiscale modelers often use homogenization techniques to construct macroscale models of complex systems that are extremely heterogeneous at smaller scales (Batterman 2013; Gerisch, Penta, and Lang 2017; Rice 2017; Wilson 2017). *Homogenization* is a technique for "describing the behavior of a material that is inhomogeneous at some lower length scale in terms of a (fictitious) energetically equivalent, homogeneous reference material at some higher length scale" (Böhm 2016, 4). In other words, while these systems are extremely heterogeneous at their microscales, they display macroscale behaviors that are similar to those displayed by a particular homogenous system. A common example is modeling an electrically insulating medium that includes several conducting inclusions as a homogeneous medium with the same overall conductance properties throughout (Milton 2002; Torquato 2002). As Batterman notes, in these cases: "*We need methods that tell us how to homogenize heterogeneous materials*" (Batterman 2013, 269).

Homogenization theory considers at least two sufficiently different scales of a system. The first scale, a, is associated with the relevant heterogeneous features at some microscale. The second scale, ξ, is associated with the size of the overall system. The key question is how the heterogeneous features at scale a are related to the homogeneous features of the system at the ξ scale. Assuming that these scales are sufficiently different, one can then introduce the following scaling parameter

$$\varepsilon = \frac{a}{\xi}$$

that is associated with the fluctuations of the heterogeneities at the microscale (Batterman 2013, 281). One then looks at a family of functions u_ε and looks for a limit $u = \lim_{\varepsilon \to 0} u_\varepsilon$, which tells us what the effective properties of the material will be at the macroscale. Introducing the appropriate scaling parameters enables one to replace the heterogeneous medium with an idealized homogeneous medium that captures the dominant properties that are important for understanding the behavior of the system at the macroscale. Moreover, like applications of renormalization, finding a homogenous description that preserves the important parameters for the macroscale behaviors can be used to demonstrate that most of the heterogeneities at smaller scales are irrelevant.

In sum, homogenization is a technique for extracting information about the dependence and independence relationships among various scales of the system. The result is an explanation of the relative autonomy of the macroscale parameters from most of the features of the system at smaller scales. What is more, homogenization enables scientific modelers to find highly idealized macroscale models that preserve the most important information from smaller scales—typically by subsuming them in a few macroscale parameters. However, rather than attempting to accurately mirror the causal or dynamical interactions occurring among the various scales of the system, Wilson notes that "Typically, applied mathematicians obtain these communication links by repeatedly blowing up their sub-models to an infinite population size, keeping the proportions of internal details intact as they go" (Wilson 2017, 218). In other words, these homogenization techniques are typically highly idealized mathematical manipulations that drastically distort the target systems in order to extract the dominant behaviors of the system at macroscales.

To see how this works in a specific case, we can consider the use of homogenization techniques to model heterogeneous landscapes in spatial ecology.

The goal of these spatial ecology models is "to *explain* observed spatial distribution patterns of populations and to predict their response to landscape alternations" (Yurk and Cobbold 2018, 171; emphasis added). The first step is to construct a model of the population at smaller (i.e., faster) scales. In this case, the modelers represented the density of the population, ρ, at time τ and location y using a reaction-diffusion model (Yurk and Cobbold 2018, 172):

$$\partial_\tau \rho(y,\tau) = \partial_y^2 \left[D(y)\rho(y,\tau) \right] + \varepsilon^2 f(y,\rho) \tag{1}$$

The growth and death dynamics of the population are modeled by function f. The movement of individuals is modeled by "ecological diffusion" by the function $D(y)$, which denotes the motility of individuals as their probability of movement in any direction.

The difficulty is that the environment to be modeled is typically patchy—that is, areas within the landscape are relatively homogeneous, but they are substantially different from their immediate surroundings (i.e., other patches). The reason for this is that natural landscapes exhibit several edges with sharp transitions (e.g., transitions between forests, grasslands, lakes, and shores).[15] This makes the motility and growth equations very different for different parts of the landscape. As a result, the modeling challenge is that equation (1) is very different for different patches in the landscape (i.e., the landscape exhibits lots of dynamical heterogeneity at smaller spatial scales). Consequently, "a fundamental problem in theoretical ecology then is to determine the extent to which individual processes on smaller scales affect population responses on larger scales" (Yurk and Cobbold 2018, 171). Fortunately, "a powerful tool to study the dynamics of model (1) is multi-scale analysis and homogenization" (Yurk and Cobbold 2018, 173). Indeed, Yurk and Cobbold's main goal is to "present a homogenization approach to the multiscale problem of how individual behavioral responses to sharp transitions in landscape features, such as forest edges, affect population-dynamical outcomes" (Yurk and Cobbold 2018, 171).

At the smaller scale, the key variables, present in equation (1), are y and τ. The issue is how these variables and equations can be scaled up to obtain a model in terms of the larger-scale variables x and t. The challenge, of course, is "to obtain an approximate model that describes the slow-scale behavior of the system by *appropriately* averaging the variation in the movement and growth parameters over the fast scale" (Yurk and Cobbold 2018, 174; emphasis added). This is done by introducing a scaling parameter ε

that describes the relationship between the large-scale variables of interest and the relevant features at smaller scales of the system. In this case, ε is a small dimensionless parameter, where $x = \varepsilon y$ and $t = \varepsilon^2 \tau$, where $\varepsilon \ll 1$.

Repeatedly applying this homogenization transformation across multiple scales of the system enabled these modelers to describe a homogenized diffusion equation and leading diffusion coefficient at the largest scale of the system that closely matches the results obtained from solving the various heterogeneous equations at smaller scales. This homogenization process is extremely useful because "the homogenized model is much easier to analyse analytically or numerically and often leads to important theoretical insights about the relationship between local individual movement behaviors and variation in growth rates and large-scale population-dynamical outcomes" (Yurk and Cobbold 2018, 174). In other words, the crucial results extracted from these homogenization techniques concern how features at smaller scales give rise to patterns at larger scales of the system.

What is absolutely crucial to remember here is that "the goal of multiscale modeling is not simply to model a system at multiple scales, but rather to conserve the information from a lower scale (modeled by high-dimensional models) to a higher scale (modeled by low-dimensional models), *so that the information from the bottom scale can be carried to the top scale correctly*" (Qu et al. 2011, 23; emphasis added). In other words, the power of renormalization and homogenization techniques isn't just that they identify a macroscale description of the system, but that they enable scientific modelers to discover the appropriate macroscale models that incorporate the relevant information from smaller scales. This is what Batterman means when he says that we need techniques that show us how to homogenize heterogeneous materials—the idealized homogeneous model must incorporate information from multiple scales of the system.

Finally, like the other cases described in this chapter, the use of these multiscale modeling techniques is routinely justified by appealing to universality classes (Morrison 2018b; Wilson 2017). Indeed, the development of renormalization techniques was motivated by the need to explain various instances of universality in physics (Kadanoff 1966, 1990, 2013). In addition, homogenization techniques are designed to account for various universal features of systems that are heterogeneous at smaller scales (Batterman 2015, 2017; Wilson 2017). What's more, both of these multiscale modeling techniques exploit the concept of universality in order to discover

Multiscale Modeling and Universality

highly idealized models at macroscales that are able to capture the universal behaviors of interest to scientific modelers. Because they enable scientists to discover universal features at macroscales and how they are coupled and decoupled from the features of the system at smaller scales, renormalization and homogenization techniques are extremely useful to multiscale modelers, who are interested in explaining and understanding the emergence of complex behaviors across multiple scales.[16]

Before moving to some more general philosophical conclusions, it is important to note that many of these multiscale modeling techniques are used simultaneously rather than one at a time. For example, the use of homogenization techniques typically requires a partial separation of scales that can be used to justify the use of various scale-dependent modeling techniques. In addition, it has long been known that systems develop scale-invariant properties when they are near their critical points. As a result, many instances of critical behavior that are analyzed using renormalization or homogenization techniques will also involve various scale-invariant relationships. For instance, the Arctic melt ponds case described in chapter 6 involves a separation of scales, self-similarity, and the use of homogenization techniques. What this shows is that typically no one of these techniques will be sufficient for confronting the challenges raised by the tyranny of scales. Instead, multiple multiscale modeling techniques will typically be required in order to uncover the myriad couplings and decouplings between various features across a wide range of scales.

In addition, I want to emphasize that in each of these cases, the dependence or independence found between two scales comes in *degrees*. For scale dependence, only some features at particular scales will be independent of (or decoupled from) some features at other scales. For scale invariance, only certain relationships and parameters will be stable across a wide, but limited, range of scales. For interscale relationships, only certain parameters at macroscales will be autonomous of many features at smaller scales, but they will also often depend on some important features at smaller scales. As a result, neither complete autonomy of one scale from another nor complete reduction of one scale to another is found in most cases of multiscale phenomena. Instead, we require far more subtle, case-by-case analysis of how multiscale modelers confront the tyranny of scales by using multiple modeling techniques to capture various kinds of relationships across spatial and temporal scales.

7.5 Philosophical Lessons: Practical Constraints, Pluralism, and Relative Autonomy

We have seen that the practice of multiscale modeling raises the problem of inconsistent models and the tyranny of scales. Fortunately, the concept of universality helps us address both of these challenges by showing how multiple conflicting models that describe various relationships within and across scales of the system can contribute to explanations and understandings of complex phenomena. Scale-dependent techniques focus on showing that certain features at particular scales are universal with respect to changes at other scales of the system. Scale-invariant techniques focus on discovering patterns that are universal across changes to the scale (and other features) of the system. Finally, interscale modeling techniques focus on showing how universal macroscale behaviors arise out of heterogeneous features at microscales. In general, these cases clearly "illustrate the very powerful notion of universality, which is of great interest to the study of complex systems in physics and biology" (Gisiger 2001, 171). Indeed, these uses of scale-dependent, scale-invariant, and interscale modeling techniques are all intimately intertwined with the concept of universality.

By naming various multiscale modeling techniques and analyzing the conditions involved in their justification, the framework offered here provides new vocabulary for constructing alternative accounts of the relationships among scales and conflicting modeling approaches.[17] In particular, I suggest that focusing on the concept of universality and the details of a plurality of multiscale modeling techniques will enable philosophers to develop better accounts of the relationships between models, theories, and scales than those provided by traditional discussions of levels, decomposition, or mereological hierarchies. Moreover, this shift will produce philosophical accounts of the relationships between scales and modeling techniques that can provide more useful normative guidance to scientific modelers that must confront the practical and representational constraints involved in multiscale modeling. By determining which features are universal at particular scales, which are universal across scales, and which are universal with respect to which perturbations at other scales, scientific modelers can begin to understand the complexity of the multiscale phenomena we observe.

In this final section I draw some additional philosophical lessons from these instances of multiscale modeling and compare those conclusions with

Multiscale Modeling and Universality 225

various views of multiscale modeling, explanation, and reduction offered by other philosophers. To begin, Winsberg (2006, 2010) focuses his discussion on instances of parallel multiscale modeling in which inconsistent modeling frameworks used for various regions of the system need to continuously pass information back and forth. The passing of this information is accomplished through what Winsberg calls "handshakes" between the models. For example, he describes how modelers of nanoscale cracks connect a continuum (finite element) model and a molecular dynamics model that are used for different regions of the system (Winsberg 2006, 590). In this case, the handshake is accomplished by looking at sets of points that can "see each other" across the boundaries of the regions of the system modeled by the various models. Both of these points are then modeled as if they were finite elements (in the continuum model). Then, both points are modeled as if they were lattice points (in the molecular dynamical model). The results are then averaged together in order to describe the energy interactions within these subregions of the system. These handshakes between models are similar to my discussion of the use of modeling techniques that aim to model the complex interscale interactions of the system. As a result, Winsberg gives us yet another example of how multiple conflicting modeling frameworks might be combined to tackle a multiscale modeling problem.

It is important to note, however, that the handshakes between conflicting modeling strategies will typically need to be very different depending on the case (Bursten 2018a). For example, Winsberg's example is specifically designed to model the energy of the total system, and various idealizations are incorporated into the models in order to make these handshakes possible. However, very different modeling algorithms and idealizations—which may be in conflict with those used to model the system's energy—will be required to model other interscale interactions. This means that the handshaking algorithms described by Winsberg will typically exacerbate the problem of inconsistent models rather than help to solve it. Indeed, as Winsberg notes, "When you include the handshaking regions, parallel multiscale models are—all at once—*models of an inconsistent set of laws*" (Winsberg 2010, 86). The problem is that Winsberg's handshaking account fails to address how such inconsistencies between the modeling frameworks used to model the same overlapping regions of the system ought to be addressed (Bursten 2018a, 162). That is, while handshakes may capture certain instances of how multiscale modelers tie various inconsistent models together, we still need answers about how to justify the use of multiple conflicting models to represent the same features of the models' target systems. In contrast, I argue that the earlier discussion of multiple overlapping universality classes shows how multiple conflicting models can be justifiably used to model the same systems (and the same regions of those systems).

Winsberg goes on to argue that these cases of multiscale modeling appear to be "at odds with some basic philosophical intuitions about the relationships between different theories and between theories and their models" (Winsberg 2006, 591). Despite noting this, he stops short of claiming that such cases require us to reject various claims that philosophers have made about levels, reduction, and emergence, and he tells us little about what philosophical claims ought to replace them (Bursten 2018a).[18] In what follows I aim to be explicit about which philosophical ideas I think ought to be rejected in light of these cases and which positive claims the cases suggest ought to supplant them.

Philosophical discussions of models and theories at different scales (or "levels") tend to focus on whether or not the laws, causes, or explanations at the macroscale are (or are not) *completely* autonomous of, or *completely* reducible to, the features at smaller scales. On the one hand, emergentists (or antireductionists) often suggest that the emergent patterns and explanations at a macroscale studied by the special sciences are completely autonomous of the features of the system at the scales studied by physicists—that is, these emergent features simply cannot be explained by examining the multiply realized, "lower-level" details of the system (Fodor 1974; Putnam 1975). Reductionists, on the other hand, seem to claim that all the features of the system described by so-called higher-level sciences can be derived (or explained) just from knowing the components and interactions at the most "fundamental" level. As Sober puts it, "reductionism asserts that physics unifies because everything can be explained, and explained *completely*, by adverting to physical details" (Sober 1999, 561).

The cases surveyed here suggests that both of these positions are far too extreme (Batterman 2017; Bursten 2018a). Instead, the real challenges facing multiscale modelers concern determining which particular features, at which particular scales, are dependent on, invariant across, or separable from one another. In other words, macroscale phenomena are rarely (if ever) completely emergent or completely reducible to features at smaller scales. In contrast,

Multiscale Modeling and Universality 227

what we find is a spectrum of cases, ranging from more to less universal, that is dependent on the particular systems under consideration, the particular phenomena of interest, and the available modeling frameworks. By focusing on the degree of autonomy involved, the framework developed here better captures how relative or partial autonomy plays a crucial role in determining the justification and success of various multiscale modeling strategies.

Moreover, these cases show that complex systems (and the models used to study them) are rarely nested into nice hierarchies in which properties at smaller scales fully determine the properties at larger scales (Batterman 2015; Potochnik 2009a, 2017). Unfortunately, as Winsberg notes, "to the extent that the literature in philosophy of science about levels of description has focused on whether and how one level is reducible to another, it has implicitly assumed that the only interesting possible relationships are logical ones—that is, intertheoretical relationships that flow logically from the mereological relationships between the entities posited in the two levels" (Winsberg 2006, 591). In contrast to this focus on hierarchical or mereological relationships, after pointing out that patterns occur across many scales of a complex system, multiscale modelers frequently remind us that "it would be misleading to think about the previous nested structure as a fully hierarchical one ... the understanding of the upper parts of the hierarchy are somewhat, but not totally, decoupled of the lower members. The influences between each component are likely to be *bidirectional*" (Solé and Bascompte 2006, 6).

Put somewhat differently, complex systems often involve "a recursive 'dialogue' between the parts and the whole: the parts create a macroscopic pattern that in turn modifies the boundary conditions in which the parts interact" (Bascompte and Solé 1995, 362). In short, things are often far more bidirectional and interwoven than is suggested by philosophical accounts that focus on decomposition into localized parts, hierarchical structuring, multiple realization, or part-whole relationships. Therefore, instead of focusing on these kinds of compositional or mereological concepts, philosophers ought to focus their attention on the unique modeling challenges that confront scientists attempting to understand various kinds of stability across multiple interacting scales of complex systems.

The previous examples also show that it cannot be assumed that the best modeling approach will be top-down or bottom-up (Bechtel and Richardson 1993). Instead, "defining the linkage between different scales poses a significant barrier to model development; in many cases, the link is bidirectional, meaning that higher- and lower-level variables, parameters, and functions characterizing the models are influenced by each other" (Deisboeck et al. 2011, 3). In light of these bidirectional interactions among models, debating whether scientific modelers ought to adopt a top-down or a bottom-up approach leaves most of the problems that actually confront multiscale modelers unresolved. What is more, as Batterman notes, in many cases "Our top-down consideration will inform the construction of models at lower scales. And our bottom-up attempts will likewise induce changes and improvements in the construction of higher scale models" (Batterman 2013, 285). Therefore, instead of trying to discover a "one approach fits all" methodology, scientific modelers require a deeper understanding of how multiple conflicting modeling techniques can be used to extract information about dependencies and independencies across a wide range of scales. The framework offered here provides the foundation for identifying a more pluralistic toolbox of modeling techniques that are simultaneously used to investigate the plethora of complex interactions exhibited by multiscale phenomena.

In addition, some philosophers have argued that the regularities found in the world tend to be found at particular scales. For example, Wimsatt (1976, 2007) organizes levels by those types of entities that tend to interact with one another. He says that "compositional levels of organization . . . are constituted by families of entities usually of comparable size and dynamical properties, which characteristically interact primarily with one another, and which, taken together, give an apparent rough closure over a range of phenomena and regularities" (Wimsatt 2007, 204). One of his main claims here is that interactions occur more frequently between objects of roughly the same scale (or level) than between objects at different scales (or levels).

While I agree with Wimsatt that many of our modeling perspectives have been designed to capture particular kinds of interactions or processes that commonly occur at particular scales, I think that these examples suggest that we should not assume that interactions within scales are more common (or stronger) than interactions across different scales. Moreover, we should not assume that features at a particular scale will often have similar dynamical properties or regularities. Because universality is not tied to any specific scale, and it comes in degrees, it helps us better capture the range of dependence and independence relations that hold between various scales of complex systems. Sometimes features at a specific scale will be coupled to other features at the same scale and largely autonomous of what is happening at other scales. But in many other cases, the features at a particular scale will be strongly coupled to features at both larger and smaller scales.

More generally, in agreement with Pincock (2012) and Potochnik (2017), I think that philosophers should be wary of directly drawing metaphysical conclusions from scientific models (and theories). A main reason for this is that I agree with Pincock that "most metaphysical discussion in contemporary philosophy is not informed by the complications of actual scientific practice" (Pincock 2012, 119). Indeed, trying to infer general metaphysical results about distinct levels, the relationships among them, or the existence of emergent features from these instances of multiscale modeling seems premature.

However, we need not adopt pure instrumentalism about these cases either. The epistemic contributions these techniques make to scientific explanations and understanding of complex phenomena depend on their telling us something important about how the real systems work. I have argued that universality gives us a way of understanding what these multiscale modeling techniques are telling us about the nature of complex systems, without depending on misguided notions such as "accurate representation of causes" or "mirroring of mechanisms or processes." These multiscale models enable scientists to capture universal patterns of behavior and extract correct modal information about the system, but this need not entail any metaphysical conclusions regarding the existence of distinct levels or well-defined entities, or the details of actual processes. Scientists know a lot about the counterfactual dependencies and independencies that hold in these complex systems, but they often do not (or at least need not) know whether the actual entities or mechanisms have the features ascribed to them by their multiple conflicting multiscale models. More generally, "we cannot move from the success of a multiscale representation to the conclusion that it reveals novel metaphysical features of the physical system" (Pincock 2012, 120). The cases presented here clearly show that the relationship between scales of a system is typically far more complex and dependent on features of particular cases than is suggested by the sweeping generalizations employed in many metaphysical debates in philosophy.

A related philosophical debate concerns whether the inclusion of more accurate details about a phenomenon always improves a scientific model. Several philosophers and scientists have suggested that more accurate and detailed models will provide better explanations. It is important to remember that there are two different claims here. The first claim is that more *details* concerning smaller-scale entities improves an explanation. While this is endorsed by some causal and mechanistic accounts (e.g., Salmon 1984), many philosophers who adopt a mechanistic approach explicitly reject this idea by suggesting that mechanisms at larger scales can provide better explanations (Craver 2007). The second claim is that more *accuracy* with respect to the relevant features of the actual mechanisms or processes always improves an explanation (Craver and Kaplan 2020). As we saw in chapter 2, this claim is widely endorsed by most mechanistic accounts and several other proponents of the standard view. However, both of these claims have been questioned by philosophers who suggest that removing details or introducing drastic distortions of relevant features can actually *improve* the explanations provided by scientific models (Batterman 2002a, 2009, 2010; Huneman 2010; Rice 2017, 2018; Weslake 2010). According to these accounts, sometimes a highly idealized model that does not accurately represent a system's relevant details "can better explain and characterize the dominant features of the physical phenomenon of interest. That is to say, these idealized models better explain than more detailed, less idealized models" (Batterman 2009, 429).

The discussion of multiscale modeling in this chapter suggests that all of these strategies have an important role to play in explaining complex phenomena, and *none of them should be privileged across all cases*. What we have seen is that many patterns are autonomous of many details at smaller scales. This enables scientific modelers to focus on constructing extremely idealized models that capture those universal features at larger scales. However, these cases also show that in many instances, some of the details of the system are particularly important for determining the patterns that occur at macroscales. As a result, it depends on the systems, our interests, and the modeling techniques available, whether or not including more accurate details about features at smaller scales will improve our explanations and understanding of the phenomena. The real questions of interest are *which* stable patterns we would like to explain and *which* features are relevant to those patterns, and at *which* scales of the system.

What's more, in most cases, when scientific modelers confront complex multiscale phenomena, the decision between an idealized (or general) model or an accurate (or detailed) model is simply not a choice that they are required to make. Instead, multiscale modelers almost always construct multiple idealized and conflicting models at a multitude of scales, and between scales. The real philosophical issue, therefore, isn't which of these types of models provides a better explanation than the others, but instead is how the insights of these conflicting multiscale models can be combined in order to produce a greater overall understanding of the phenomenon. I have argued that this can be done by recognizing how the kinds of multiscale modeling techniques used by scientific modelers can be related to complex phenomena by universality classes. The next step is to show how the modal information extracted from these models enables human beings to develop a better overall understanding of natural phenomena. I take up this task in the next chapter.

8 Understanding, Realism, and the Progress of Science

Much of this book has focused on the nature of scientific explanation and how idealizations make positive and essential contributions to those explanations. This focus is warranted due to the central role played by explanation and idealization within scientific practice. However, explanations are not the only epistemic achievement for which holistically distorted models are leveraged. In fact, one of the primary reasons that scientific explanations are so valuable to us is that discovering explanations enables science to produce an ever-growing body of understanding of natural phenomena (de Regt 2017; Khalifa 2017; Potochnik 2017; Strevens 2008). What's more, in addition to the understanding produced by explanations, I argue in this chapter that scientific inquiry produces understanding in many ways that do not pass through explanation (Lipton 2009; Rice 2016, 2018; Rohwer and Rice 2013). That is, even in cases where we fall short of having a complete explanation, science can greatly increase our understanding of a phenomenon.

Scientists frequently refer to the degree to which they understand a phenomenon and how scientific understanding confers understanding onto everyday observations. For example, one physics textbook says, "Understanding adiabatic processes allows you to understand why popping the cork on a cold bottle of champagne or the tab on a cold can of soda causes a slight fog to form at the opening of the container" (Halliday, Resnick, and Walker 2011, 526). Also, there has recently been more of a focus on the degree to which the public and politicians understand the phenomena studied by scientists (e.g., evolution and climate change). Furthermore, the epistemology literature has shown an increasing interest in the nature of scientific understanding (e.g., Elgin 2007, 2017; Grimm 2012; Khalifa 2012, 2013; Kvanvig 2003; Zagzebski 2001), and several philosophers of science have recently focused on how idealized models can be used to produce scientific understanding (e.g., de Regt 2009b, 2017; Khalifa 2017; Morrison 2015; Potochnik 2017; Rice 2016; Strevens 2013). In fact, some philosophers have gone so far as to argue that the ultimate epistemic aim of science is the production of understanding (Elgin 2017; Potochnik 2017). As de Regt succinctly puts it, "We value understanding, and we value science because it provides us with understanding of the world" (de Regt 2017, 44). Therefore, providing an account of understanding and how idealized models contribute to the production of understanding is necessary for showing how holistic distortion contributes to the epistemic aims of science.

In this chapter I use examples from physics, biology, and economics to motivate a modal account of scientific understanding. I begin by surveying several cases of understanding without explanation in section 8.1. In section 8.2 I use these cases to develop a factive account of scientific understanding (and its relationship to the counterfactual account of explanation presented in chapter 4), and then I argue that this account of scientific understanding is compatible with recognizing that our best scientific models and theories are holistically distorted representations. In section 8.3 I argue that this compatibility gives rise to a view that I call *Understanding Realism*. According to this view, science aims at, and often achieves, a factive understanding of the natural world—even if the models and theories used to acquire that understanding are holistic distortions. This version of realism provides new ways of thinking about scientific progress across theory change (section 8.4) and highlights the role of diversity in promoting science's epistemic aims (section 8.5). Section 8.6 concludes the chapter.

8.1 Understanding without Explanation

As I noted in chapter 4, many philosophers recognize that there is a strong link between explanation and understanding. In fact, I argued there that a requirement for any satisfactory account of explanation is for it to show how explanations produce understanding. However, several philosophers have taken this idea further, arguing that the *only* way to scientifically understand a phenomenon is to grasp a true or correct explanation of that phenomenon (de Regt 2009b, 2017; Khalifa 2012, 2017; Strevens 2008, 2013; Trout 2007). For example, Trout claims that "scientific understanding is the state produced, and only produced, by grasping a true explanation" (Trout 2007, 585–586). Following Trout, Strevens argues, "An individual has scientific understanding of a phenomenon just in case they grasp the correct scientific explanation of that phenomenon" (Strevens 2008, 3; Strevens 2013, 1). Khalifa also says, "S has minimal understanding of why p if and only if, for some q, S believes that q explains why p, and q explains why p is approximately true" (Khalifa 2017, 126). Finally, de Regt claims that "understanding a phenomenon [is equivalent to] having an adequate explanation of the phenomenon" (de Regt 2009b, 25). While explanations are certainly a key source of understanding, overemphasizing the relationship between explanation and understanding has led to a neglect of myriad cases where scientific models are able to produce understanding of a phenomenon without providing an explanation (Lipton 2009; Rohwer and Rice 2013, 2016).[1] In this section I briefly survey some examples that show how understanding can be produced by models that fail to explain. Recognizing both sources of understanding is key to capturing the variety of ways that science produces an ever-expanding body of understanding regarding natural phenomena.

8.1.1 Investigating Necessity Claims

One way in which models produce understanding without providing an explanation is by aiding in the investigation of claims about what is necessary or unnecessary for a phenomenon to occur. An example of this is Schelling's checkerboard model, used to investigate segregation in cities (Grüne-Yanoff 2009; Rohwer and Rice 2013; Schelling 1978; Weisberg 2013). The dominant belief at the time was that segregation must be the result of explicit (and strong) racist preferences. Schelling's model aimed to show that this was not necessarily the case.

In Schelling's model, dimes and nickels are used to represent two types of individuals, *A* and *B*. An individual's "neighborhood" is represented by a set of nine adjacent squares on a chessboard. The model assumes that individuals prefer to have neighbors that are at least 30 percent of the same type (e.g., *A*s want at least 30 percent of their neighbors to be *A*s). The agents then take turns determining if their preferences are met. If so, the agent remains in the same location; if not, the agent moves to the nearest unoccupied location. The model is run until all of the agents are satisfied with their locations.

I think that it is rather obvious that this extremely simple model drastically distorts most of the relevant features of any real-world segregated city.

Agents are not nickels and dimes, and they do not continuously decide whether to move to the nearest unoccupied location based on the same degrees of preference for like neighbors. Indeed, the model does not even aim to accurately represent the features of any real-world target system (Weisberg 2013). Despite these distortions, Schelling's results showed that across a wide range of changes to the dynamics of the model, including the use of various utility functions, rules for updating, neighborhood sizes, and spatial configurations, segregation is the equilibrium point of the model. Consequently, Schelling's model shows how minor preferences for like neighbors make it extremely hard to avoid segregation. As a result, explicit and strong racist preferences are not necessary for segregation to evolve. Given the context in which it was formulated, Schelling's checkerboard model allows us to learn something important because, "Before the models' publication, it seems, many people believed that segregation was necessarily a consequence of explicitly racist preferences. Schelling's model showed that there were plausible settings in which this was not so" (Grüne-Yanoff 2009, 96). As Schelling himself puts it, his goal was to investigate "some of the individual incentives and individual perceptions of difference that *can* lead collectively to segregation" (Schelling 1978, 138). What he showed, through the use of a highly idealized model, is that this outcome is possible even in cases where every individual acts on a small preference for similar neighbors.

In this case, the modeler uses a highly idealized model that does not aim to explain the observed pattern of interest. For one thing, simply grasping that this is a possible route to segregation fails to provide a complete explanation of how any actual segregated city has arisen (Weisberg 2013). Merely knowing that it is possible for mild preferences for like neighbors to produce segregation falls short of being able to explain why cities are actually segregated. However, by showing that a particular set of preferences could possibly give rise to the phenomenon, Schelling's model is able to produce some information that enables us to better understand the phenomenon of interest. Specifically, this model is able to justify the true belief that a neighborhood can become segregated even if there are no strong racist preferences (i.e., individuals acting on strong racist preferences is *not* a necessary condition for segregation to occur). This information is enlightening even if this fact is not part of the actual explanation of why cities are segregated. Therefore, Shelling's checkerboard model produces some understanding by undermining a formerly accepted claim about what was necessary for segregation to occur, but it does not even aim to provide an explanation of that phenomenon.

8.1.2 Modeling Hypothetical Scenarios

In a second kind of case, a model is constructed to better understand the general behavior of a small set of related features in order to determine what kinds of behavior could possibly result from those features. This is typically accomplished by constructing a model of a hypothetical scenario that isolates the features of interest in order to determine what behaviors those features are able to produce (Rice 2016; Rohwer and Rice 2013). A hypothetical scenario is not intended to accurately represent any particular features of a real-world system (i.e., the model has no real-world target system whose features it aims to capture). Instead, the modeler constructs a hypothetical system in order to explore the counterfactual situation in which only a few key features are present. The crucial difference between these kinds of cases and the Schelling case has to do with the historical context and goals of the model builders (Giere 2010; Plutynski 2004; Weisberg 2013). In Schelling's case, because it was believed that strong racist preferences were necessary for segregation to occur, the goal of the modeler is to investigate a possible system that could demonstrate that such a condition was unnecessary. This is why the key adjustable parameter in Schelling's model is the degree to which the agents prefer to have similar neighbors. In modeling hypothetical scenarios, in contrast, the goal is typically to show what can possibly be produced with a few minimal features in order to determine what *is* necessary for the phenomenon to occur. Thus, while both types of modeling clearly aim at investigating merely possible systems, they differ in the ways in which the context and goals of the modelers influence which of the possible (or hypothetical) systems the model is intended to capture.

The biologist Joan Roughgarden and colleagues describe this kind of hypothetical modeling as the construction of a "minimal model for an idea":

> A minimal model for an idea is intended to explore a concept without reference to a particular species or place. An example is the evolution of sex. Biologists do not understand why sexual reproduction has evolved so ubiquitously in nature. Many conjectures have been stated about the costs and benefits of sexual reproduction as compared with asexual reproduction. These conjectures have been explored theoretically to see simply if they make sense. Does a hypothetical population with recombination, mutation, and selection really evolve faster than

a hypothetical population subject to the same mutation and selection but lacking in recombination. The answer turns out to depend on the initial condition, and on the statistical nature of environmental change. Thus, models about the evolution of sex explore an idea, they are not really intended for testing, nor to apply to specific systems. Most of the early textbook models in ecology and evolutionary biology are of this type. (Roughgarden et al. 1996, 27)

By building a related set of such models, scientists can begin to understand how various sets of features might contribute to the overall behavior of a system in different contexts (e.g., in conjunction with different sets of assumptions or other features). Understanding the possible contributions of these features to overall system behavior often enables the modeler to answer how-possibly questions (Forber 2010; Odenbaugh 2005; Resnik 1991). What's more, these background beliefs (e.g., about what is possible) are often *true* and can contribute to our understanding of the phenomenon of interest, even if they are not part of any actual explanation of that phenomenon (Rohwer and Rice 2013).

An example of this kind of modeling is Maynard Smith's original use of the Hawk-Dove game (Maynard Smith 1978; Maynard Smith and Price 1973). Several species exercise restraint in combat instead of fighting to the death. The Hawk-Dove game is intended to show how individual selection could *possibly* produce this behavior in a wide range of populations.

In the Hawk-Dove game, two organisms compete for a resource that will increase their fitness by V. The basic game allows only two strategies. Hawks (H) escalate until injured or until the opponent retreats; Doves (D) display and then retreat if their opponent escalates. This results in three kinds of interactions: (1) *Hawk versus Hawk,* where each player has a 50 percent chance of obtaining the resource, V, and a 50 percent chance of receiving some cost, C, of being injured; (2) *Hawk versus Dove,* where the Hawk obtains the resource and the Dove retreats; and (3) *Dove versus Dove,* where the resource is shared equally. These interactions lead to the following payoff matrix:

	H	D
H	½(V−C), ½(V−C)	V, 0
D	0, V	V/2, V/2

where $V > V/2 > 0 > \frac{1}{2}(V-C)$.

The Hawk-Dove game also employs several idealizing assumptions that drastically distort the features of real biological populations. These idealizations include (1) infinite population size, (2) random pairing of players, (3) asexual reproduction, (4) symmetric contests, (5) pairwise contests, (6) constant payoff structure across individuals and across iterations of the game, and (7) perfect correlation between winning the resource and reproductive success (Maynard Smith 1982). In addition, the model represents the available strategies, interactions, and payoffs in a highly distorted way (i.e., the model misrepresents or ignores most of the features of real-world populations). Indeed, as Maynard Smith and Price say of their model, "real animal conflicts are vastly more complex than our simulated conflicts" (Maynard Smith and Price 1973, 17). In other words, the Hawk-Dove game holistically distorts most of the features of any real-world biological population in which the target phenomenon occurs.

Despite these distortions, given this payoff matrix, the model shows that neither Hawk nor Dove is an evolutionarily stable strategy (ESS). A stable equilibrium does occur, however, when the average payoffs for Hawks are equal to the average payoffs for Doves. This can occur in one of two ways: (1) the population could consist of a mixture of some Hawks and some Doves or (2) the population could consist of individuals who all adopt a mixed strategy of playing Hawk with probability x and Dove with probability $(1-x)$. Either way, the model predicts that individual selection will lead to restraint in combat in some instances. Consequently, the Hawk-Dove game shows us how it is possible for individual selection and a particular kind of payoff structure to produce the phenomenon of interest (Rohwer and Rice 2013).

While this is an important result, merely knowing that it is possible for individual-level selection to produce the phenomenon of interest is not enough to provide an explanation of why this phenomenon occurs. In general, merely knowing that a *type* of explanation could possibly account for the explanandum is insufficient to provide that explanation. Nonetheless, the Hawk-Dove model does enable us to understand that our observations could possibly be produced by selection operating only at the level of individuals. As Maynard Smith and Price put it: "A main reason for using [the model] was to test whether it is *possible even in theory* for individual selection to account for 'limited war' behaviour" (Maynard Smith and Price 1973, 15; emphasis added). Therefore, despite its failure to provide an explanation on its own, the Hawk-Dove game allows us to answer a key how-possibly question concerning the compatibility of individual-level selection with the observed behavior by investigating a hypothetical scenario. Furthermore, this modal information is true of the phenomenon of interest. Therefore, by providing true modal information, this hypothetical model produces some scientific understanding of the phenomenon, despite the fact that it does not explain the phenomenon.

8.1.3 Exploring Possibility Space

Another way that understanding is produced by models that fail to explain is by exploring a wide range of possible systems so as to better understand the overall space of possibilities. This is extremely common in biological modeling (Odenbaugh 2005). As Beckner notes: "Selectionists have devoted a great deal of effort to the construction of models that are aimed at demonstrating that some observed or suspected phenomena are possible, that is, that they are compatible with the established or confirmed biological hypotheses" (Beckner 1968, 165). Indeed, in many cases, only by exploring a wide range of possibilities can scientists come to understand certain features of the possibility space that might include the real systems of interest (see Kennedy 2012 for some interesting examples from astrophysics). As Weisberg puts it, "Theorists ultimately aim to partition the space of possibilities. They aim to understand what is possible, what is impossible, and why" (Weisberg 2013, 128). Moreover, as Weisberg notes, this is typically accomplished by providing counterfactual information about systems that go well beyond the actual system. The key difference with modeling hypothetical scenarios such as the Hawk-Dove game is that, in exploring possibility space, scientists are often interested in a wide range of possibilities that are likely to include the actual systems. That is, instead of focusing on one hypothetical system that is known to be nonactual, the goal in these cases is to model a wide range of possible systems in order to understand the space of possibilities that is likely to include the actual cases of interest.

As an example, astrophysicists Daniel, Heggie, and Varri (2017) use a mathematical model of possible star orbits to demonstrate that even if a star has an energy that exceeds the energy required to escape from a cluster, the star may be unable to escape. These "potential escapers" play an important role in the density and kinematics of stellar systems near the outskirts of star clusters. The problem is that the "growing body of numerical and observational information is not matched by comparable progress in the theoretical understanding of the phase space." Moreover, "none of the analytic models which are currently available include the contribution of the potential escapers" (Daniel, Heggie, and Varri 2017, 2). In order to explore this phase space, these physicists needed to construct a new idealized mathematical model that could be used to explore the possible orbits of these potential escapers (figure 8.1).

By building an idealized model that could explore possible orbits and energies, these modelers showed that "even star clusters on simple circular orbits possess a population of stars with energies above the escape energy which are none the less confined within the stellar system itself" (Daniel,

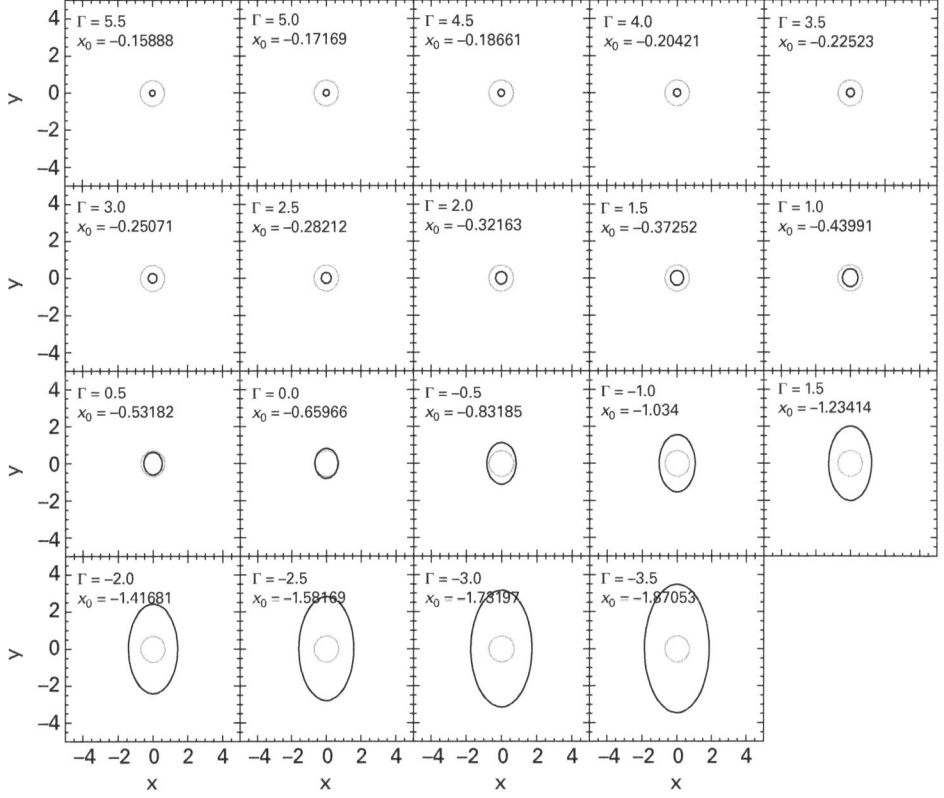

Figure 8.1
Several examples of circular and elliptical f-orbits. Each star's orbit is shown by the darker (black) line, while the tidal (Jacobi) radius is shown by the lighter (gray) line. The tidal radius is the radius where the gravity from the cluster becomes dominated by the gravity from the galaxy around which the cluster orbits (from Daniel, Heggie, and Varri 2017, 5).

Heggie, and Varri 2017, 2). In other words, by exploring a wide range of possibilities, these physicists were able to demonstrate something about the phase space of possible orbits. As these modelers put it, "For our purposes, the importance of these results is that they show us in simple terms that *it is possible* for a star to remain inside the cluster even though its energy exceeds the energy of escape" (Daniel, Heggie, and Varri 2017, 4; emphasis added).

After demonstrating this possibility by modeling some simple circular orbits, these modelers then explore the phase space in more detail by simulating a much wider range of possible orbits in order to investigate which kinds of initial conditions could give rise to escapers and nonescapers (figure 8.2).

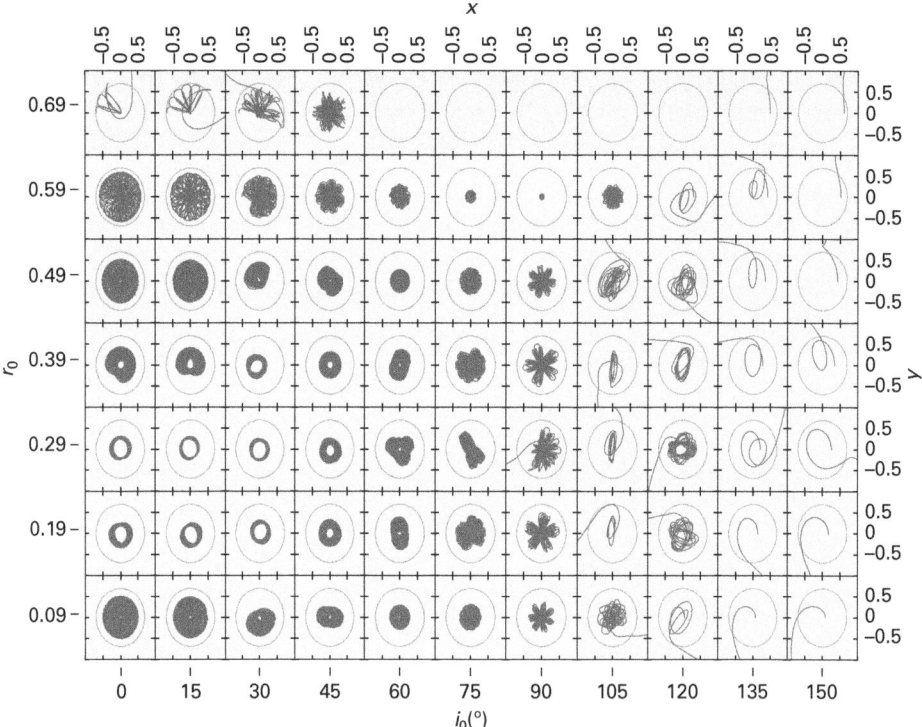

Figure 8.2
Various three-dimensional orbits for stars with different initial conditions. Each step down increases r_0 by 0.1 and i_0 is increased by 15 degrees with each step to the right. The initial parameters (r_0 and i_0) are given on the lower-left axes. The orbital trajectory is plotted in black, with the tidal (i.e., escape) radius shown as a light gray circle (from Daniel, Heggie, and Varri 2017, 6).

These idealized models do not aim to accurately represent the features, causes, or mechanisms that produce the occurrence of any particular star orbits. Instead, the goal of these simulations is to explore the space of possibilities that are likely to include some actual cases and to demonstrate the need to incorporate potential escapers into models of stellar systems. I contend that these models produce some understanding of stellar systems by providing modal information concerning which orbits are possible (and which are impossible) and showing how changes in various parameters (e.g., the initial conditions of the system) result in changes to the overall behavior of the system. In other words, these models produce understanding by exploring the phase space of possible orbits rather than explaining any particular occurrences of escaper or nonescaper orbits.

8.1.4 Summing Up

The cases discussed in this section raise a serious challenge to recent accounts that have claimed that the *only* way to produce scientific understanding is by grasping a correct explanation (Khalifa 2012, 2013, 2017; Strevens 2008, 2013; Trout 2007). The first problem with such claims is that, according to most accounts of explanation, a correct explanation will need to accurately represent difference-making features of a real-world target system. In contrast to these accounts, I argue that a highly idealized model can produce understanding even if it is an inaccurate representation of most (or perhaps even all) of the features of real-world systems—including known difference-makers for the phenomenon of interest. In fact, holistically distorted models can still produce understanding, even when the model does not even aim to be an accurate representation of the features of a target system.

However, given that I have argued that holistically distorted models can still be used to explain, focusing on the inaccuracy of these models is not enough to establish that they can produce understanding without providing an explanation.[2] In order to support this conclusion, we need to look at what counterfactual dependence and independence information is provided by these models. In chapter 4 I argued that explanations are constituted by a set of information about the counterfactual dependencies and independencies that hold between features of the system and the explanandum. Moreover, I argued that the set of counterfactual dependence information must be sufficient to account for why the explanandum occurred.

The question, then, is whether these models provide a set of modal information that is sufficient to account for the occurrence of the phenomenon.

In the first example, Schelling's checkerboard model shows that it is possible to produce segregation from mild preferences for like neighbors (within a highly idealized model system). This might be used to justify our belief in the following counterfactual claim: even if people did not have explicitly racist preferences (and everything else remained the same), then segregation would still occur in many cities. While this might be useful information for how *not* to intervene when the goal is to eliminate segregation, grasping the truth of this single counterfactual is woefully inadequate for providing a complete explanation of why cities are actually segregated. For one thing, this piece of information tells us nothing about which features of real cities the explanandum does counterfactually depend on— it merely shows that one feature that was thought to be necessary is not required. Furthermore, while I have argued that information about counterfactually irrelevant features can sometimes contribute to an explanation, citing one counterfactually irrelevant factor is insufficient to explain. Now perhaps Schelling's model can be interpreted as showing us that segregation depends on there merely being *some* kind of preferences for like neighbors in the population.[3] I am not confident that Schelling's model succeeds in showing how actual segregation depends on this feature. But, even if we grant that this is so, the model tells us almost nothing about how this single factor gives rise to segregation in actual cities. In addition, given that segregation counterfactually depends on a whole host of systematic factors beyond individuals' preferences concerning like neighbors, citing this single feature fails to provide enough dependence information to account for why actual cities are segregated.

In our second example, the Hawk-Dove game shows how individual-level selection could possibly produce restraint in combat. This information might be used to support the counterfactual claim that even if there is no group selection operating, restraint in combat could still (possibly) evolve. While this is certainly true, and it is interesting given the context of the group selection debate, this information is inadequate to explain why real-world species often show restraint in combat. For one thing, the model fails to show that individual-level selection *is* the process responsible for the phenomenon in actual populations (i.e., the phenomenon could still be due to group-level selection). In addition, while the model is capable of showing that individual-level selection is a possible type of explanation for the phenomenon, it tells us little about how the features of real-world systems actually give rise to the phenomenon. That is, the model tells us very little about how the phenomenon counterfactually depends on the features of real biological populations. For these reasons, the Hawk-Dove game fails to provide enough information about the counterfactual dependence and independence relations that hold in real populations to be able to provide a complete explanation of the restraint in combat we observe in many real biological populations.

In our third example, the goal is to explore various possible orbits of stars and show how those orbits depend on some of the initial conditions within the model system. While this exploration of possibility space does provide some counterfactual dependence information, merely citing these initial conditions is insufficient for showing why (or how) those initial conditions give rise to those orbits in real stellar systems. Moreover, the purpose of this kind of modeling is not to tell us what the actual initial conditions were for any real star cluster in which the phenomenon (of being an escaper or nonescaper) occurs, but instead to explore the range of conditions that could possibly give rise to those observations. Therefore, while it does provide extensive modal information about the space of possible systems, these models do not provide sufficient counterfactual dependence (and independence) information to be able to explain why there are escapers or nonescapers in any real-world stellar systems.

One possible objection here is to suggest that, while these models may produce some understanding, they are unable to produce an understanding of *why these phenomena occur*. For example, both Strevens (2013) and Khalifa (2017) argue that grasping a correct explanation is necessary only if we are concerned with understanding why a phenomenon occurred. However, following Lipton, I think these cases show that, "even on a narrow conception of understanding as understanding why, we may nevertheless get understanding without actual explanation" (Lipton 2009, 54). The reason for this is that these cases provide the same kind of modal information that is provided by explanations about how different possible states of the system would change the phenomenon of interest. Providing only *some* of this information will often fall short of being able to provide a complete explanation, but such information can still improve our understanding of why the phenomenon occurs in the cases it does. In response, one might suggest that we only have understanding why when we can completely answer a why question about the phenomenon, but this begs the question about whether there could ever be understanding why without having an explanation. These cases show how the same kind of information (i.e., modal information) provided by explanations can be provided independent of explanation. Despite this, like Lipton, I also see no reason to restrict the understanding produced by scientific inquiry to understanding why. Understanding how to produce a phenomenon, alternative ways that a phenomenon could have occurred, what is necessary for a phenomenon to occur, and what is possible or impossible all seem to be legitimate epistemic aims of scientific inquiry, even if they do not ultimately lead to an explanation of the phenomenon.

Another possible objection would be to argue that, while these models do not provide how-actually explanations, they each provide *how-possibly* explanations (Brandon 1990; Craver 2006; Forber 2010; Plutynski 2004). This would mean abandoning Trout's and Strevens's claim that understanding requires us to grasp a *correct* explanation. Along these lines, Sober (2011, 577–578) suggests that many idealizations in biological models can be removed by interpreting the claims of the model merely as suggesting what selection would promote in the nonactual situation. While there certainly can be how-possibly explanations, in order for the concept to go beyond mere prediction, there must be additional criteria for what makes something capable of explaining how something is possible. Presumably these criteria should be at least partially drawn from our account of how-actually explanations, without the requirement that the explanans be true. For example, Brandon draws on Hempel's account of explanation in order to argue that "how-possibly explanations are potential explanations, none of whose explanatory premises contradict or conflict with 'known facts'" (Brandon 1990, 178–179). Independent of Brandon's use of Hempel's account, we might require that a how-possibly explanation provide an account of how our world could possibly produce the phenomenon. Of course, the cases presented here fail to meet this kind of criteria because many of their idealizing assumptions are in conflict with known facts. For example, while Schelling's model answers a how-possibly question (concerning a merely possible system), it does not plausibly explain how the actual world could possibly have given rise to the phenomenon because we know that there is systematic structural racism and that some individuals in the population do have strong racist preferences. Moreover, we know that both of these features have made significant contributions to why actual cities are segregated.

In contrast to Brandon's approach, Plutynski suggests that models in population genetics "are explanatory in that they are effective at providing proofs of possibility" (Plutynski 2004, 1208). However, if all that was required to be a how-possibly explanation is that a model provide a set of conditions that could give rise to the phenomenon, then any model that makes an accurate prediction would provide a how-possibly explanation. In order to avoid this result, we require additional criteria for providing an explanation of how something is possible. What I have argued here is that if these criteria are drawn from a more general account of scientific explanation (e.g., the account defended in chapter 4), then these models produce understanding (by providing true modal information) without providing a how-possibly explanation because they fail to show how the features of real systems could give rise to the phenomenon. Indeed, even if we ignore the truth requirement for explanations, these models fail to provide sufficient modal information to be able to explain.

I conclude that these highly idealized models are able to contribute to scientific understanding without being able to provide a complete explanation. The modal information they provide is simply too limited to be able to explain why the phenomenon actually (or possibly) occurs. Consequently, accounts that require a model to provide an explanation in order to produce understanding miss a large amount of the understanding produced by the idealized scientific models used to investigate necessity claims, hypothetical scenarios, and the range of possible systems.

8.2 An Account of Understanding

In light of these examples, we need an account of scientific understanding that can show how understanding can be produced via routes that do not pass through explanation (Lipton 2009; Rohwer and Rice 2013, 2016). In this section I provide a factive account of scientific understanding that meets this requirement and accounts for the central role of holistically distorted models in science's production of understanding.

Before embarking on this project, however, I should note that while I suspect that there will be some continuity between scientific understanding and the understanding produced outside of science (e.g., in history, art, or literature), my focus will be on the features required for *scientific* representations to produce understanding. In addition, I will focus on how scientific representations can be used to understand the phenomena we observe, rather than an individual's ability to understand a particular model or theory, such as Willa's understanding of quantum mechanics. That is, I assume that the understanding that is an epistemic aim of science is concerned with understanding real phenomena in the world rather than simply understanding our theories or models of those phenomena (de Regt 2017, 23; Kvanvig 2009; Schurz and Lambert 1994, 68). Of course, understanding a scientific model or theory might be necessary for understanding the phenomenon (de Regt 2009a, 2017), but this does not mean that understanding the model or theory is *sufficient* to understand the phenomenon itself.

With those caveats out of the way, we can start to look at the particular features of scientific understanding that distinguish it from other epistemic achievements like knowledge and explanation. The first feature that distinguishes understanding from knowledge is that it is widely accepted that understanding requires the ability to grasp relationships among the pieces of a larger body of information (Grimm 2012, 103). This is part of what makes understanding more valuable than just having a set of knowledge. As Kvanvig puts it: "what does understanding add that knowledge can lack? . . . Understanding requires the grasping of explanatory and other coherence-making relationships in a large and comprehensive body of information" (Kvanvig 2009, 192). In particular, the grasping of these interconnections "systematizes our thinking on a subject matter in a way beyond the mere addition of more true beliefs or even justified true beliefs" (Kvanvig 2009, 205). For example, one who understands planetary motion not only justifiably believes many true things about planets, but also grasps how those pieces of information are connected in various systematic ways. Elgin puts the point in this way: "Understanding is primarily a cognitive relation to a fairly comprehensive coherent body of information. The understanding encapsulated in individual propositions derives from an understanding of larger bodies of information that include those propositions" (Elgin 2007, 35).

Philosophers of science echo this idea, but in slightly different terms: "to understand a phenomenon P is to know how P fits into one's background knowledge" (Schurz and Lambert 1994, 67). In other words, in order to understand a phenomenon, not only must an agent grasp various relations among the components of a body of information, but those connections must enable the agent to systematically incorporate the phenomenon into their background knowledge. Following these views, on the account I present here, in order to understand a phenomenon, an agent must (1) grasp the relations between a fairly comprehensive body of information about the phenomenon and (2) see how that information is related to various pieces of their background knowledge. Grasping these systematic relationships is the "something further" that must be grasped in order to genuinely understand.

These features of understanding are also crucial for seeing how communities can gain understanding of a phenomenon that goes beyond the understanding of any individual (Zagzebski 2001). For one thing, communities (e.g., the community of scientists studying a phenomenon) can have a much larger set of background information about the phenomenon (Woody 2015). Moreover, as new information about the phenomenon is acquired through various scientific activities such as experimentation, modeling, or theorizing, that new information can be incorporated into the community's overall body of information about the phenomenon of interest. The interconnections among all those pieces of information may not be grasped by any single individual in the group, but those individuals can each contribute overlapping sets of information and connections that will constitute the scientific community's overall understanding of the phenomenon. Determining precisely how we should think about how communities can understand will have to be worked out in more detail later (see Boyd 2019 for a discussion of some of the challenges here). For now, the important point is that our account of scientific understanding should not be restricted to the understanding had by individuals (i.e., the account should at least *allow* for communities of scientists to have understanding that goes beyond the understanding had by any individual scientist). Allowing this possibility enables our scientific understanding of a phenomenon to expand beyond the cognitive limitations of individual human agents.

Before focusing on the particular kinds of connections that need to be grasped in order to scientifically understand a phenomenon, we need to address the issue of whether the understanding produced by scientific inquiry is *factive*. To say that understanding is factive is to claim that at least some of the information or beliefs within one's body of understanding must be true (Grimm 2006, 2012; Mizrahi 2012; Strevens 2008, 2013). For example, Young Earth creationists believe that a great flood formed the Grand Canyon in about a year, but it seems that they do not genuinely understand this phenomenon because their story is incorrect in crucial ways (Strevens 2013). Indeed, there is a strong intuitive pull to say that understanding natural phenomena cannot involve believing falsehoods because "what we are trying to understand is how things actually stand in the world" (Grimm 2006, 518).[4] No matter how confident one is that they understand a topic, if their beliefs are mostly false, they do not genuinely understand.

One might think this suggests that all the propositions or elements that contribute to one's understanding ought to be true. However, this requirement is far too strong (Elgin 2007; Kvanvig 2003; Zagzebski 2001). For example, it appears that someone might be said to understand evolution even if he or she has some false beliefs about the details of how it operates. Moreover, such a standard cannot do justice to the epistemic contributions of science because scientific understanding typically depends in essential ways on the use of highly idealized models and theories (Elgin 2007, 2017). As a result, the widespread use of idealizations within our best scientific models and theories suggests that requiring all of the propositions that contribute to one's understanding be true is too high a standard.

In response to these observations, Elgin (2007, 2017) argues that we ought to abandon a factive conception of scientific understanding. However, contrary to Elgin, I think that allowing that not all the propositions that contribute to one's understanding must be true does not require us to adopt a nonfactive account of scientific understanding. For example, Kvanvig (2003) and Mizrahi (2012) both argue for "quasi-factive" accounts of understanding by distinguishing between central propositions and peripheral ones. On Kvanvig's view, all of the central propositions of one's understanding must be true, but a few false beliefs about peripheral propositions does not undermine one's understanding. He suggests:

> Suppose that the false beliefs concern matters that are peripheral rather than central to the subject matter.... When the falsehoods are peripheral, we can ascribe understanding based on the rest of the information grasped that is true and contains no falsehoods ... in this way, the factive character of understanding can be preserved without having to say that a person with false beliefs about a subject matter can have no understanding of it. (Kvanvig 2003, 201–202)

This is similar to Strevens's (2008, 2013) suggestion that scientific understanding requires the grasping of an explanation that correctly describes the difference-making causes of the phenomenon. This allows some falsehoods that distort peripheral (or non-difference-making) causes to play a role while still requiring a factive component to understanding.

While this is a step in the right direction, as I've argued extensively in previous chapters, idealizations are often central to many of the scientific models and theories used to understand in science. Indeed, as Elgin notes, in scientific practice, "elimination of idealizations is not a desideratum. Nor is consigning them to the periphery of a theory" (Elgin 2007, 38).[5] In addition, as I argued in chapter 5, the idealizations used in scientific practice routinely result in representations that holistically distort contextually salient difference-making features (i.e., they distort precisely those features that are taken to be relevant and are essential to the way in which the phenomenon was produced). As a result, requiring that the idealizations within our best scientific models be resigned to the periphery of our understanding is the wrong way to argue that scientific understanding can still be factive.

Instead, my view is that scientific understanding is factive because in order to genuinely understand a natural phenomenon, *most* of what is believed about that phenomenon—especially about certain contextually salient propositions—must be true. To be clear, I am not claiming that simply having a majority of what an individual or community believes (or accepts) about the phenomenon be true is sufficient for understanding. In addition, I don't mean to suggest that determining whether an agent or group understands ought to involve counting up all the propositions within their understanding and determining whether the percentage of true propositions meets some universally applicable threshold (e.g., 65 percent). Rather than counting propositions, I recommend a case-by-case approach that allows for a plurality of context-sensitive ways that scientific understanding might meet this factive requirement (Rice 2016). For one thing, in various contexts of scientific inquiry, some pieces of information or connections will be more salient, and so their truth-value or accuracy ought to carry more weight in determining whether one's understanding is factive. In light of this fact, I argue that the factive component of scientific understanding is somewhat flexible and is highly context sensitive.

Despite this context sensitivity, the context of inquiry will specify a particular question of interest, contrast class, and set of features that are thought to be relevant and irrelevant to the phenomenon. As a result, merely having many true beliefs about the phenomenon of interests will typically be insufficient for understanding because the agent who understands will need to grasp certain truths and relationships that are made particularly salient by the context. Precisely how these contextual factors interact with the factive requirement on scientific understanding will have to be uncovered by analyzing particular cases because they will likely be different for different areas of science that are focused on understanding different kinds of phenomena.

The important point is that requiring that most of the contextually salient components of one's understanding be true is able to accommodate the central contributions of idealizations to scientific understanding. For one thing, these idealizations are rarely believed to be true by the scientists who use them. But, even when scientists unwittingly accept them as truths, the centrality of the contributions made by these falsehoods is not enough to undermine scientists' ability to factively understand the phenomenon. The challenge for factive accounts of understanding isn't to show that idealizations *never* play a central role in scientific understanding but rather to show how particular ways of distorting centrally important features, causes, and mechanisms can make positive contributions to our overall understanding of the phenomenon. In short, allowing idealizations to make central contributions to scientific understanding is compatible with requiring that most of the contextually salient components of that understanding must be true or accurate (to the degree dictated by the context).

Therefore, while I maintain a factive requirement for understanding, I have no universal account to offer about how to determine precisely across all cases when most of what scientists believe (or what the community accepts) about the phenomenon is true in a way that is sufficient for understanding the phenomenon (Rice 2016). Instead, I suggest that the body of information that constitutes the scientific community's understanding must simply be "true enough" for our purposes in particular contexts of inquiry (Elgin 2007, 2017; Potochnik 2017). Still, although the factive requirement suggested here is admittedly vague, I think we can make clear judgments in many cases. For example, the Grand Canyon case seems to clearly fail this requirement because these agents fail to grasp the particularly salient facts that the cause of the Grand Canyon was the Colorado River (not a great flood), and that the process took approximately 6 million years to complete. As a result, the contextually salient beliefs that these agents have about the phenomenon and why it occurred are mostly incorrect. We can contrast this case with physicists' understanding of the movements of planets, which seem to clearly meet the factive requirement because—although it may contain some false beliefs or depend on essential idealizations—it is mostly constituted by true beliefs, and those beliefs are precisely those that are salient in the context of inquiry (e.g., where the Earth is, which bodies orbit which others, and the approximately elliptical shape of planetary orbits).

What's more, I think this vagueness regarding the factive requirement for understanding is due to the inherent vagueness in our judgments about when the falsity of one's beliefs will undermine one's understanding (i.e., vagueness in how we apply the concept of understanding). Indeed, "we typically acknowledge that people can have a measure of understanding even if the contentions making up the bodies of information they endorse diverge somewhat from the truth" (Elgin 2017, 62). Therefore, while requiring that most of the contextually salient propositions within one's understanding be true is admittedly vague and context sensitive, I suggest that this tracks the way that we attribute scientific understanding to individuals and communities. We don't require that experts have accurate information about *all* of the central propositions regarding a phenomenon (in fact, few experts would be able to meet this high standard). Instead, we want a scientific community whose understanding contains mostly true information about the important features of a phenomenon—even if their route to that information, or the way they transmit or represent that information, depends on idealizations. Despite allowing idealizations to contribute in essential ways to the production and transmission of understanding, I see no reason not to call this a factive notion of understanding because the truth of most of the information continues to play a key role in our judgments about whether one genuinely understands.

Applying this account to the representations used by scientists, I claim that a scientific model or theory is able to produce factive scientific understanding of a natural phenomenon if it enables an agent or a community to grasp some true information about the phenomenon and the agent or community grasps how that information can be systematically incorporated into a larger body of information in which most of their contextually salient beliefs about the phenomenon are true. This account of the factive nature of scientific understanding leaves open the possibility that many (perhaps central) propositions or representations that contribute to the understanding produced by science might be false or inaccurate. In this way, my account maintains that scientific understanding is factive, while accommodating the central contributions made by holistically distorted scientific representations.

At this point, it is important to recall that the explanations and understanding provided by idealized models ought to be distinguished from the assumptions of the models themselves (see section 4.6.2). In contrast with this kind of view, Elgin argues that any kind of *veritism* that takes truth to be necessary for epistemic success is unacceptable because "if we accept it, we cannot do justice to the epistemic achievements of science" (Elgin 2017, 9). Specifically, Elgin argues that "the more serious problem comes with the laws, models, and idealizations that are acknowledged not to be true but that are nonetheless critical to, indeed at least particularly constitutive of, the understanding that science delivers" (Elgin 2017, 14). In a similar way, Potochnik argues that understanding is not factive because it is often produced by idealized representations that do not aim to be true (Potochnik 2017, 113). As Potochnik puts it, "If the epistemic aim of science is truth, then science has fallen short of this aim again and again" (Potochnik 2017, 121).

While I am sympathetic with many aspects of Elgin's and Potochnik's views, I disagree with the claim that the understanding produced by scientific inquiry must be *partially constituted* by the myriad idealizations used in science. Although theories, models, and idealizations are certainly the tools with which scientists produce understanding of natural phenomena, it does not directly follow that the assumptions involved in those tools must be included in the explanations and understanding that scientists extract from those tools. Furthermore, as we saw in the previous discussion, a factive conception of scientific understanding can allow that some of the central propositions in our body of information about a phenomenon be false or inaccurate. In short, not all the idealizations of science must be incorporated into the understanding produced by scientific inquiry, and even those that are incorporated will often be insufficient to undermine the facticity of the understanding produced. If this distinction between the holistically distorted representations used by scientists and the understanding provided by scientific inquiry can be maintained, then recognizing the essential role of multiple, conflicting, highly idealized models in science need not force us to adopt a nonfactive conception of scientific understanding.

While I have elsewhere (Rice 2019a) phrased this in terms of "separating" scientific understanding from the assumptions of idealized models, I now think that this way of putting it is not quite right because it suggests a sharp product-process distinction that misrepresents how science provides explanations and understanding. As Sullivan and Khalifa (2019) correctly note, if idealizations were only part of the process by which we discover understanding, then they don't seem to play an essential role within the resulting understanding itself. The problem with this stark separation, I suggest, is that explanation and understanding are dynamical processes of scientific inquiry. These epistemic achievements are not simply end points that float free of the processes by which they are produced. More specifically, scientific explanations and understanding depend in essential ways on the models, theories, methods, and scientists that are involved in their construction, communication, and application. As a result, the epistemic justification for believing the truths involved in scientific explanations and understanding *requires* the models and idealizations from which they are derived—if we divorce the explanations and understanding from the models, then we lose the justification that supports believing them (Elgin 2017).[6] As an illustration of this point, the explanations provided by the ideal gas law still appeal to the model that holistically distorts the properties of real gases—including all the idealizations (Woody 2015). Just as we cannot isolate the contributions of the true parts of our scientific models from the contributions of their idealizations, we cannot isolate the resulting explanations and understanding from the representations that scientists use to develop, communicate, and provide them.

Despite this fact, I think we can and should maintain a clear *distinction* between the claims of an idealized model and the claims included in the explanation or understanding produced by appealing to that model because not all the claims in the model will necessarily be in the explanation or understanding and vice versa (because multiple models might be involved in providing that understanding). Factive accounts of these epistemic achievements need not claim that there is some epistemic product that includes only truths. What is required is a clear sense of which things need to be true within our explanations and understanding in order for them to count as genuine. Put differently, just because these epistemic achievements cannot be completely separated from the holistically distorted representations used to achieve them does not mean that those achievements must be nonfactive.

The last piece of my account involves specifying the kind of relationships that need to be grasped to scientifically understand a phenomenon.

Following my account of explanation in chapter 4, I argue that the crucial kind of information involved in scientific understanding is information about counterfactual dependence and independence relations. That is, scientific understanding is constituted by a body of mostly true beliefs about how changes to the features of the system would (or would not) alter the phenomenon of interest (see Le Bihan 2017 and Saatsi 2019 for similar views). As I argued in chapter 4, this clear link between explanation and understanding is one of the main reasons for adopting a counterfactual account of scientific explanation. By focusing on how modal information produces understanding, we can clearly see why explanations are such a good source of understanding: they provide a large amount of the modal information required to understand.

In addition, the examples reviewed earlier in this chapter show that understanding can be produced in ways besides providing explanations. In these cases, we learn something about the possible states of the system (i.e., we acquire true modal information about the phenomenon), but we may not grasp how the phenomenon counterfactually depends on the features that account for its occurrence in any real-world system. This enables understanding to be produced (or improved) via routes that do not go through explanation.

My focus on modal information aligns with an idea originally proposed by Nozick: "I am tempted to say that explanation locates something in actuality, showing its actual connections with other things, while understanding locates it in a network of possibility showing the connections it would have to other nonfactual things or processes. (Explanation increases understanding too, since the actual connections it exhibits are also possible.)" (Nozick 1981, 12). In other words, like explanations, understanding is produced by providing true modal information about the phenomenon of interest—in Nozick's terminology, it "locates it in a network of possibility." This modal information then improves science's understanding of the phenomenon by telling scientists about the range of possible states of the system that would (and would not) alter the phenomenon of interest. More recently, Grimm has expanded on Nozick's idea by suggesting that when we don't have the understanding provided by an actual explanation, we might have what he calls "proto-understanding":

> By an agent's proto-understanding, I mean an agent's convictions *about* the sorts of possibilities that are live or relevant, relative to the situation in question. [This is] a further specification of Nozick's notion of a "network of possibility"; it is

something like a person's "modal sense" of the various alternatives that might have obtained, relative to the fact in question. (Grimm 2008, 491)

I suggest that Grimm's changes in proto-understanding amount to an improvement of actual scientific understanding that, although it may be provided by an explanation, may be produced in other ways as well. In the cases surveyed in section 8.1, the modal information provided produces true beliefs about the network of possibilities and can answer how-possibly questions despite the failure of those models to explain. Moreover, these beliefs about the network of possibility are often true and can be systematically incorporated into larger bodies of information that contain mostly true information about the phenomenon. I argue that such adjustments in our "modal sense" of the possible alternatives improves our understanding of the phenomenon. In addition, I suggest that the more true modal information that one grasps about the target phenomenon, the better one understands the phenomenon—that is, more modal information deepens one's understanding (Saatsi 2019, 68).

It is important to remember, however, that some pieces of modal information might improve our understanding more than others.[7] For example, many counterfactual situations will be of little interest to scientists or directly follow from more interesting counterfactuals (e.g., the fact that a uranium sphere cannot be larger than a mile in diameter also means that it cannot be larger than two miles in diameter). As with explanations, I suggest that the context of inquiry will play an absolutely crucial role in determining which counterfactual situations are most interesting to us, and therefore which pieces of modal information will deepen our understanding in the ways we desire. Nonetheless, I suggest that more modal information always improves our understanding to some (perhaps very small) degree—even if the counterfactual situation is particularly distant or not that interesting within certain contexts.

This highlights another important feature of understanding: unlike knowledge or explanation, understanding seems to clearly come in *degrees* (Elgin 2017; Khalifa 2017; Kvanvig 2003).[8] For example, some people understand a subject better than others. This feature of understanding is important for two reasons. First, it implies that there might be degrees of understanding that fall below the understanding provided by having a complete explanation. In science, this allows for the improvement of our degrees of understanding of a phenomenon even if scientists are unable to provide an explanation of that phenomenon. Second, understanding of a phenomenon might be improved beyond the understanding provided by an explanation. One way that this can happen is when scientists provide multiple explanations for the same phenomenon, each of which contributes to our overall understanding of the phenomenon (Longino 2013; Massimi 2018; Morrison 2015; Potochnik 2017). In other cases, models that fail to explain might deepen our understanding of a phenomenon beyond what our current explanations can provide. The general point is that the degree of understanding that the scientific community has about a phenomenon neither starts nor stops with the understanding provided by a single explanation (Rice 2019a).

My account also allows us to see why there can be multiple factive ways to understand the same phenomenon without making understanding into a purely subjective feeling. As Trout (2002) argues, someone can certainly be mistaken about whether they understand and the subjective feeling of understanding can be positively misleading about whether one actually understands. To accommodate these facts, my account requires not just that one incorporate the information into one's own conception of the world, but also that the modal information grasped and the body of information it is incorporated into must be mostly true. Which features are counterfactually relevant and irrelevant to the occurrence of a phenomenon is objectively determined by the world, not by our cognitive states, background beliefs, or feelings. This is what ensures that genuine understanding, on my account, is factive—and thereby at least partially objective—rather than just a subjective achievement or feeling. As de Regt (2017) convincingly argues, understanding can be objective even if it pragmatically depends on features of agents and communities, as well as the representations they have available.

However, there is a plethora of true modal information about any given phenomenon—so much information that no single individual could have *all* the modal information that would be involved in an ideally complete or deep understanding of any phenomenon. As a result, different individuals can scientifically understand the same phenomenon in different ways by having different (but perhaps overlapping) sets of modal information about the possible states of the system and how they would change the phenomenon. This possibility is easily illustrated by the fact that there can be multiple explanations of the same phenomenon, which each produce understanding of the phenomenon, but they do so by citing different facts. For example, an equilibrium explanation of a gas's behavior might appeal to statistical features of the overall system, while a causal process explanation might trace the trajectories of individual particles. Each of these ways of understanding the phenomenon might be constituted by sets of beliefs that are mostly true and are incorporated into sets of background information that are largely correct. The fact that there are multiple ways this can occur need not lead us to accept inconsistent claims about "the way the world really is." Instead, different ways of understanding will simply involve different sets of modal information and incorporate that information into different sets of background beliefs, experiences, and assumptions. Because different individuals can have different sets of mostly true background information and different true beliefs about which states of the system are genuine possibilities, different individuals can legitimately and factively understand the same phenomenon in different ways.

Furthermore, because factive understanding can tolerate some central falsehoods, the overall body of understanding produced by a scientific community need not be entirely consistent (i.e., a community of scientists can understand a phenomenon even if some of its members hold beliefs that are inconsistent with beliefs held by other members). This is compatible with each of those scientists (and the overall scientific community) having sets of beliefs about the modal space of possibilities that are mostly true (even if they are mostly true in different ways). This is also what enables multiple conflicting scientific models or theories to all contribute to the community's overall understanding of natural phenomena (Rice 2019b). Like the different sets of values, background beliefs, and interests of individual agents, different modeling perspectives that use different representational frameworks, idealizations, and abstractions can each contribute some true modal information to science's overall understanding of a phenomenon. Rather than aiming to directly represent the ways that scientists believe the world to be, the development of scientific understanding is typically a process in which scientific communities extract various kinds of modal information by interacting with multiple representations that they know are inaccurate in different ways.

8.3 Understanding Realism

This account of scientific understanding (in combination with the other views defended in this book) has important implications for the ongoing debate concerning scientific realism. Traditional scientific realism is typically supported by some version of the "no miracles" argument, which claims that the accuracy of scientific theories is the only (or at least the best) explanation of the predictive successes of science (Putnam 1975). Given this argument, several philosophers have noted that science's widespread use of idealizations raises a serious challenge for the realist (Cartwright 1983; McMullin 1985; Odenbaugh 2011; Psillos 2011; Suárez 1999). In particular, given that we know that our models and theories include intentional and drastic distortions of reality, even if they make accurate predictions, we have no reason to believe that they, or the theories that employ them, are true (or accurate).

As we saw in chapter 2, a common response—coming from defenders of the standard view—has been to argue that highly idealized models can still provide *partially* accurate representations (Bueno and Colyvan 2011; Kitcher 1993; Peters 2014; Pincock 2011; Strevens 2008; Vickers 2017; Weisberg 2007a, 2007b; Worrall 1989). That is, many philosophers are quick to point out that not every part of a model is intended to accurately represent. As a result, idealizations that distort irrelevant, negligible, or nonsalient features can be made compatible with the realist's claim that models accurately represent the relevant features of real systems. This "selective confirmation" response requires us (1) to distinguish the relevant parts, features, or aspects required for the success of the model (or theory) from those that are idle or irrelevant; and (2) to show that the features relevant to the success of the theory or model are just those features that the theory or model accurately describes (Stanford 2003, 2006). However, as I argued in chapter 5, given the kinds of holistic distortion that we observe throughout scientific theorizing, I am not optimistic that these conditions will routinely (or ever) be met.

Fortunately, this decompositional strategy is not the only response available to the realist. In particular, I argue that the realism debate needs to move beyond focusing on the accuracy or truth of scientific models or theories themselves and instead analyze the factive body of understanding produced by scientists' use of those holistically distorted representations (Rice 2016, 2019a, 2019b). Unfortunately, the realism debate has been structured around determining whether our current best theories are accurate descriptions of the unobservable world. Given the pervasive use of holistic distortions in scientific theorizing, I think the answer to this question is clearly no. However, there are many other important questions surrounding the factive epistemic achievements of science that are compatible with our best models and theories being inaccurate. For example, as Morrison notes regarding the Hardy-Weinberg law: "The Hardy-Weinberg law enables us to understand fundamental features of heredity and variation.... Hence, the claim that the law is false in some sense misses the point if our concern is understanding" (Morrison 2009, 133). Indeed, the overemphasis on accurate representation (derived from the standard view) has made traditional versions of realism incompatible with the pervasive distortion of relevant features that we find in scientific practice. But this only spells trouble for realists if they are committed to the standard view's assumption that the only (or best) way to achieve the epistemic aims of science is to construct (partially) accurate representations of reality.

Throughout this book I have argued against this core assumption of the standard view. As a result, an alternative approach is needed. Therefore, in this section I use the account of factive scientific understanding defended in the previous section to argue for a view that I call Understanding Realism. According to Understanding Realism, one of the primary epistemic aims of science is to produce factive understanding, and this aim can be (and often has been) accomplished by the use of holistically distorted models and theories (Rice 2016, 2019a). This version of realism focuses on the accuracy (or facticity) of the body of understanding produced by scientific inquiry rather than the accuracy of scientific models or theories themselves.

In chapters 4 and 5 we saw how holistically distorted models can be used to explain—and thereby understand—phenomena by enabling the extraction of true information about the counterfactual dependencies and independencies that hold between features of the system and the explanandum. In chapters 6 and 7 we saw how multiple, conflicting holistic distortions can be used to extract true modal information about the universality of various patterns across changes to the features of the systems. Finally, in this chapter we have seen that holistically distorted models can produce understanding by investigating necessity claims, exploring hypothetical scenarios, or surveying the range of possible systems. Each of these holistically distorted scientific representations is able to contribute to the scientific community's factive scientific understanding of natural phenomena without requiring the representations that they provide to accurately describe the observable or unobservable features of real systems. In these ways, as well as others, scientific models can provide a wide range of true information about the counterfactual relevance and irrelevance of various features of real systems to the occurrence of natural phenomena (Batterman and Rice 2014; Le Bihan 2017; Massimi 2018; Rice 2013, 2016, 2017; Saatsi 2019).

Recognizing that holistically distorted representations can be leveraged in epistemically fruitful ways reveals a large corpus of systematized factive understanding that scientific inquiry has produced about the natural world. Once again, it is crucial to note that the factive understanding provided by scientists' use of idealized models (or theories) ought to be distinguished from the assumptions of the models and theories used to produce that understanding. As I argued in the last section, factive understanding of natural phenomena is provided by the accurate modal information *extracted from* scientists' use of idealized representations. This modal information can then be incorporated into the scientific community's overall corpus of mostly true beliefs about the phenomenon of interest. Consequently, realists can claim that science is able to produce a wide range of factive understanding, despite the fact that scientific models are typically highly idealized and conflict with one another.[9] In fact, my account suggests that realists ought to encourage such a proliferation of conflicting models because they will typically provide access to a wider range of modal information and therefore will provide more overall understanding than any single model (or perspective) (Rice 2019a, 2019b).

Given that Understanding Realism denies one of the central tenets of traditional realism—that our scientific models and theories are accurate descriptions of real-world phenomena—it is worth taking a moment to consider why this position ought to be considered a kind of realism. Indeed, one might think that Understanding Realism has more in common with instrumentalism than realism. The key difference is that for the instrumentalist, the products of science are either accurate prediction or empirical adequacy regarding observable features of the phenomenon (van Fraassen 1980). In contrast, Understanding Realism claims that scientific models and theories can be used to produce the epistemic achievement of factive understanding, which requires that the model enable scientists to see how various changes to the observable and unobservable features of the real systems would (or would not) result in changes in the phenomenon of interest. This goes well beyond merely "saving the phenomenon" or being predictively accurate regarding only observable aspects of real systems. In short, while scientific models and theories are instruments, according to Understanding Realism, they are instruments for producing factive understanding, which requires much more than empirical adequacy (Rice 2019a; Saatsi 2019, 80).

In addition, the view defended here has much in common with traditional forms of realism. First, my account argues that one of the primary aims of scientific inquiry is *factive* understanding (Rice 2016, 2019a). This means that science still aims at truth in an important way. Some philosophers have interpreted the reliance on essential idealizations as requiring the claim that science does not aim for truth or that science only aims for nonfactive understanding (de Regt 2017; Elgin 2017; Potochnik 2017). In contrast, I argue that science aims to acquire factive understanding of natural phenomena by discovering correct modal information about real-world phenomena. In fact, this means that science aims at a larger set of truths than is typically assumed by accounts of realism. If this account is correct, then an achievable aim of science is not only to tell us what the world is like, but also about the ways the world would be different in various counterfactual situations. Second, this account holds that the factive understanding that science aims at ought to include extensive accurate information about unobservable entities and their dependence and independence relations with observable phenomena. Indeed, in line with other realists, I maintain that much of science's factive understanding is produced via the construction of factive explanations that must "latch on to" the features of unobservable reality (Saatsi 2019, 66). This means that the epistemic aims of science often include providing accurate information about unobservables much like traditional accounts of realism. Finally, Understanding Realism maintains that we are justified in believing much of what science tells us about the operations of unobservable entities and their features, even if we are not justified in believing the models and theories used to generate that understanding. That is, we are still justified in believing much of the information that science has provided about the nature of unobservable entities, processes, and features of real systems—that information is just contained in scientists' understanding of the phenomenon rather than being accurately reflected by the representations used to discover that information. In sum, Understanding Realism counts as a kind of realism because (1) it makes a claim about the epistemic aims of science involving truth, (2) it includes unobservable entities and their features within those truths that science aims at, and (3) it asserts that science has been epistemically successful in terms of achieving factive understanding of many natural phenomena (Rice 2019a).

The mistake of traditional accounts of realism has been assuming that these factive epistemic aims must be achieved via (or within) true or accurate models and theories. By showing that these factive epistemic achievements can be accomplished by other means, we can preserve the realist's commitments regarding the aims and achievements of science without having to show that our models and theories accurately represent reality. More generally, we ought to be realists about the factive epistemic achievements of science without requiring that our scientific models and theories accurately represent the entities, interactions, or causes of real systems (Rice 2019a, 2019b). Just as a false story can produce justified true beliefs in the listener (e.g., a fable with the moral that vegetables are good for you) known to be inaccurate scientific representations can be used to develop factive explanations and understanding. However, just as the author of a fable strategically uses distortion to make their point (rather than just hoping that the story will luckily produce true beliefs in the listener), scientists strategically use distortion of the relevant features of real systems in order to reveal the modal information of interest. Although scientific theories and models will always be pervasively distorted and conflict in various ways, we can still be justified in believing that much of the information that scientists strategically extract from them is accurate.[10]

8.4 Understanding and Scientific Progress

I have already noted that a key feature of understanding is that it comes in degrees. This enables the understanding that we have about a phenomenon to both fall short of, and exceed, the understanding produced by an explanation of that phenomenon. This fact also entails that our overall understanding of a phenomenon can be improved over time—even as our theories and models of that phenomenon change. Most realist accounts have focused on demonstrating the continuity of reference of key theoretical terms or showing that our theories are getting progressively more accurate across theory change. But I contend that neither of these conditions needs to be met in order for our past theories and models to contribute to our current factive understanding of natural phenomena. Instead, theories and models that are known to be false or have been rejected can still contribute some degree of understanding of natural phenomena by providing true modal information about the possible states of real systems.

Furthermore, we can improve our degrees of understanding of natural phenomena (e.g., by exploring additional possibilities) without having to construct models or theories that are getting progressively more accurate. For example, a central tenet of Copernicus's theory is that the Earth travels around the sun in a circular orbit, but this claim is false—the Earth's orbit is, in fact, elliptical. Nonetheless, I argue that Copernicus's theory contributed—and continues to contribute—to our understanding of planetary motion. In particular, Copernicus's theory provides some true modal information about what the universe would be like if the Earth's orbit were circular rather than elliptical (along with some other changes), and thus answers a range of "What if things had been different?" questions. In short, a model or theory need not tell us how the world actually is in order to produce factive understanding—often a model or theory contributes to our understanding by revealing what would be different if various features of the world were different in various ways (Rice 2019a). The problem with most realist accounts derived from the standard view is that they suggest that this advancement in understanding (or explanation) must be captured by showing that the later theories are more accurate than the previous theories—or that the later theories can capture what was accurate about the previous theories.

In contrast, I argue that our past theories of planetary motion—despite their distortions—enabled the scientific community to uncover true modal information about the possible states of the universe. Moreover, scientists have incorporated some of this modal information into the community's overall body of information about planetary motion that is constituted by mostly true beliefs. Grasping this kind of modal information about the possible states of the universe has increased our understanding of planetary motion, regardless of whether the theories and models used to study those phenomena are becoming progressively more accurate representations of reality. What's more, because the truth of this modal information is separate from the accuracy of the models and theories themselves, our justifications for believing this modal information can survive radical changes to the theories and models used in science.

As another example, Bokulich points out that, "Without knowledge of the classical orbits, our understanding of quantum spectra and wavefunction morphologies is incomplete. In sum, although reference to classical structures is in some sense eliminable from the explanation of phenomena such as diamagnetic Rydberg spectra and wavefunction scarring, such an elimination *comes at a rather high cost of understanding*" (Bokulich 2008, 155; emphasis added). This is the case even though classical physics has been replaced by quantum mechanics. The same holds true in biology. Although central claims of Darwin's theory of evolution have been replaced by conflicting assumptions (e.g., during the modern synthesis), our understanding of many biological phenomena is still constituted by numerous insights from his original theory. As a result, we see that—despite their inaccuracy with respect to relevant (e.g., difference-making) features—previously rejected theories and models can still contribute in significant ways to our current factive understanding of natural phenomena. The key to seeing how this is possible is to reject the requirement that factive understanding be produced by accurate models or theories and focus on the true modal information about possible states of real systems that can be extracted by the strategic leveraging of holistically distorted scientific representations. Just as there are many ways that our current (conflicting) holistically distorted models and theories can contribute to the production of factive understanding, so can our past (conflicting) holistically distorted models and theories.

Unfortunately, given the accurate representation requirements derived from the standard view, many philosophers seem to suggest that past theories could produce genuine understanding only *if they were true* (or if we could isolate the true parts responsible for their successes). For example, Strevens argues that, "To have 'understanding with' Newtonian physics is to be able to construct or grasp an array of Newtonian explanations that are good in the sense that they are internally correct—they would be correct if only Newtonian physics were true" (Strevens 2013, 513). However, Strevens also says that understanding can only be produced by grasping a correct explanation, which for Strevens requires that the model or theory provide an accurate representation of difference-making causes. As a result, it seems that Strevens's view entails that Newtonian physics provides no actual understanding of physical phenomena. To me this seems to be a mistake. The widespread use of idealizations in science shows us that our current theories and models are no more likely to be accurate representations of the relevant features of reality than Newtonian physics is. So, if we follow Strevens in requiring that genuine understanding is produced only by representations that accurately represent difference-makers, then the scientific community doesn't understand nearly as much as we thought.

Fortunately, the arguments provided here show how theories and models can produce factive scientific understanding without being accurate representations of real-world systems—including their difference-making features. Understanding Realism capitalizes on this possibility to argue that the scientific community can have a large body of factive understanding (which they intuitively have), while allowing our past and current theories to be pervasively inaccurate descriptions of real systems. I don't think philosophers of science should be in the business of doubting whether the pervasively idealized representations used in science have enabled us to genuinely explain and understand natural phenomena. However, I do think that the widespread use of holistically distorted representations in science shows that we cannot justifiably believe that science accomplishes these epistemic achievements via the construction of partially true or accurate theories or models. Moreover, we ought to acknowledge that these idealized scientific representations are the best epistemic tools we have for understanding our world. My goal throughout this book has been to show how this could be so by providing factive accounts of explanation and understanding that do not require the accurate representations assumptions involved in the standard view.

Recently, Potochnik has similarly suggested that if we focus our attention on scientific understanding, then we can adopt a more cumulative picture concerning past theories:

> Many superseded scientific theories posit radically different views of the world.... But taking these theories to represent causal patterns enables a subtler accounting of what they had right. Of, course, some scientific conjectures are simply wrong: they posit causal patterns that do not exist. But many previously accepted theories latched on to causal patterns that *are* embodied in phenomena, even if they were later replaced by other theories that depicted other patterns.... My interpretation of science's epistemic aim enables this and many other episodes of scientific change to be interpreted in a cumulative way. (Potochnik 2017, 121)

In other words, Potochnik suggests that past scientific theories can contribute to our current understanding by focusing on causal patterns that are embodied in real phenomena but are different from the causal patterns targeted by our current theories. While I certainly agree that focusing on understanding enables us to interpret the achievements of science in a more cumulative way, there are some crucial differences between this kind of proposal and my own. First, Potochnik's account works only for causal patterns, but, as I argued in chapter 3, many of the patterns of interest to scientists are noncausal, and many scientific theories are focused on noncausal patterns of behavior. Second, Potochnik's account enables past scientific theories to contribute to understanding only if they track causal patterns that actually occur (or are embodied) in real systems. However, many of the patterns described by past (as well as current) scientific theories and models will be patterns that could never occur (even approximately) in real systems—that is, many of the models that produce understanding do so by displaying patterns are not embodied in any real phenomena, but rather describe only merely possible states of real systems. In contrast, my account shows how models and theories that are used to investigate possible systems (or patterns) that are not, and never will be, instantiated can still contribute true modal information about the space of possibilities.

In addition, as a consequence of focusing only on patterns that are embodied in real phenomena, Potochnik's account—like Elgin's use of exemplification—struggles to deal with cases where multiple models and theories make contradictory assumptions about the same patterns or features of real systems. Specifically, the only way that past theories can contribute on Potochnik's account is if they target different patterns, or aspects, of real phenomena than those scientists are currently interested in. Saatsi (2019) adopts a similar strategy by arguing that scientific understanding is cumulative because past theories focused on different aspects of the target phenomenon.[11] Consequently, these accounts enable scientific understanding to be cumulative only if we assume that each of the scientific representations contributing to that understanding is designed to investigate different features or patterns.

But this is patently not the case in the instances of theoretical replacement that are problematic for the scientific realist. The issue is to show how our (known to be inaccurate) theories and models that have been, or will be, replaced by later theories about the very same features of the very same phenomena can nonetheless contribute to our current understanding. In contrast to these accounts, my account of understanding based on modal information shows how past scientific representations that describe the same features, aspects, and patterns in contradictory ways can all still contribute to our current understanding of natural phenomena. This is accomplished by showing how each of those theories or models contributes a different set of modal information about a phenomenon by exploring how things would have been different in different ranges of possible systems (which may or may not include any of the actual cases).

8.5 The Role of Diversity in Promoting Understanding

This account of understanding also shows why one of the best ways to improve science's overall understanding is to increase the diversity of the scientists (and nonscientists) investigating a phenomenon. No single individual can grasp all the information available regarding a phenomenon, or all the ways that the phenomenon would (or would not) have been different in various counterfactual situations. However, diverse individuals' different ways of understanding (and explaining) the same phenomenon can all contribute different pieces to the overall body of information that constitutes the understanding of the entire scientific community.

The main reason that a diverse group of scientists is likely to produce a greater degree of understanding is that understanding is deeply connected to the cognitive features of the agents who understand (Potochnik 2017). Different scientists will incorporate new information into different bodies of background information and experiences, which can reveal novel connections that would go unnoticed by individuals with different background knowledge or experiences. In addition, different background assumptions, experiences, and values will lead different scientists to see the investigation of different possibilities as being relevant to understanding the phenomena (or to other goals) (Hrdy 1981, 1986; Okruhlik 1994). This is important because certain pieces of modal information about the phenomenon of interest may be accessible only by individuals with certain skills (e.g., an expert's ability to interpret the data), embodiments (e.g., physical appearances that provide access to experiences inaccessible to others), or experiences and values (e.g., that enable some scientists to see various biases involved in others' assumptions about what is relevant or irrelevant) (Collins 1986, 2000; Longino 1990, 2002; Okruhlik 1994; Pitts-Taylor 2017). Each of these features of individual scientists shapes the kinds of understanding they are able to contribute to the community's overall understanding.

Additionally, these features will influence the kinds of modal information that a scientist is able to extract from a scientific model because "understanding through models comes typically by way of building them, experimenting with them, and trying out their different alternative uses"

(Knuuttila and Merz 2009, 154). Consequently, the skills, experiences, and embodiments that individuals draw on in their interactions with scientific models—and the data used in their construction—will heavily influence the understanding that they are able to discover about the phenomena via the model or theory (de Regt 2017; Harding 2004; Keller 1982, 1983, 1985; Pitts-Taylor 2017; Wylie 2004). As a result, individuals with different beliefs, experiences, skills, and embodiments will be able to extract different (but often overlapping) sets of modal information about the phenomenon of interest (de Regt 2009a, 2017; Knuuttila and Merz 2009; Leonelli 2009).

In addition to diversity with respect to types of scientists, scientific understanding is often improved by the use of multiple conflicting modeling and theoretical approaches. Understanding Realism accounts for this observation by suggesting that using multiple incompatible models can provide access to a wider range of modal information—a body of modal information that is inaccessible from any single research perspective. Unfortunately, because of the traditional assumption that explanation and understanding are best produced by accurate or true theories or models, this possibility has been largely overlooked within the realism debate. However, feminist epistemology (see in particular Longino 1990, 2002, 2013) and the growing literature on model-based science shows us that often the best (or only) way to understand complex phenomena is by employing a set of representations or perspectives that we know are idealized and incompatible. Moreover, given that different scientific models or theories will often be responsible for different empirical successes, scientific communities typically make progress by distributing their resources across a range of incompatible theoretical approaches (Solomon 2001).

Let me be a bit more specific about how both these kinds of diversity—of individuals and modeling approaches—directly contribute to the epistemic aims of explanation and understanding. In chapter 4 I argued that scientific explanations are constituted by a set of counterfactual dependence and independence information about how the contextually salient features of the system are related to the explanandum. What this means is that a large set of counterfactual information is required to provide a complete explanation of a phenomenon. In some cases individual scientists might grasp the required set of modal information, but in many other cases no single individual, or modeling approach, will have access to all the counterfactual information required to explain a complex phenomenon. When this occurs, different individuals might grasp different pieces of the scientific community's explanation of the phenomenon (e.g., because those features are of particular interest to them). Moreover, different modeling approaches might be used to acquire and communicate different pieces of the explanation (e.g., by modeling different features on which the explanandum counterfactually depends). What's more, by combining the overlapping counterfactual dependence and independence information provided by different individuals, models, and approaches in various ways, scientists can perhaps provide multiple explanations for the same phenomenon.[12] In short, by providing access to a larger body of modal information, more diverse groups of scientists using a more diverse set of modeling approaches will provide more (and better) explanations of the phenomena we observe.

In addition to expanding the range of explanations provided, broadening the diversity of scientific researchers and modeling approaches expands our overall understanding of a phenomenon by expanding the range of possibilities that are considered. For example, different individuals will often consider different states of the system as live possibilities and therefore will investigate different counterfactual situations. If the features altered in those counterfactual situations turn out to be relevant to the phenomenon of interest, then we will have increased our overall understanding of the phenomenon. But even if those features or changes turn out to be irrelevant, that information is often important to our understanding of the phenomenon as well. This is because we are often interested in understanding why certain patterns are repeated across systems that are very heterogeneous. Understanding the stability of these patterns often requires showing why the differences between those systems are irrelevant to the occurrence of the pattern. Because of this, exploring a wide range of possibilities and discovering that various features are irrelevant to (or unnecessary for) the phenomenon can also deepen science's understanding of the phenomena we observe. Because different individuals and modeling approaches will be capable of (or interested in) exploring different possible states of the system, a diverse group of scientists using a diverse group of modeling approaches will produce more accurate modal information about the possible states of the system than any single scientist or modeling approach. Therefore, more diverse communities using a wider range of conflicting modeling methodologies will typically understand a phenomenon better than more homogenous communities.

Unfortunately, most epistemological accounts (of both knowledge and understanding) have focused exclusively on the cognitive achievements of individuals.[13] This is fine for answering some questions about how individuals acquire understanding, but it is at odds with the way that science produces understanding. In the case of understanding in science—and in many nonscientific fields as well—the crucial questions concern how a diverse community of individuals using conflicting approaches can interact to produce a better overall understanding of exceedingly complex phenomena. Fortunately, the account of understanding (and realism) provided here can easily be scaled up to encompass the understanding of the overall scientific community. Specifically, we can think of the scientific community's overall understanding of a phenomenon as the sum total of the accurate modal information regarding the possible (and actual) states of the systems of interest that are had by the members of that community. The more of this kind of information the scientific community has about a given phenomenon, the better the community understands that phenomenon. There are certainly several additional questions about how communities understand that need to be answered here (e.g., how do the social structures of communities allow or inhibit the incorporation of modal information from different perspectives into the community's understanding)—but those additional questions will have to be addressed elsewhere (e.g., see Rubin and O'Conner 2018; Sledge and Rice manuscript; or Woody 2015). My aim here is only to argue that an account based on the accurate modal information grasped by a community of scientists better captures how science actually understands complex phenomena and provides a clear case for increasing the diversity of the types of scientists, methods, and modeling approaches involved in our attempts to understand natural phenomena.

These benefits provide clear, epistemic arguments for the value of diversity and pluralism in science. These conclusions also have serious implications for how we ought to fund and pursue scientific research as a society. Rather than seeing conflicting values, assumptions, methods, models, and theories as barriers to the epistemic aims of science, Understanding Realism recognizes these kinds of diversity as absolutely essential to the expansion of scientific understanding beyond the understanding that can be achieved by any one perspective. If science aims to factively explain and understand complex phenomena, which I think they intuitively do, these epistemic goals will be best accomplished by diversifying the scientific community, supporting the use of multiple incompatible modeling approaches, and creating social structures that yield epistemic authority for multiple conflicting approaches to studying the same phenomenon (Longino 1990, 2002, 2013).

8.6 Conclusion

This chapter began by arguing that scientific understanding could be provided without having an explanation of a phenomenon. I then provided an account of scientific understanding in which understanding is constituted by the scientific community's grasp of a body of accurate modal information that contains mostly true beliefs about the phenomenon of interest. This account shows how we can be realists about the factive understanding produced by scientific inquiry without having to show that our best scientific models or theories are accurate representations of the (unobservable) features of real systems.

In addition, this account shows how past scientific models and theories that conflict with our current beliefs can still contribute to the scientific community's understanding of phenomena. Finally, while most accounts of scientific realism suggest that we should aim for the discovery of a single best theory or model, Understanding Realism makes a clear case for expanding and maintaining the diversity of individuals, methods, models, and theories whose conflicting assumptions, idealizations, and values directly and positively contribute to the epistemic aims of science.

9 Leveraging Holistic Distortions: The Positive Contributions of Idealizations to Explanation and Understanding

The goal of this book has been to move philosophical accounts of science away from focusing on the concepts of causation, decomposition, and accurate representation toward the concepts of modal information, holistic distortion, and universality. When combined, the arguments presented in previous chapters recommend a significant shift in how philosophers conceive of explanation, modeling, idealization, understanding, and scientific progress. In this final chapter I combine various pieces of these views into a more general account of scientific theorizing, explore some of its philosophical implications, and address some of the outstanding questions raised by this alternative picture of science. Along the way, I also review what the cases discussed in this book have revealed about the variety of ways that holistic distortions make positive contributions to scientific explanations and understanding.

9.1 Some Other Questions about Explanation

The first part of this shift, which occurred in chapters 3 and 4, was to argue for the adoption of an account of explanation based on counterfactual dependence and independence relations that can replace (and subsume) current accounts that focus only on causal explanations. However, this counterfactual account goes well beyond only answering this traditional philosophical question concerning the kind of dependence (and independence) relations involved in explanation. In order to illustrate the breadth of this account of explanation, I will show how the views defended in previous chapters address a set of additional questions about explanation that Potochnik (2018) has correctly suggested that philosophers of science need to consider.

Potochnik's first question is: what is the priority of communicative features of explanation versus the ontological features of explanation? (Potochnik 2018, 59).[1] Proponents of the ontic approach suggest that the important features of scientific explanation are largely independent of human influences—for instance, they are independent of who is doing the explaining (Craver 2014; Salmon 1989; Strevens 2008). In contrast, the communicative approach, adopts the view that "human explanatory practices must be the starting point for any account of explanation" (Potochnik 2018, 59). This approach is defended by Achinstein (1983), Bromberger (1966), Potochnik (2017), and van Fraassen (1980).

I have argued that an account of explanation is not required to—and indeed should not—choose between these two options. The communicative sense of explanations is crucial because our account of explanation ought to focus on how the scientific community actually provides explanations, and what representations they use to do so (Bokulich 2016; Potochnik 2017). This means that the influence of pragmatic features goes beyond merely specifying the target explanandum, contrast class, and features of interest. It also includes the numerous pragmatic considerations that influence the practice of scientific modeling (e.g., the practical constraints that must be faced by multiscale modelers confronting the tyranny of scales).

However, in contrast with other communicative accounts of explanation (e.g., van Fraassen 1980), I have argued that there is one kind of dependence relation that explains—counterfactual dependence (as well as independence). Moreover, I have argued that once the pragmatics of explanation have specified the explanandum, the contrast class, and a set of contextually salient features, only particular sets of features on which the explanandum counterfactually depends will be able to account for the occurrence of the explanandum. This puts important limits on the influence of pragmatic features. No amount of pragmatic influence can make dependence and independence relations appear or determine whether the features cited in the explanans are sufficient to account for the occurrence of the explanandum—pragmatics can only alter what we want to explain, which alternatives we would like our explanation to account for, and which modeling techniques scientists have available to develop and communicate those explanations.

Communicative accounts overemphasize the pragmatic influences of context when they begin to suggest that what ought to be included in an explanation is merely whatever best serves our purposes of communication, whatever is easiest for our minds to grasp, or whatever the community of scientists decides is an explanation. Most important, while scientific explanations are certainly formulated for and by humans, they succeed only by providing accurate information about counterfactual dependencies and independencies that are made true by the world. In other words, the communicative aspects of explanation are constrained by the way the world really is—this is what enables scientific explanation to be factive.

Despite these ontic constraints, ontic accounts of explanation (e.g., Craver 2014) overemphasize the metaphysical aspects of explanation when they suggest that explanations are completely independent of the representations that scientists actually use to develop and communicate explanations. While explanations provide information about real-world phenomena, the widely cited idea that explanations exist out in the world before any human being discovers them has led many accounts to neglect important issues about how idealizations, values, and practical limitations influence the process of how science actually provides explanations of natural phenomena.

Consequently, this historical divide in the explanation literature has led to accounts of explanation that either make it unclear how explanations can be provided by the holistically distorted representations used by scientists (which typically leads to limited normative guidance for practicing scientists), or make it unclear how scientific explanations connect to the real world (which typically leads to the rejection of realism). If the goal is to provide an account of how our science provides factive explanations, then we require both the pragmatic and ontic aspects of explanation to play a role. I think the accounts of explanation, idealization, and modeling defended in earlier chapters provide one example of such a view by focusing on how the context specifies the set of counterfactual relationships of interests, and showing how holistically distorted models and theories can provide the desired information about those relationships.

The next question that Potochnik raises is: how are explanation and understanding related? I have argued that explanation is neither necessary nor sufficient for understanding. Explanation is not sufficient for understanding because some explanations might fail to produce understanding if they are grasped incorrectly or if acquiring understanding from the explanation requires background knowledge that the agent lacks. Moreover, I argued in chapter 8 that explanation is not necessary for understanding because science's understanding can be improved by idealized models that fail to explain. These claims distinguish my account from Strevens's (2013) view, on which explanation is necessary for understanding, and from Potochnik (2017)'s view, on which all explanations produce understanding. Despite drawing these important distinctions between explanation and understanding, my view also shows why explanations are one of the primary ways that science produces understanding. Because explanations provide a wide range of the kind of modal information that must be grasped in order to understand, grasping an explanation will greatly improve one's understanding. This provides a clear link between the information that constitutes an explanation and the achievement of understanding while still maintaining a clear separation between the concepts of explanation and understanding.

Relatedly, Potochnik asks: what is the relevance of the psychology of explanation? On my account, the psychology of explanation is relevant to the pragmatic aspects of explanation (e.g., what we find interesting), to how different individuals conceive of different relevant possibilities (e.g., due to different past experiences or background beliefs), and to the ways in which understanding is produced by grasping explanations (e.g., in the incorporation of new information into a larger body of background knowledge). Unfortunately, research regarding the psychology of explanation has tended to focus exclusively on the benefits of seeking explanations for individual agents (e.g., see Lombrozo 2011). While there are certainly interesting questions to ask about the role of explanations in individuals' cognition, this research neglects many important questions about how communities of scientists interact to explain and understand complex phenomena (Woody 2015). Although the ways that individuals explain and understand in their daily lives will surely have some similarities to the ways that the scientific community explains and understands, the role of interactions between diverse individuals, conflicting modeling approaches, and background theories needs to be addressed by any account of how science explains and understands our world (Longino 1990, 2002, 2013; Solomon 2001). Therefore, while the psychology of explanation and understanding is important, there are several key questions about how these epistemic achievements are accomplished by science that require us to expand our perspective beyond the psychological aspects of individuals—for instance, we must also consider the social structures of science and how they enable or inhibit individuals' ability to contribute to the epistemic aims of the community (Longino

1990, 2002; Rubin and O'Conner 2018; Sledge and Rice manuscript; Solomon 2001).

Potochnik next asks about the priority of accurate representation to explanation. Throughout this book, I have argued against a significant role for accurate representation relations in our accounts of explanation, understanding, and modeling. Indeed, in chapter 5, I argued that the models used by scientists to explain and understand ought to be characterized as holistically distorted representations of their target systems. Moreover, I have argued that idealizations play positive and ineliminable roles within the explanations and understanding produced by scientific practice. Instead of appealing to accurate representation relations between scientific models and difference-making causes, mechanisms, or processes, I have argued that scientists leverage holistically distorted models that are within the same universality classes as the real-world systems in which the phenomena of interest occur (chapter 6). This relationship between idealized models and real-world systems enables holistically distorted representations to be justifiably used by scientific modelers to extract the modal information required to explain and understand.

This relates to Potochnik's next question: what are the representational aims of explanations? In particular, when should explanations idealize or abstract? My answer here is: *often!* Given the myriad ways that scientists are able to leverage distortions of real-world systems in order to accomplish their epistemic aims, I suggest that scientists do and *should* continue to strategically use myriad idealizations and abstractions of relevant features in order to access modal information that would otherwise be difficult—if not impossible—to acquire. What the examples discussed throughout this book show is that drastic distortion (even of difference-making or contextually salient features) is one of the most fruitful epistemic tools used in scientific practice. By removing the requirement that the epistemic achievements of science be produced via models or theories that accurately represent, we can see how distortion of relevant features makes positive contributions to extraction of the modal information required to explain and understand.

Potochnik also asks how explanation relates to some of the other aims of scientific inquiry (e.g., prediction or manipulation). I've already discussed the relationship between explanation and understanding. In addition, while explanations often improve our ability to predict and manipulate the world around us, these goals can often come apart. For example, in many instances of noncausal explanation, we are able to explain why the phenomenon occurs, but we may not have (or even be interested in) the ability to manipulate, or intervene on, the explanatorily relevant features. Unfortunately, too many accounts tie the value of explanation directly to the ability to intervene on, or manipulate, features of the system (Douglas 2009; Woodward 2003). This results in accounts that struggle to capture cases in which we are interested in explaining, without the possibility of intervening on the features cited in the explanans. We may not have any interest in making the planets move differently from the way they do, and it may be impossible to intervene on things like the universe's space-time structure, but we would still like an explanation of planetary motion.

Moreover, the exclusive focus on these kinds of practical applications of explanations for purposes of intervention misses the intrinsic epistemic value of having an explanation. In many cases, scientific explanations are pursued and considered valuable simply because they enable us to better understand the world and our place in it (i.e., explanations and understanding are also intrinsically valuable to us). Indeed, many students of science and practicing scientists are driven by the desire to explain and understand the phenomena that we observe, independent of our ability to manipulate those outcomes. Accounts of explanation ought to capture this kind of wonder about the universe and the satisfaction that we achieve when our curiosity about why things happen is satisfied. I suggest that focusing on the grasping of modal information about possible states of the world (beyond just those that result from interventions) and on the incorporation of that information into a larger body of information about natural phenomena captures much of the intrinsic value of these epistemic achievements.

Finally, Potochnik raises the question of what the proper levels of explanation are. This question connects issues of explanation with various debates regarding reduction, emergence, and the unity or disunity of scientific fields. The views argued for in this book are certainly antireductionist, in the sense that I reject the claim that only lower-level or smaller-scale details explain, or that reduction of macroscale explanations to explanations at microscales is typically (or even often) possible. However, the more important conclusion is that philosophical accounts of science need to move beyond the discussion of levels of explanation or disciplinary ontology entirely, and instead focus on how the relevant features of complex phenomena change with changes in spatial and temporal scales. For example, in chapter 7 I argued that the concepts of scale dependence, scale invariance, and interscale interaction are all essential to how scientists investigate multiscale phenomena. In most cases, the phenomena, our explanations, and our understanding cannot be tied to particular levels, modeling perspectives, or disciplines. As a result, asking questions such as "is the macroscale explanation or the microscale explanation better?" is a mistake (Potochnik 2009a).[2] It is also a mistake to ask whether a given scale (or theory) is completely autonomous or reducible to another scale (or theory). Instead, we find a spectrum of cases with varying degrees of autonomy with respect to what is happening at other scales of the system. This is why the concept of universality is so important. Instead of restricting our discussion of how scales relate to questions about how microscale features give rise to macroscale features of the system, scientists are often more interested in identifying which of the system's features are stable with respect to changes to which other features of the system—regardless of the "level" or discipline of the stable features or the "level" or discipline of the features being changed. I have argued that the concepts of universality and universality classes free our discussion of the importance of stability across changes from the requirement that this be tied to particular levels, hierarchies, or other mereological pictures of levels of explanation.

Before moving on, I want to raise three additional sets of questions for future investigations of explanation that are not on Potochnik's list (which she admits is not exhaustive). The first question is: how are explanations and understanding constructed over time by incorporating information from multiple individuals, modeling approaches, and theories? Unfortunately, philosophical discussions of explanation and understanding have tended to focus only on the finished epistemic products of science that show up in textbooks or articles. I suggest that there are many important questions to be addressed about the process of explaining and understanding natural phenomena via the strategic use of idealization and other modeling techniques. In other words, perhaps it would be better to think of explanation and understanding as dynamical epistemic schema that stretch across time and often depend on multiple models, theories, and scientists (Rice, Rohwer, and Ariew 2018). Some of the pieces of this kind of dynamical view can be found in earlier discussions of the process of modeling and the justifications given for the introduction of distortions to construct explanations and understanding. Only by focusing on the processes, decisions, constraints, and techniques involved in the practice of scientific modeling can we get a complete account of how science explains and understands natural phenomena.

Investigating these modeling practices has also raised several questions concerning how multiple conflicting representations across a wide range of scales can be used to explain and understand complex phenomena. Unfortunately, there has been relatively little work done on the details of these multiscale modeling techniques and their philosophical implications, although Batterman (2013), Bursten (2018a), Green and Batterman (2017), Wilson (2017), and Winsberg (2006, 2010) have started to address some of these cases. Yet these cases focus our attention on important new questions, such as: What justifications can be provided for using multiple conflicting models of the same phenomenon? How can the insights of multiple conflicting models be integrated into a coherent explanation or understanding of the phenomenon? How do idealized models enable us to explain universal patterns across distant spatial and temporal scales? Given the implications of these cases for a variety of philosophical debates (see chapter 7), additional investigation of instances of multiscale modeling focused on these questions is called for.

As I mentioned above, asking questions about the dynamical processes by which science explains and understands natural phenomena raises additional questions concerning how these epistemic aims are accomplished by communities rather than individuals. Although individuals can certainly explain and understand many things, often the information is distributed across several individuals, such that no single person has all the information involved in science's various explanations or overall understanding of the phenomenon. This brings up important issues concerning how the diversity of a community in terms of individuals, cognitive styles, modeling approaches, and theoretical backgrounds might produce a *better* overall understanding of our world than is possible from within any single perspective or approach. While I have attempted to address some of the issues regarding the use of multiple conflicting modeling approaches, and feminist philosophers have discussed the role of diversity in improving overall understanding (Harding 2004, 2015; Keller 1985; Longino 1990, 2002; Okruhlik 1994; Solomon 2001; Wylie 2004), these areas could use additional attention in philosophical discussions of the nature of explanation, idealization, and modeling. Addressing these questions will help clarify how the dynamical process of science, performed by limited agents using limited representational tools, is able to construct explanations and understanding of the phenomena we observe.

9.2 Holistic Distortion Is Necessary for the Modeling Techniques Used to Explain

In addition to providing answers to several questions concerning scientific explanation and understanding, this book has argued for a drastic shift in the ways that philosophers and scientists conceive of the role of idealization within scientific theorizing. Drawing on the examples discussed throughout this book, I enumerate here several of the ways in which scientific modelers leverage holistic distortions to explain and understand natural phenomena. I do not claim that the following list is exhaustive, but the myriad contributions extracted from these cases do show that leveraging holistic distortions is central to the way in which science explains and understands our world.

As I argued in chapters 3 and 5, the idealizations used in scientific theorizing are often constitutive of the core mathematical frameworks used in scientific models that explain. These idealizations typically result in pervasive distortions of relevant and irrelevant features because they make use of mathematical frameworks that represent the system as a fundamentally different kind of system that is amenable to particular kinds of mathematical treatment. The motivation for introducing these pervasive distortions is that they are necessary for applying the modeling techniques required to provide the explanation of interest (Batterman 2002b; Cartwright 1983, chap. 7; Morrison 2015; Wimsatt 2007). As a result, removal of these distortions through some process of deidealization would result in the inapplicability of the necessary modeling techniques and the loss of the very explanations and understanding that motivated their introduction. In short, a science that did not employ the central use of holistically distorted models would be far less epistemically successful than actual science—this is why the idealizations are epistemically ineliminable.

We have seen this kind of contribution play out in numerous examples. Biological optimality models make use of a number of mathematical modeling techniques to demonstrate how the equilibrium point of the evolving population counterfactually depends on various constraints and trade-offs involved in the trait's selection (Rice 2012, 2013). Bringing these results to bear on the evolution of phenotypic traits requires these modelers to make several idealizing assumptions concerning the entities, interactions, and dynamics of the target system—including contextually salient difference-makers. In addition, we've seen how statistical modeling techniques can demonstrate that many phenomena depend on certain statistical features of the population, and that most of the physical details of the system are irrelevant to those macroscale statistical patterns (Ariew, Rice, and Rohwer 2015; Walsh, Ariew, and Matthen 2017; Walsh, Lewens, and Ariew 2002). In order to use these statistical modeling techniques, scientists routinely make idealizing assumptions about the size of the population and the independence of various features. These idealizations (and others) enable the application of statistical theorems and modeling tools that would otherwise be inapplicable. Finally, we have seen that when physicists explain various behaviors, they often require the introduction of limits, in which the idealized mathematical description functions as a necessary condition for explaining and understanding the phenomenon (Batterman 2002b; Morrison 2015, 30–31). In chapter 5 I argued that similar considerations apply to explanatory uses of the ideal gas law and various population genetics models. Indeed, examples can be found throughout scientific modeling. These examples show that pervasive distortion of real-world systems is often absolutely necessary for the application of the modeling techniques used to extract the explanations that science has provided for many puzzling phenomena.

As a result of these cases, I have argued that a primary goal of philosophical accounts of idealized modeling should be to justify scientists' use of these holistic distortions in terms of the explanations and understanding that they enable them to achieve that would otherwise be unobtainable. In chapters 6 and 7 I argued that this can be accomplished by showing that the idealized models, which are conveniently amenable to the available modeling techniques, are within the same universality class as the real-world systems in which the phenomenon occurs. This enables scientists to justifiably extract modal information about universal patterns despite the fact that many idealized models are mere caricatures of real systems and distort most of their difference-making causes, mechanisms, and features. Unfortunately,

> most accounts of idealization and abstraction involve adding back properties that were left out of the description or deidealising in order to make the system more realistic or concrete. However, there are many instances where this type of deidealization isn't possible (e.g. the thermodynamic limit) and standard accounts say

nothing about how we should understand cases were the abstract mathematical representation is *necessary* for representing the target system. (Morrison 2015, 21)

In contrast, my leveraging holistic distortions approach has focused on uncovering precisely these kinds of positive contributions that idealizations make to scientific explanations by allowing the application of various mathematical modeling techniques that reveal or demonstrate the information required to explain.

What is more, given that scientific explanations are constituted by a set of information about counterfactual dependencies and independencies (chapter 4), we can see just how these necessary idealizations make positive contributions to scientific explanations. This can occur in at least two ways:

1. The idealizations are constitutive of a modeling framework that allows the discovery of counterfactual dependencies between features of the system and the explanandum.
2. The idealizations are constitutively involved in the modeling techniques that demonstrate the counterfactual irrelevance of various features to the explanandum.

Consequently, idealizations are not introduced merely as a way of distorting what is already believed to be irrelevant. Neither are idealizations innocent bystanders that merely focus our attention on the accurate representation of relevant features. Instead, the use of idealizations and abstractions that result in holistically distorted representations of real systems is typically justified by their being absolutely necessary for applying the theoretical and modeling techniques required to discover the modal information required to explain. The holistic distortion view gives us a way to understand the reasons that these idealizations are typically introduced long before scientific modelers have been able to identify which features are relevant and which are irrelevant to the explanandum.

This view of the necessity of holistic distortion has arisen out of my focus on the process of scientific modeling and on the various representational/mathematical limitations that scientific modelers must confront when attempting to explain and understand complex phenomena. Unfortunately, as Wilson notes, "Such attitudes strongly contrast with those that prevail within philosophical circles today, in which considerations of 'metaphysics' are pursued without concern for the issues of mathematical capacity and linguistic representation" (Wilson 2017, 239). Focusing on these representational limitations also gives us a very different picture of the role and justification of idealizations than focusing only on how our cognitive limitations motivate the use of idealization. While many idealizations and abstractions simplify in ways that are useful for dealing with complexity (Mitchell 2009; Potochnik 2017; Wimsatt 2009), I have focused on the justifications that can be provided for the introduction of idealizations as a means to using the limited theoretical modeling tools that scientists have available. Not only are there limited modeling tools available, but each of these modeling techniques will come with their own representational limitations and affordances. For example, as we saw in chapter 7, the practice of scientific modeling must often confront the tyranny of scales problem, in which no single modeling approach can capture all the features that are relevant to the phenomenon of interest. As a result, often the challenge faced by scientific modelers is not to decide how to distort the system in order to make the representation graspable by human beings, but instead to decide which of the available modeling techniques will enable for the extraction of explanatory information, and which idealizations are required to use those modeling tools.

9.3 Holistic Distortions Often Provide Better Explanations

In addition to being necessary for the modeling techniques used to explain various phenomena, holistically distorted models can contribute to the epistemic aims of science by producing *better* explanations. This is because many explanatory virtues can be improved by introducing idealizations and abstractions of relevant features. For example, by distorting various features of real systems, an idealized and abstract model will often be able to apply to more real and possible systems (Levins 1966; Matthewson and Weisberg 2009; Potochnik 2017; Strevens 2008). This in turn often means that highly idealized models that holistically distort their target systems are better able to reveal counterfactual information about why the phenomenon would or would not occur in other heterogeneous systems. Moreover, by allowing the patterns of interest to scientists to include patterns that hold across systems whose relevant causes are heterogeneous, the accounts of modeling defended in this book expand our conception of the importance of investigating stable patterns beyond only focusing on causal patterns. While generality is not always valuable to us, in many contexts, more general explanations that idealize or abstract away from the features of particular systems will be preferable to explanations that aim to accurately represent those heterogeneous features.

Holistically distorted models are also often better for extracting information about why many features of the system are irrelevant to the explanandum. Minimal model explanations and the use of the renormalization group provide extreme examples of this, but many other highly idealized models are also used to extract information about the irrelevance of various features of the system. Because this information about irrelevance often plays crucial roles within scientific explanations of patterns across different systems (see chapters 3 and 4), holistically distorted models within the same universality class as the real system will often provide better explanations than a model that accurately represented the system's relevant features. Accurately representing relevant features tells us little about why the other features are irrelevant to the explanandum. Only by investigating classes of systems in which those features are different can we extract the required information about irrelevance. Because holistic distortions enable scientists to investigate a range of nonactual systems, and because idealizations are often essential to the application of modeling techniques that demonstrate irrelevance (e.g., renormalization or homogenization), holistic distortion can often improve our explanations of patterns that occur across heterogeneous systems.

Finally, given that many of the phenomena that we would like to explain depend on noncausal structural, statistical, mathematical, or topological features of the system, in many cases, models that holistically distort the causes or mechanisms operating within a system will enable us to better appreciate the explanatorily relevant features involved in noncausal explanations. In other words, holistically distorting causes and mechanisms is often what enables us to focus our attention on the noncausal features of the system on which the explanandum counterfactually depends. This isn't to say that noncausal explanations are always better than causal explanations. Instead, the point is that when a noncausal explanation is preferred or required, introducing drastic distortions of the system's causes or mechanisms often will improve our ability to uncover the dependence relations between the system's noncausal features and the explanandum.

9.4 Holistic Distortions Improve Understanding

Explanations are the primary way in which science produces understanding. Therefore, the epistemic contributions that holistic distortions make to explanations are contributions to scientific understanding as well. However, as I argued in chapter 8, holistically distorted models also make epistemic contributions to understanding beyond their contributions to explanations.

One way in which distortions contribute to human understanding is that they often make complex phenomena intelligible for limited beings such as ourselves (Giere 1988; Mitchell 2009; Potochnik 2017; Sober 2015; Wimsatt 2007). The world is complicated, and our minds simply cannot process all the dizzying details of the plethora of entities, features, and interactions that contribute to many natural phenomena. While simplicity does not necessarily make for a better explanation (or for an explanation that is more likely to be true), simplification almost always aids in our ability to grasp information. Holistically distorted models often result in greatly simplified representations by abstracting away many features, using parameters to stand in for multiple factors, or using idealizations that simplify our representations in a variety of ways. Consequently, holistically distorted representations are often much easier to grasp and can be communicated more efficiently than more complex realistic descriptions.

Beyond helping limited beings deal with overwhelming complexity, idealized scientific representations are useful for focusing our attention on the features that are most important for our epistemic aims (Potochnik 2017). One way that this can happen is that idealizations might indicate that various features are irrelevant to the pattern of interest by distorting irrelevant, insignificant, or nonfocal features (Elgin and Sober 2002; Potochnik 2017; Strevens 2004, 2008; Weisberg 2007a, 2013). As defenders of the standard view have noted, this process can then focus our attention on other relevant features.

Of course, I think the epistemic contributions of distortion also go well beyond the identification of irrelevant features as a means to focusing on relevant features (see chapter 5). For one thing, idealizations can draw attention to the very features that they distort by either exaggerating those features or producing drastically distorted representations in which those features are particularly salient. As Elgin (2017) notes, this is often how caricatures work. By exaggerating certain features, we can draw attention to them. Moreover, many mathematical modeling frameworks make particular features more salient, even though the model drastically distorts how those features are realized in actual systems. In short, as Elgin argues, "Effective models afford an understanding of their targets because their simplifications, idealizations, elaborations, and distortions make salient important features of the targets" (Elgin 2017, 249).

However, as I argued in chapter 8, I disagree with Elgin that being able to instantiate or approximate important features via exemplification is the only way that models make features salient. This is because, as Elgin herself suggests, "Many scientific models such as equations and diagrams, are incapable of instantiating the properties they apparently impute to their targets. If they cannot instantiate a range of properties, they cannot exemplify them" (Elgin 2017, 258). Elgin's response to this objection is to argue that, "Where their divergence is negligible, the models, although not strictly true of the phenomena they denote, are true enough of them.... Where a model is true enough, we do not go wrong if we think of the phenomena as displaying the features that the model exemplifies" (Elgin 2017, 261). However, I have argued throughout this book that scientific models can afford access to salient features of real phenomena, even when they drastically distort those features in ways that make the model incapable of instantiating or approximating those features. Models (and the modelers who use them) routinely emphasize and deemphasize various features of the phenomenon, but they often do so by drastically distorting those features. As a result, while a scientific model certainly "equips us to see the target differently than we otherwise might" (Elgin 2017, 263), this is often accomplished with no regard to whether the salient features are exemplified, instantiated, or accurately represented within the model.

Another way that scientific representations focus our attention on certain features is by choosing models that focus our attention on particular spatial and temporal scales. Many philosophers have recognized that different patterns become recognizable at different scales (Bursten 2018b; Garfinkel 1981; Potochnik 2017). This is one of the main motivations for abstracting away from details at smaller scales in attempting to study patterns at more macroscales, but it is also more generally applicable to any selection of scales to be represented within a scientific model. Idealizations, too, often enable us to limit our representations to particular spatial or temporal scales. For example, using idealizations that enable the application of the central limit theorem can enable scientific modelers to ignore the underlying features of individuals within the population so as to focus on the macroscale statistical patterns at the level of the overall population. In short, idealization and abstraction are decisions to distort that routinely contribute to our understanding by focusing our representations, observations, and inferences on particular scales of the system. Moreover, as we saw in chapter 7, distortions can also focus our attention on various relationships across scales (e.g., power law relationships that are stable across changes in scale). By strategically using distortions to investigate particular universal features at or across particular scales of the system, drastic distortion routinely contributes to scientists' ability to grasp the modal information required to understand.

Next, as we saw in chapter 7, multiple conflicting idealized models are often used to study the same phenomenon. Such a situation looks epistemically problematic if we adopt the standard view's focus on accurately representing relevant features. Fortunately, the views defended in previous chapters show how such a situation can promote a better overall understanding of the phenomenon than could be provided by any single model, approach, or perspective. Specifically, different models will often be in different universality classes, each of which includes the systems in which the target phenomena occur. Discovering these classes of systems enables scientific modelers working from within different modeling frameworks and perspectives to extract different pieces of modal information about the system—even if their background assumptions and models are in conflict with one another. Moreover, multiple conflicting models can be used to explore different counterfactual situations with respect to the same patterns or features. Indeed, it is often by virtue of being holistically distorted and in conflict with each other that a set of inconsistent idealized models enables the scientific community to extract a wider range of modal information about the phenomenon than those models could provide individually.

Finally, another key contribution of holistically distorted models is that they aid in scientists' exploration of the possibility space of natural systems. That is, scientific models of merely hypothetical systems often improve our understanding of real systems by representing merely possible systems that may never be realized. The examples (from chapter 8) of Schelling's checkerboard model, the Hawk-Dove game, and the modeling of possible stellar orbits all contribute to understanding in this way. This is because understanding is constituted by a set of true modal information, including modal information about what would occur in fairly distant possible systems. Precisely how we ought to compare the closeness of various possible systems and how the understanding produced by investigating very distant possibilities compares with that produced by investigating closer possible systems will have to be specified more exactly elsewhere. The point that I've tried to make here is that constructing various kinds of holistically distorted models enables scientists to explore a wider range of possibilities. Exploring those possibilities can then enable scientists to justify various claims about what is possible, impossible, necessary, or unnecessary for the phenomenon to occur. In doing so, these holistically distorted models directly contribute to the scientific community's overall understanding of various phenomena.

In sum, holistically distorted models aid in the extraction of modal information about our world, and this modal information enables us to better understand real phenomena. Only by recognizing a wide variety of contributions do we get a more complete picture of how holistically distorted representations contribute in positive ways to the understanding provided by scientific inquiry. Pervasively distorted models are not merely way stations on the path to more accurate models that produce the real epistemic fruits. Instead, holistically distorted models directly and positively contribute to the discovery of the modal information required to factively understand natural phenomena.

9.5 Intertwined Idealizations and Holistic Distortion

In addition to highlighting distortion's epistemic contributions, the holistic distortion view helps us avoid some of the misleading ways that philosophers have sought to analyze the role of idealizations in science. First, some philosophers—myself included—have attempted to distinguish between various kinds of idealization. This is typically done by looking at the motivation for introducing the idealization and the ultimate representational goals of the model builders (Rohwer and Rice 2013; Weisberg 2007a, 2013). The value of demarcating various kinds of idealization lies in showing that there is no univocal process or justification for idealization in science (Weisberg 2007a). The problem with this sort of project is that these kinds of idealization are rarely used in isolation and are often difficult to separate in actual scientific practice. While I have tried to uncover a wide range of the kinds of epistemic contributions that idealizations make within scientific practice, enumerating an exhaustive list of types of idealization or their contributions seems intractable.

Instead, we should note that there are many overlapping and inseparable reasons to idealize scientific representations (Potochnik 2017). For example, a single scientific model may include both Galilean idealizations introduced for computational tractability and hypothetical pattern idealizations introduced in order to investigate a necessity claim via a hypothetical scenario. Moreover, a single idealizing assumption might have multiple reasons for its introduction (e.g., it may simultaneously enable the application of various modeling techniques and simplify the model's representation of the target system). This lends further support to the idea that most of the idealized models used in science are holistically distorted representations of their target systems. Because there are multiple overlapping motivations for introducing distortions into scientific representations, we have good reason to expect most scientific models to be pervasively idealized. What is more, this observation also helps explain why so many idealizations are deeply entrenched in our best scientific models and theories in ways that make them ineliminable. Because they serve so many epistemic and nonepistemic functions, idealizations of relevant features are absolutely crucial to scientific practice. Moreover, given the multiple overlapping reasons that motivate the introduction of distortions into scientific representations, removing one of those motivations (e.g., by overcoming computational tractability in the future) will often be insufficient to motivate the removal of an idealizing assumption because it will typically serve several other aims of the modeler as well.

In addition, the holistic distortion view blocks philosophers' attempts to describe singular idealizing assumptions and their contributions in isolation. Several philosophical accounts begin by describing a particularly egregious idealizing assumption and then suggesting that the motivation for, and contributions of, that assumption can be described in isolation of the rest of the scientific representation (e.g., suggesting that the only role of assuming an infinite population is to claim that drift makes no difference). However, as we have seen throughout this book, idealizing assumptions rarely act in isolation. Instead, idealizations typically play crucial roles within the overall scientific representation that is constituted by many intertwined assumptions. In short, it is a mistake to conceive of idealizing assumptions as isolable propositions within scientific models that can be easily removed, replaced, or changed independent of the rest of the model. Conceiving of idealized models as holistic distortions emphasizes both the multiple intertwined reasons for introducing distortions and the ways that those distortions are intertwined with multiple other assumptions and features of scientific models.

9.6 Universal Patterns and the Epistemic Goals of Science

The view of science that I've presented here has also focused on the explanation and understanding of universal patterns in science, rather than on the explanation and understanding of particular events. This is because the very pursuit of science depends on there being stable patterns across space and time such that we can predict, explain, and understand what we observe in different places and times. Although Hume's (1739, 1999) problem of induction has troubled philosophers for centuries, what it makes clear is just how central the discovery of projectible patterns across different cases are to the practice of science. I have offered little that might be used to directly respond to his concerns with inductive inferences or confirmation; what I have done is construct a philosophical approach to science that takes the discovery and explanation of universal (and therefore projectible) patterns across a range of real, possible, and model systems to be central to the epistemic aims of science.

While several philosophers have noted the importance of general patterns in science (e.g., Garfinkel 1981; Kitcher 1981; Potochnik 2017; Strevens 2008; Woodward 2003), many accounts of explanation—such as many causal or mechanistic accounts—begin by giving a detailed analysis of how explanations are provided for singular events.[3] I suspect that this is largely in response to the need to make explanation a bit more local than is allowed by covering-law models that focus on universal and exceptionless generalizations. However, most causal and mechanistic accounts of explanation begin by providing a few simple examples of how a particular causal story (e.g., the poisoning, shooting, and drowning of Rasputin), a token causal claim (e.g., the flagpole causes its shadow), or a particular mechanism (e.g., how a bike's drivetrain can explain its motion) can be used to generate an account of which events, properties, or causes are required to explain a singular event (Glennan 2017; Salmon 1984; Strevens 2008). That is, the foundation of many (if not most) causal or mechanistic accounts has been the explanation of singular events.

After using these cases to clarify how causes or mechanisms are able to provide an explanation, these accounts then scale up to macrocauses or mechanisms that are the same across multiple systems in order to explain why various patterns occur. For example, Strevens adopts what he calls the "causal-mechanical" approach to explaining patterns, "according to which the end point for regularity explanation is, as for event explanation, a causal model representing—however abstractly—facts about fundamental-level causal influence" (Strevens 2008, 221). Moreover, Strevens suggests that the fundamental idea behind this approach is that "the explanation of a causal generalization and the explanation of any instance of the generalization invoke the same causal mechanism" (Strevens 2008, 223). This macrocausal strategy has been widely employed in the explanation literature.[4] The general idea is that whenever we want to explain a pattern, we simply need to abstract away from the details of particular causal mechanisms until we abstractly describe what the causal mechanisms that produce the various instances of the pattern have in common (Strevens 2008, 289–296). Several mechanistic accounts of explanation account for the explanation of general patterns in a similar way (Craver 2007; Millstein 2006). Likewise, Potochnik's (2017) causal patterns are presumably just those patterns that are produced by similar causes across different cases.

This approach works well if we can assume that patterns are always generated by the same (of fairly similar) causes or mechanisms across different systems. However, given the extreme causal heterogeneity of complex systems, this will often not be the case. Causal accounts repeatedly make the mistaken assumption that simply abstracting to a larger scale of the system (or away from details) will transform causal and mechanistic heterogeneity into causal or mechanistic homogeneity. I do not deny that there can be causal and mechanistic explanations at macrolevels or macroscales—there certainly can. But it cannot be assumed that systems that are extremely heterogeneous in most of their physical components, causes, and mechanisms at smaller scales will always (or even frequently) involve the same macrocausal relationships or mechanisms that produce patterns at larger scales or higher levels of abstraction.

In contrast to this approach, I have offered a view of science that begins with the explanation and understanding of universal patterns across systems that are often extremely heterogeneous in their difference-making causes or mechanisms. These types of patterns are widespread across the sciences. For example, physicists routinely discover universal patterns that are stable across many fluids and magnets that have almost nothing in common in terms of their physical causes or mechanisms. Similarly, biologists repeatedly depend on the discovery of stable patterns across species, populations, and ecosystems that are extremely diverse with respect to the causes or mechanisms that generate those patterns. Even if these systems have some similarities in their causal relationships or mechanisms, those minimal similarities will rarely be enough to account for the occurrence of the pattern.

Moreover, in many cases, scientists are particularly interested in providing an explanation for why a particular pattern continues to occur despite various changes to the physical features of the system. That is, beyond just noting the scope of a pattern, scientists often want to explain why a pattern has the particular scope it has. Indeed, several other philosophers have noted the need to account for scientific explanations that capture explanations of universality rather than focusing on the explanation of singular instances (Batterman 2002b; Morrison 2015). For example, Nicolas Fillion and Robert Moir recently argued that:

> In accordance with sound scientific theorizing, our objective is not only to account for explanations of particular phenomena, but also for explanations of robust patterns emerging in the phenomena. . . . In order to provide a general strategy that explains patterns, it is necessary to have a framework that is rich enough to capture universality of physical phenomena. (Fillion and Moir 2018, 738)

Focusing on these universal patterns instead of token causal explanations results in an account of explanation that can accommodate explanations of stable patterns across causally heterogeneous systems. Indeed, the counterfactual account of explanation offered in chapter 4 was built out of three types of noncausal explanations that each aim to explain patterns that are stable across myriad changes to the causes or mechanisms that produce the phenomenon in real systems. In each of these kinds of cases, I have argued that the explanation requires information about noncausal features of the system to show why these patterns hold across systems that are extremely heterogeneous in their causes or mechanisms (at any level or scale of the system). What is more, I have argued that explaining the stability of these patterns across different systems requires information about why various features of the systems (e.g., all their causal features) are irrelevant to the patterns' occurrence. While other philosophers have recognized the role of stability through concepts such as invariance, generality, or multiple realizability, the concept of universality captures these ideas without biasing our philosophical approach toward the stability of only causal features, stability at only macroscales, or ontological commitments regarding supervenience or mereological relationships.

9.7 Realism about Epistemic Products

Finally, the views defended in this book suggests that we must break the tie between representational accuracy and the production of factive explanations and understanding. This approach has been driven by two basic observations about scientific practice:

1. Scientific models and theories are pervasively distorted representations of real systems.
2. Scientific inquiry intuitively has produced a plethora of factive explanations and understanding of natural phenomena.

Given the seeming tension between these two claims, most philosophical accounts of science have tried to give up, or drastically alter, one of them. For example, as we saw in chapter 2, many philosophers suggest that claim (1) ought to be altered by arguing that scientific models and theories, when successful, only distort features that are irrelevant, insignificant, or otherwise not of interest (Elgin and Sober 2002; Potochnik 2017; Strevens 2008; Weisberg 2007a, 2013). However, in chapter 5 I argued that scientific practice routinely involves the direct and drastic distortion of contextually salient difference-making features that we know are relevant to the occurrence of the phenomenon. Other philosophers have suggested that we ought to instead alter claim (2) by suggesting that the epistemic achievements of science are nonfactive (de Regt 2017; Elgin 2007, 2017; Potochnik 2017). However, most of these accounts motivate the alteration of claim (2) by pointing to the pervasive use of idealization in science. That is, they follow the standard view in assuming that accurate representation is required for the epistemic achievements of science to be factive. Then, after convincing us of the centrality of idealization to scientific practice, these accounts argue that the epistemic achievements of science, therefore, must be nonfactive.

What we see, then, is that both of these responses adopt the standard view's assumption that, if the epistemic aims of science are going to be factive, then scientific models and theories must provide accurate representations of the relevant features of real systems. In contrast, I have argued for a third option. We can and should consistently hold that both claims (1) and (2) are true. The key to recognizing this option is to break the link between representational accuracy and the production of factive epistemic achievements. The arguments provided in this book show how holistically distorted scientific representations can nonetheless be used to produce factive explanations and understanding. In particular, factive explanations are provided not by constructing models or theories that accurately represent relevant features, but instead by extracting a set of true counterfactual dependencies and independencies via the use of various modeling processes that drastically distort the explanatorily relevant features of real systems. Moreover, factive scientific understanding is produced by grasping a set of true modal information about the possible states of the system, not by providing a scientific representation that accurately represents the features or relationships found in real systems. Therefore, we can be realists or factivists about the epistemic achievements of science, even if we reject the claim that our best models and theories are accurate descriptions of the relevant features of real systems.

9.8 Conclusion

In conclusion, let me lay out the main claims I have defended throughout the book:

1. Idealization is central, essential, and ineliminable from scientific practice due to science's widespread reliance on multiple conflicting idealized models and theories to explain and understand natural phenomena (chapter 1).
2. Most philosophical accounts argue that causation is essential to explanation, models that explain must accurately represent relevant (e.g., difference-making) causes, and scientific representations can be decomposed into their accurate and inaccurate parts (chapter 2).
3. There are many noncausal explanations in science. These explanations cite noncausal features on which the explanandum counterfactually

depends, often demonstrate that many features of the system (e.g., its causes) are irrelevant to the explanandum, and routinely make essential use of idealizations that distort difference-making causes (chapter 3).

4. Causal and noncausal explanations can be unified under a counterfactual account of explanation in which explanations are constituted by a set of information about how the explanandum counterfactually depends on, and is counterfactually independent of, the contextually salient features of the systems of interest (chapter 4).

5. Instead of trying to decompose models into an accurate representation of relevant features and the distortion of irrelevant features, idealized models ought to be characterized as holistic distortions of their target systems whose use is justified by the explanations and understanding they enable that would otherwise be inaccessible (chapter 5).

6. Holistically distorted models can be justifiably used to explain and understand when they are in the same universality class as the systems in which the explanandum occurs (chapter 6).

7. Appealing to universality classes can also justify the use of multiple conflicting models for the same phenomenon and various multiscale modeling techniques used to confront the tyranny of scales (chapter 7).

8. Scientific understanding of a natural phenomenon is constituted by a community's or agent's grasp of a set of modal information about possible states of the system that is incorporated into a body of information that is mostly true. This understanding can be produced via routes that do not go through explanation, is improved by various kinds of diversity, and shows how a kind of realism is compatible with granting that scientific representations are pervasively inaccurate (chapter 8).

9. These views reveal a wide variety of ways that holistically distorted representations are strategically leveraged in order to produce better scientific explanations and improved understanding of natural phenomena (this chapter).

Collectively, these claims provide essential pieces of an answer to the fundamental question presented at the beginning of the book: how can the drastic distortions of relevant features contribute to the epistemic aims of science? My response to this question has required us to rethink the three main tenets of the standard view. Instead of requiring all explanations to be causal, we require a counterfactual account of explanation that

Leveraging Holistic Distortions 299

can show what causal and noncausal explanations have in common. Instead of requiring idealized models to be decomposed into their accurate and inaccurate parts, we require a view that recognizes how multiple, intertwined assumptions, idealizations, and abstractions result in scientific representations that holistically distort the features of real systems. Finally, instead of focusing on accurate representation relations between models and difference-makers, we require an account of how holistically distorted models within the same universality classes as real systems can be justifiably used to discover modal information. Adopting this leveraging holistic distortions approach accomplishes two important tasks: (1) it captures a much wider range of the kinds of explanations, understandings, idealizations, and modeling techniques found in scientific practice; and (2) it enables us to maintain factive accounts of the epistemic achievements of science while recognizing the central and ineliminable use of distortion in scientific practice.

This alternative approach has also highlighted several new questions for future philosophical research concerning scientific explanation, understanding, idealization, and modeling. For example, instead of (only) asking, "What are the norms of mechanistic and causal explanation?" we also need to ask, "What are the norms that govern the inclusion of information about counterfactual dependence and independence in causal and noncausal explanations?" Instead of asking, "How can the distortions of idealized models be relegated to features that are irrelevant, insignificant, or not of interest?" we need to ask, "How can the distortion of relevant, significant, and difference-making features be leveraged in order to extract explanatory information?" Instead of asking, "What features must models accurately represent in order to be justifiably used to explain?" we should ask, "How can universality and other model–world relations be used to justify the appeal to idealized models within an explanation?" Instead of asking, "How can the ontic features of explanation be isolated from the pragmatic aspects of explanation?" we should ask, "How do the limited representational tools used by scientists reveal the ontological dependencies (and independencies) required to explain?" Instead of asking, "Can explanations at one level be reduced to the explanations at another level?" we ought to ask, "How can multiple conflicting models across a range of spatial and temporal scales be used to explain and understand multiscale phenomena?" Instead of asking, "How can we be realists about scientific models and theories?" we need to ask, "How can we be realists about the understanding that scientists extract by strategically using holistically distorted representations?" I have tried to offer some answers to these alternative questions, but they have unfortunately received comparatively little attention in the philosophical literature. Yet we require answers to these questions if we want to account for the variety of ways that scientists leverage pervasively distorted models and theories to explain and understand the phenomena we observe.

The claims listed above present a philosophical account of science that explicitly recognizes science's pervasive distortion of relevant features as a *virtue*, rather than an obstacle that needs to be overcome or corrected. Indeed, I have argued that a science that did not make use of holistically distorted representations would be far less epistemically successful in terms of the factive explanations and understanding that it would be able to provide. By moving beyond the foundational assumptions set in place by the standard view, we can instead focus our attention on the variety of ways that holistically distorted models are strategically leveraged by scientists to explain and understand natural phenomena. In order to lay the foundation for this project, this book has provided alternative accounts of explanation, idealization, modeling, and understanding that show how it is possible for science to produce factive epistemic achievements despite its essential and ineliminable reliance on holistically distorted models and theories. I do not pretend to have provided definitive accounts of any of these concepts, nor have I answered all the questions that may arise concerning them. However, recognizing the possibility of consistently maintaining that science both depends on holistically distorted representations and produces factive epistemic achievements reveals an important set of new questions for philosophers of science to investigate going forward. Answering those questions will result in philosophical accounts of science that more accurately describe scientific practice and will hopefully provide useful normative guidance for scientists' continual attempts to leverage holistically distorted representations in order to explain and understand our world.

Notes

Chapter 1

1. Here, *inaccessibility* refers to the idea that drastic and deliberate distortion of the very features that scientists are interested in is essential to much of the knowledge, understanding, and explanations that science has provided for natural phenomena (i.e., if we were to remove these distortions, we would lose those epistemic achievements).

2. Even those accounts that suggest that the understanding produced by science is nonfactive (e.g., Elgin 2017) suggest that this nonfactive understanding is produced by having the scientific representation instantiate or exemplify salient features.

3. The existence of these dissenting views is precisely why I will be careful to claim that only "most" of the prominent views in the literature have adopted these claims.

4. In addition, while there are several differences among the views that I will group under the standard approach, such differences are already well worked through elsewhere in the literature, and where those differences matter to the arguments that I wish to make, I've tried to offer more detailed discussions of the particular features of these views.

5. Throughout this book, I adopt Weisberg's general description of models as interpreted concrete, mathematical, or computational structures that specify model systems via the use of equations, propositions, and other representational devices (Weisberg 2013, chap. 1). In what follows, I will refer to the system specified by these interpreted structures, along with the consequences, predictions, or dependence relationships that the assumptions of the model generate, as the "model system." Thanks to Chris Pincock for pressing me to be clearer about what I take a model system to be.

6. Thanks to Chris Pincock for suggesting that I be clearer on this point.

7. My views concerning idealization in science also have much in common with the views of Nancy Cartwright (1983). However, unlike Cartwright, I will move away from the focus on causation and forces as being important for explanation.

Moreover, my views will derive from the analysis of examples from a wide range of disciplines, rather than focusing primarily on explanations in physics.

8. This book also focuses on numerous topics that are not part of Potochnik's discussion (e.g., universality, the strategies used in multiscale modeling, and the realism debate).

9. As I have noted in this chapter, there are several exceptions to this focus on accurate representation of causes that will be discussed throughout the book, including Batterman (2002a, 2002b), Bokulich (2008), Lange (2013a, 2013b), Pincock (2012), and Walsh, Ariew, and Matthen (2002).

10. It is a bit difficult to know exactly where to put Potochnik's views in this schema because she argues that scientific representations must accurately describe the causal factors of interest to the research program, and also that the resulting understanding must be nonfactive because it depends on idealizations. I will argue against both of these claims.

11. As de Regt (2017) argues, this adeptness at using models and theories for the accomplishment of our epistemic aims often requires various skills.

Chapter 2

1. In particular, "having students construct explanations (and/or arguments) using evidence or scientific principles and reasoning is supported by a large body of evidence that shows improved student learning" (Cooper 2015, 1274).

2. Hempel understood that not all explanations were of the sort described by the standard DN model. Since explanations can also be probabilistic or statistical, Hempel also included an account of *inductive-statistical* (IS) explanation (Hempel 1965). The crucial difference is that, where DN explanations establish deductive certainty, IS explanations confer a high inductive probability on their explanandum.

3. Ideal intervention captures Woodward's idea that while an actual intervention may not be possible due to practical limitations, "causally explaining an outcome involves the identification of factors and relationships such that *if* (perhaps contrary to fact) manipulation of these factors *were* possible, this would be a way of manipulating or altering the phenomenon in question" (Woodward 2003, 10).

4. In fact, the literature on mechanisms rarely addresses the relationship between mechanistic models and mechanisms themselves, because it is routinely assumed that "the parts and organization of the model typically map directly to parts and organization of the object being investigated" (Matthewson and Calcott 2011, 738).

5. A similar strategy is adopted by various accounts that attempt to use robustness analysis to show that the results of scientific models do not depend on their idealizations (Kuorikoski, Lehtinen, and Marchionni 2010, 551).

6. Similarly, idealizations can facilitate understanding only when they stand in for features "that are unimportant to the focal causal pattern" (Potochnik 2017, 107).

7. Thanks to an anonymous reviewer for raising this possible response.

8. One might distinguish between a feature's being irrelevant and a feature's being relatively unimportant. While this distinction can certainly be made, I don't think this subtle difference in degree has much of an impact on the arguments that follow. Nonetheless, to avoid any confusion, my use of irrelevance (and relevance) here is meant to capture claims concerning causal irrelevance (e.g., difference-making), relative unimportance (e.g., having a negligible impact), or contextual nonsalience.

9. Given the range of views, from a number of debates, it is impossible to find the perfect wording that captures all the subtle details of every version of the general decompositional strategy. I think putting these assumptions in terms of relevant/irrelevant and accurate/inaccurate does the best job of capturing what these views have in common. Indeed, however these accounts determine relevance/irrelevance, the claim is that the model will be successful/explanatory, just in case it accurately (to some degree) captures the relevant or most important features and uses idealizations only to distort irrelevant (or largely insignificant) features.

Chapter 3

1. Optimality modeling also includes game-theoretic models, which are often referred to as *frequency dependent* optimality models because the payoffs to different strategies depend on the frequency of strategies in the population.

2. Although I believe many of the claims made here can be applied to optimality explanations within other disciplines (e.g., economics or physics), for purposes of space, I will restrict my discussion to examples of optimality explanations from biology.

3. In biological contexts, fitness, or inclusive fitness, is the ideal currency, but often a more easily measured currency is used (e.g., average energy intake).

4. These are often referred to as the strategies' *payoffs*.

5. Of course, many optimality models are not used to provide an explanation. However, these alternative uses, though important, are outside the focus of this chapter.

6. For instance, if most of the population is foraging in one area, it might be better for an individual to forage elsewhere where there will be less competition.

7. Of course, equilibrium and evolutionary stability get defined in several ways. However, I will not be concerned with those differences here because my arguments rely only on the general fact that these models explain by showing that a particular strategy (or set of strategies) is a system's equilibrium point.

8. This kind of analysis was initially used in economics to determine the optimal strategy for investing limited resources.

9. Indeed, Sober (2000) cites Parker's model as meeting a more rigorous standard for testing optimality models.

10. In addition, the male-female trade-off remains essential even when various underlying assumptions of Fisher's model are changed (Hamilton 1967).

11. In addition, Strevens argues: "What the equilibrium model contains above and beyond the mechanism, I suggest, is a basing generalization that guarantees the existence of the appropriate filigreed surface for any given release point" (Strevens 2008, 271). Specifically, the basing generalization in the equilibrium case guarantees the existence of the appropriate parts of the basin for each of the possible paths that the ball might take to the bottom of the basin.

12. Later, Strevens makes it explicit that the key to explaining stable ecological patterns is to fill in the causal mechanisms (Strevens 2008, 159). According to him, an ecological model that represents features that are multiply realizable without specifying the causal mechanisms of the system merely "functions as an explanatory template, to be filled out in different ways to obtain deep explanations of stability in different ecosystems" (Strevens 2008, 160).

13. Potochnik also proposes a weak use of optimality models, in which "an optimality model must accurately represent selection, but if selection is only part of the evolutionary story, then observed trait values may vary from the values the optimality model predicts" (Potochnik 2009b, 186). In particular, she argues that these models must be "fully accurate" with respect to their fitness functions and the range of possible trait values (Potochnik 2009b, 185).

14. Indeed, one difficultly in constructing biological optimality models is that it is often impossible to tell what phenotypic strategies were available in the history of the population.

15. To borrow a weighted phrase from economics, a model's optimization assumptions are usually taken to be adequate, so long as the system behaves "as if" it were optimizing the model's criterion (Appiah 2017; Vaihinger 1952).

16. An interesting thing to note about the assumption of infinite population size is that many philosophers have defended a statistical interpretation of fitness that entails that selection and drift cannot be pulled apart as two separate causal processes (Matthen and Ariew 2009; Walsh 2007, 2010; Walsh, Lewens, and Ariew 2002). If selection and drift are inseparable in this way, then one cannot assume infinite population size without consequently distorting the representation of the entire evolutionary process.

17. Indeed, as Odenbaugh notes, "It is extremely difficult to manipulate ecological systems in systematic and controlled ways" (2005, 233).

18. In addition, given that optimality models provide very limited accurate information about the causal dynamics that led to the explanandum, they will usually provide almost no information about how the system would behave under an intervention of this kind.

19. Moreover, it is important to remember that the contribution of food availability to the fitness of these foraging strategies is *drastically* distorted in models that assume that they are directly correlated, given that a wide range of other causal factors contribute to the fitness of these foraging strategies. So even if we restrict the model's representation to this simple causal relationship, the model fails to accurately describe those causes.

20. Huneman has recently used similar language to describe "topological explanations" that "abstrac[t] away from causal relations and interactions in a system, in order to pick up some sort of 'topological' properties of that system and draw from those properties mathematical consequences that explain the features of the system" (2010, 214).

21. Galton is, of course, a rather controversial figure, given his advocacy of eugenics and the applications of his work as part of that movement and in the development of IQ testing (Hacking 1990, 183). The ethical components of this history should certainly be kept in mind and investigated further, but my focus here will be on the statistical regression techniques that Galton developed rather than on his particular applications of those techniques in support of problematic ends.

22. Galton also attempted to demonstrate the reasonableness of these statistical assumptions using his quincunx to demonstrate how individual events aggregate to form the curve of error (Ariew, Rohwer, and Rice 2017).

23. Matthen and Ariew (2009) defend this claim by first granting that trait differences, such as variation in camouflage, often are causes of evolutionary change. These same tests, however fail to establish causation when we consider the claim: natural selection causes evolutionary change because the relationship between variation-in-advantageous-traits, and natural selection is purely mathematical. This entails that one cannot manipulate natural selection independent of variation-in-advantageous-traits.

24. In addition, contrary to Sober (2011), many counterfactual dependencies will cite features that do not promote or produce the phenomenon to be explained. For example, a statistical fact concerning the distribution of traits might be said to promote the selection of a particular trait, even if that statistical distribution is not a cause of the trait. Moreover, it can be the case that if X had been different, then Y would have been different even if X has no role is causally promoting or producing Y. In many cases, this will occur because changing X would require changing other features that *are* responsible for the production of Y. For example, overcrowding in Philadelphia schools might counterfactually depend on the average number of children per household, even if the only features that actually causally contribute to

that overcrowding are individual children in those classrooms. In addition, there are many forms of counterfactual dependence that are grounded in noncausal relationships. For example, a shape being a triangle depends on its having three sides, but it makes little sense to say that the shape's three-sidedness promotes or produces it being a triangle. More generally, a phenomenon can *depend* on the values taken by various variables within a system without those variables (or the properties they refer to) being causally involved in producing or promoting the phenomenon.

25. As we will see in chapter 5, these features of population genetics connect it with similar statistical modeling approaches used in physics. Indeed, "the birth of population genetics resulted from the application of mathematical techniques that invoked infinite populations and ideas from statistical physics. To that extent, then, the methods of population genetics look very much like the methods of physics" (Morrison 2015, 33).

26. In particular, the LGA model is able to reproduce the parabolic profile of momentum density that we find in (incompressible laminar) flow in a pipe.

27. What's more, Lange's framing of this objection completely leaves out the role of renormalization within the minimal model explanation.

28. While some authors have recently suggested that Batterman (2002b) suggests that renormalization group explanations are noncausal simply because they eliminate details, I do not think he ever makes such an argument. Instead, he argues that the explanations of universality provided by the renormalization group are noncausal because they do not reference the features that would be provided in a causal-mechanical explanation of the system.

Chapter 4

1. This is because I think philosophical debates concerning explanation need (or ought) to move beyond the never-ending cycle of proposing universal necessary conditions and counterexamples to those conditions (Rice and Rohwer 2020).

2. Consequently, I will explore a middle ground between analyzing isolated cases and the Hempelian tradition of attempting to provide a universally applicable theory of explanation (Rice, Rohwer, and Ariew 2018).

3. Ideally, the account also ought to tell us what distinguishes these kinds of explanations (e.g., causal versus statistical) from one another.

4. There are surely other requirements for an account of explanation (i.e., this list is not exhaustive). However, this list is sufficient to provide the main arguments for adopting the counterfactual account.

5. Of course, like causal relevance, this kind of relevance and irrelevance can come in degrees.

6. Thanks to two anonymous reviewers for suggesting that I clarify this point here.

7. While I have some sympathies with Humean motivations for making this distinction, they are not driving the separation here. Rather, the distinction is motivated by the need to account for noncausal explanation in science.

8. I won't worry about the details of those cases here, but I think the account that I defend can take on much of what Woodward (2003) says about these cases. All that is required for my purposes is that in many cases, we can evaluate various counterfactual situations, and this dependence relation is typically asymmetric in the way that is required to explain.

9. It is also worth noting that rather than universally privileging dependence relationships at any particular level of description, the counterfactual approach allows that relationships of counterfactual dependence can exist at multiple levels.

10. Thanks to an anonymous reviewer for encouraging me to address this objection earlier in the chapter.

11. In the same paper, Woodward seems to agree that in other cases, irrelevance information is playing a more substantial explanatory role (Woodward 2018, 130). Thus, the disagreement here may simply concern the requirement that some explanations must explicitly include information about the irrelevance of certain features to the occurrence of the explanandum.

12. In addition, it is worth noting that in most cases that involve a common cause, the counterfactual situations of interest to us will be those in which there is a hypothetical intervention on some variable of the system that influences the explanandum (Woodward 2003, 14–15). That is, in cases in which there is a common cause, the relevant set of counterfactual information typically will be only the set of counterfactuals involved in evaluating Woodward-style interventions. There is no reason why a more general counterfactual account of explanation cannot adopt the kinds of holding-fixed solutions proposed by various causal accounts, given that in these cases we are clearly interested in counterfactuals grounded in causal dependence. However, given that not all counterfactual situations of interest to scientists can be produced by a hypothetical surgical intervention, the set of counterfactuals involved in the explanation will not always include the counterfactuals that hold under interventions. Consequently, while additional counterfactuals about what would happen under interventions will often help us address instances of causal explanation where there is a common cause, this does not mean that all explanations will need to involve interventionist counterfactuals.

13. However, even if we grant that some preemption cases are genuine counterexamples to a counterfactual account of explanation, I think the arguments given here still provide a strong motivation for adopting the counterfactual account over other existing accounts, which are subject to their own sets of counterexamples (e.g., Rice and Rohwer 2020).

14. In doing so, the why question and contrast class can also pick out a particular kind of counterfactual explanation (e.g., causal versus statistical) that is required to explain the explanandum.

15. See Rice (manuscript) for a more detailed discussion of how this feature of my view goes beyond what other accounts have said about the role of irrelevant features in explanation.

16. This also allows us to make some sense of talk involving the discovery of explanations that enable an agent to acquire understanding. Explanations are constituted by sets of true modal information, but an agent who grasps the workings of a model that provides an explanation can discover this modal information.

17. Kitcher (1981) uses Darwin's theory as an example.

18. Thanks to an anonymous reviewer for raising this possible objection.

19. I will be considering understanding of a real-world phenomenon only because I assume that this is the understanding provided by scientific explanations. Insofar as one can understand theories, models, examples, and people, that will be outside the scope of this discussion.

20. This is often taken to be a key feature that distinguishes understanding from knowledge because the unit of knowledge is typically thought to be individual propositions.

21. These dependence relations can hold between events, states, and properties within the system (Kim 1994).

22. For my purposes, the modifier *scientific* here is only intended to imply that the understanding is produced by scientific inquiry.

23. Moreover, perhaps idealizations of those features would make the explanation worse because the very act of including irrelevant factors might misleadingly suggest that those features *do* matter.

24. As further evidence of the problem, only requiring a single counterfactual dependence leaves Reutlinger's account vulnerable to counterexamples such as the counterfactual dependence of storms on barometer readings.

25. In fact, some have argued that explanations of universality itself are counterexamples to accounts that require an explanation to cite a relationship of counterfactual dependence (Khalifa, Doble, and Millson 2019).

26. Thanks to an anonymous reviewer for pressing me on this point.

Chapter 5

1. I also think that the realism debate should avoid having philosophical accounts of realism imposing metaphysical structures on the world a priori. Instead, our

metaphysical claims about the world ought to be epistemically justified by first looking at the nature of our models/theories and how they are able to relate to the world.

2. The law of large numbers states that as the sample size increases, the average of the quantities sampled will be closer and closer to the expected outcome.

3. It is possible, however, that removing or replacing the idealizing assumptions would result in a modified version of the model, a different kind of model, or a different explanation. The most important point here is not whether the idealizations can be eliminated from the model, but rather the pervasive nature of the distortions that they introduce due to their role in the foundational mathematical frameworks of the model. This is enough to demonstrate the failure of the decompositional assumption.

4. Thanks to an anonymous reviewer for raising this possible objection.

5. Potochnik's (2015) account in which idealizations can be used to frame the system in different ways that make different causal patterns accessible is similar in that the reframing of the system make positive epistemic contributions. The key difference between our views is that I would resist the claim that this is always done by enabling the model system to display a causal pattern.

6. This is not intended to suggest that no philosophers have discussed some of the positive contributions of idealizations (e.g., Potochnik 2017; Strevens 2008), but only that we should focus on the development of those accounts in line with the holistic distortion view, which allows for the distortion or difference-makers that are contextually salient instead of accounts that require the assumptions of the decompositional strategy.

7. I do not intend to suggest that these are the only goals of scientific inquiry. They are merely representative of the kinds of goals that scientists have been able to achieve with holistically distorted models.

8. Of course, which modeling tools we have available might very well be a function of our cognitive limitations. However, in many cases, I suggest that the problem is the limitations of mathematical tools themselves rather than our inability to comprehend a model.

9. Holistically distorted models can also allow for predictions that otherwise would be inaccessible, but that kind of contribution is outside the scope of my discussion here.

Chapter 6

1. Here it is also important to remember that scientists' interactions with scientific models can reveal counterfactual information, even if that information is not directly reflected in the model (see section 4.6.2).

2. Indeed, scientists often explain by appealing to idealized models that are in the same universality class as their target systems before attempting any explicit delimiting or explaining of universality.

3. This response to Lange (2014) on behalf of Batterman and Rice (2014) comes from an excellent paper by Travis McKenna (forthcoming). Thanks to that paper I was able to see how this response to Elgin and Sober (2002) also applied to the objections raised by Lange.

4. This is important because explanations ought to specify the scope of the dependencies they cite and demonstrate that the relationships being cited are stable with respect to changes to features of the system that are irrelevant (Potochnik 2017; Woodward 2003).

5. There are two main versions of the Eden Growth Model. The first assumes that all perimeter sites have the same probability of being occupied, while the second assumes that all open bonds are occupied with equal probability.

6. These growth rules can be modified in various ways. I use Eden's original growth rules for simplicity, but none of what follows depends on the particular patterns obtained from those rules.

7. The Eden model could be considered a minimal model. However, just because it is minimal does not mean that the explanation provided is a *minimal model explanation*. For details, see chapter 3 of this book and Batterman and Rice (2014).

8. This discussion of Gaussian universality can be generalized to many other instances of statistical modeling in which we learn about the universality class by investigating limit theorems (Frank 2009, 1565). For example, Benford's law and Zipf's law might be treated in the same way (Tao 2012, 26).

Chapter 7

1. For example, the characteristic time scale for models of nuclear inactions will be on the order of microseconds, but if scientists want to model glacial movement, the characteristic time scale will be years or longer. Similar considerations apply to spatial scales that can range from the size of individual particles to the entire universe.

2. Thanks to an anonymous reviewer for pressing me to be clearer about how to individuate these classes from one another.

3. Moreover, some of these models aim to represent dynamical interactions of the system with well-defined functions, while others investigate the same behaviors by simply assigning idealized computational algorithms to the elements of the system and studying the emerging behaviors (Qu et al. 2011).

4. In addition, however, the inconsistencies among those models often result in substantial modeling challenges within the practice of multiscale modeling (Wilson 2017, 204). That is, the problem of inconsistent models also often generates practical methodological challenges for multiscale modelers.

5. For example, some models incorporate noise in conflicting ways that can give rise to what is called *Keizer's paradox* (Qu et al. 2011, 26).

6. This kind of modeling technique was recently discussed by Hillerbrand (2015). One of her simpler examples is the use of a two-body model to study the motion of the Earth around the Sun.

7. It is also important to note that universality sometimes occurs in the opposite direction because sometimes features and processes at smaller scales are stable with respect to changes to features at larger scales. By determining which of the features of the system are dependent only on a particular (or a few) scales of the system, modelers can use models designed specifically for those scales and processes to model these scale-dependent features.

8. In cases where this occurs across discrete scales of the system, these features are called *fractals*, which are often defined as structures without characteristic scales. In biology alone, fractal structures have been identified in bones, the circulatory system, and the lungs (Gisiger 2001, 165).

9. Some philosophers have also noted that "the question of how complex systems are formed and evolve via pattern formation requires multi-scale descriptions using mathematical tools such as scaling laws" (Morrison 2014, 1144).

10. In these cases, the phase transitions characterized by the power law are typically continuous, similar to the transitions observed in ferromagnets and water when it is raised above its critical point.

11. This idea is echoed by Stanley et al. as well: "It is beginning to appear that a number of ecological phenomena obey regular laws which are scale invariant, and that these laws are universal in the sense that they do not depend on details concerning the actual species" (Stanley et al. 2000, 63).

12. What is more, the features or patterns that are stable or dependent within particular subsystems are often not the same features or patterns that are stable after those subsystems are aggregated together by our modeling techniques and allowed to change over time.

13. For example, Qu et al. (2011) discuss the use of mean field theory, coarse graining, homogenization, self-organization theories, and systems biological approaches.

14. Rather than working through another detailed example, I refer interested readers to the earlier discussions of minimal model explanations in chapter 3, the use of renormalization to extract counterfactual information at the end of chapter 4, or Batterman's (2002b) excellent discussion of renormalization group methods.

15. In order to model these features, these modelers make another idealizing assumption: the landscape is one-dimensional and can be partitioned into interval patches (y_{i-1}, y_i).

16. While not all multiscale modeling involves emergence, in many cases the presence of universality is a hallmark of emergent behavior that is used to justify the use of idealized multiscale modeling techniques to understand the emergence of complex behaviors (Morrison 2018b).

17. Thanks to Julia Bursten for pushing me to emphasize this point.

18. Bursten (2018a) argues for what she calls a conceptual strategies account of how to make sense of these relationships. What I say here is compatible with Bursten's account, I think, although I focus on the role of universality in justifying mathematical modeling strategies, whereas she focuses on the role of our conceptualizations of the problem and the representational features of the models in order to connect different theoretical frameworks. However, in many ways, my account provides the details of what some of the conceptual strategies used in multiscale modeling are and so builds on Bursten's account.

Chapter 8

1. While I agree with Lipton (2009) on this point, I find some of his examples unconvincing. Moreover, I think Lipton's most convincing examples are focused on showing that a merely potential explanation (i.e., something that would be an actual explanation if it were true) can produce understanding. In contrast, I do not think we should restrict scientific understanding to understanding why something is the case; rather, I think that there are numerous cases from scientific practice that show how understanding is possible without providing a potential explanation.

2. Potochnik (2017) raises this objection to the arguments provided in Rohwer and Rice (2013). I agree that we were a bit too focused on what the model accurately represents in that paper, but the point that these models fail to be accurate representations is still important, given that many accounts have required such accurate representation for explanation (see chapter 2 for details).

3. Thanks to an anonymous reviewer who suggested this possibility.

4. In addition, Grimm (2006), Kvanvig (2003), and Pritchard (2009) all argue for a factive notion of understanding that requires at least some of the propositions in one's understanding be true.

5. These are importantly different claims because an idealization might be permanent but be restricted to the periphery of the theory. I agree with Elgin (2017) that many of the idealizations are central to our best models and theories, and that these idealizations are often not eliminated, or even cannot be eliminated without destroying the understanding provided by the model or theory.

6. Thanks to Catherine Elgin for helping me to clarify and emphasize this crucial point.

Notes to Chapter 9

7. Thanks to Catherine Elgin for raising this concern and pushing me to clarify my thoughts here a bit more.

8. While explanations might be better or worse, or perhaps can be deepened, whether an explanation has been provided is typically treated as a threshold phenomenon.

9. Juha Saatsi (2019) has defended a similar kind of view, according to which scientific realism is compatible with science aiming at understanding.

10. The extraction of this modal information will require scientists to have various skills. I will not discuss these skills in any real detail here, but I think the account here is consistent with most of the skills that de Regt (2017) argues are important for using a model, theory, or explanation in order to understand a phenomenon.

11. For example, Saatsi argues that the theories of Descartes, Newton, Young, and Fresnel all provide understanding because they target different aspects of the rainbow and consequently can answer different sets of "What if things had been different?" questions (Saatsi 2019, 72).

12. These different explanations might include or emphasize different pieces of counterfactual information.

13. This is the case even within many parts of social epistemology that only focus on placing individuals into a larger social context.

Chapter 9

1. This actually combines two of Potochnik's questions, but my answer addresses both, so I have combined them here.

2. I also think that it is a mistake to try and clearly demarcate fields of science such that there are distinctively biological or chemical phenomena, but those arguments will have to be provided elsewhere.

3. Potochnik (2015, 2017) is an important exception here.

4. For example, Strevens suggests that the explanation of the pattern "All ravens are black" first cites the causal mechanism used to explain a single instance that connects some property or cluster of properties, P, with blackness. The pattern explanation then simply adds the basing-generalization that "All ravens have P" (Strevens 2008, 229). He then goes on to argue that this same line of reasoning applies to pattern explanations in higher-level sciences such as biology, psychology, and economics.

References

Achinstein, P. 1983. *The Nature of Explanation.* New York: Oxford University Press.

Agyingi, E., Wakabayashi, L., Wiandt, T., and Maggelakis, S. 2018. Eden model simulation of re-epithelialization and angiogenesis of an epidermal wound. *Processes 6*: 1–17, doi: 10.3390/pr6110207.

Alexandrowicz, Z. 1980. Critically branched chains and percolation clusters. *Physics Letters A 80*: 284–286.

Appiah, K. A. 2017. *As If: Idealization and Ideals.* Cambridge, MA: Harvard University Press.

Ariew, A., Rice, C., and Rohwer, Y. 2015. Autonomous statistical explanations and natural selection. *British Journal for the Philosophy of Science 66*(3): 635–658.

Ariew, A., Rohwer, Y., and Rice, C. 2017. Galton, reversion and the quincunx: The rise of statistical explanation. *Studies in History and Philosophy of Biological and Biomedical Sciences 66*: 63–72.

Baker, A. 2005. Are there genuine mathematical explanations of physical phenomena? *Mind 114*: 223–238.

Baker, A. 2009. Mathematical explanation in science. *British Journal for the Philosophy of Science 60*(3): 611–633.

Barenblatt, G. I. 1996. *Scaling, Self-similarity, and Intermediate Asymptotics.* Cambridge: Cambridge University Press.

Baron, S., Colyvan, M., and Ripley, D. 2017. How mathematics can make a difference. *Philosophers' Imprint 17*: 1–19.

Barzel, B., and Barabási, A. 2013. Universality in network dynamics. *Nature Physics 9*: 673–681.

Bascompte, J., and Solé, R. V. 1995. Rethinking complexity: Modeling spatiotemporal dynamics in ecology. *TREE 10*(9): 361–366.

Batterman, R. W. 2000. Multiple realizability and universality. *British Journal of Philosophy of Science 51*(1): 115–145.

Batterman, R. W. 2002a. Asymptotics and the role of minimal models. *British Journal for the Philosophy of Science 53*(1): 21–38.

Batterman, R. W. 2002b. *The Devil in the Details: Asymptotic Reasoning in Explanation, Reduction, and Emergence*. Oxford: Oxford University Press.

Batterman, R. W. 2005. Critical phenomena and breaking drops: Infinite idealizations in physics. *Studies in History and Philosophy of Modern Physics 36*: 225–244.

Batterman, R. W. 2009. Idealization and modeling. *Synthese 169*(3): 427–446.

Batterman, R. W. 2010. On the explanatory role of mathematics in empirical science. *British Journal for the Philosophy of Science 61*(1): 1–25.

Batterman, R. W. 2011. Emergence, singularities, and symmetry breaking. *Foundations of Physics 41*: 1031–1050.

Batterman, R. W. 2013. The tyranny of scales. In *The Oxford Handbook of Philosophy of Physics*, ed. R. Batterman, 255–286. Oxford: Oxford University Press.

Batterman, R. W. 2015. Autonomy and scales. In *Why More Is Different: The Frontiers Collection*, ed. B. Falkenburg and M. Morrison, 115–135. Berlin: Springer.

Batterman, R. W. 2017. Autonomy of theories: An explanatory problem. *Noûs 52*(4): 858–873.

Batterman, R. W. 2019. Universality and RG explanations. *Perspectives on Science 27*(1): 26–47.

Batterman, R. W., and Rice, C. 2014. Minimal model explanations. *Philosophy of Science 81*(3): 349–376.

Beatty, J. 1980. Optimal-design models and the strategy of model building in evolutionary biology. *Philosophy of Science 74*(4): 532–561.

Beatty, J. 1995. The evolutionary contingency thesis. In *Concepts, Theories, and Rationality in the Biological Sciences: The Second Pittsburgh-Konstanz Colloquium in the Philosophy of Science*, ed. G. Wolters and J. Lennox, 45–81. Pittsburgh: University of Pittsburgh Press.

Bechtel, W. 2015. Can mechanistic explanation be reconciled with scale-free constitution and dynamics? *Studies in History and Philosophy of Biological and Biomedical Sciences 53*: 84–93.

Bechtel, W. 2017. Analysing network models to make discoveries about biological mechanisms. *British Journal for the Philosophy of Science 70*(2): 459–484.

Bechtel, W., and Abrahamsen, A. 2005. Explanation: A mechanist alternative. *Studies in History and Philosophy of Biological and Biomedical Sciences 36*: 421–441.

Bechtel, W., and Abrahamsen, A. 2010. Dynamic mechanistic explanation: Computational modeling as an exemplar for cognitive science. *Studies in History and Philosophy of Science 41*(3): 321–333.

Bechtel, W., and Richardson, R. C. 1993. *Discovering Complexity: Decomposition and Localization as Strategies in Scientific Research.* Princeton, NJ: Princeton University Press.

Beckner, M. 1968. *The Biological Way of Thought.* Los Angeles: University of California Press.

Belmonte-Beitia, J., Woolley, T. E., Scott, J. G., Maini, P. K., and Gaffney, E. A. 2013. Modelling biological invasions: Individual to population scales at interfaces. *Journal of Theoretical Biology 334*: 1–12.

Benguigui, L. 1995. A new aggregation model: Application to town growth. *Physica A: Statistical Mechanics and Its Applications 219*: 13–26.

Bishop, R. C. 2008. Downward causation in fluid convection. *Synthese 160*: 229–248.

Böhm, H. J. 2016. A short introduction to basic aspects of continuum micromechanics. *Institute of Lightweight Design and Structural Biomechanics (ILSB) Report 207.* http://www.ilsb.tuwien.ac.at/links/downloads/ilsbrep206.pdf.

Bokulich, A. 2008. *Reexamining the Quantum-Classical Relation: Beyond Reductionism and Pluralism.* Cambridge: Cambridge University Press.

Bokulich, A. 2011. How scientific models can explain. *Synthese 180*: 33–45.

Bokulich, A. 2012. Distinguishing explanatory from nonexplanatory fictions. *Philosophy of Science 79*(5): 725–737.

Bokulich, A. 2016. Fiction as a vehicle for truth: Moving beyond the ontic conception. *The Monist 99*: 260–279.

Bokulich, A. 2018. Searching for non-causal explanation in a sea of causes. In *Explanation beyond Causation: Philosophical Perspectives on Non-causal Explanation*, ed. J. Saatsi and A. Reutlinger, 141–163. Oxford: Oxford University Press.

Bonachela, J. A., Nadell, C. D., Xavier J. B., and Levin, S. A. 2011. Universality in bacterial colonies. *Journal of Statistical Physics 144*(2): 303–315.

Boyd, K. 2019. Group understanding. *Synthese*, https://doi.org/10.1007/s11229-019-02492-3.

Brandon, R. 1990. *Adaptation and Environment.* Princeton, NJ: Princeton University Press.

Brigandt, I. 2013. Systems biology and the integration of mechanistic explanation and mathematical explanation. *Studies in History and Philosophy of Biological and Biomedical Sciences 44*: 477–492.

Bromberger, S. 1966. Questions. *Journal of Philosophy* 63(20): 597–606.

Buchanan, M. 2014. Equivalence principle. *Nature Physics* 10: 543.

Bueno, O., and Colyvan, M. 2011. An inferential conception of the application of mathematics. *Noûs* 45(2): 345–374.

Burnston, D. C. 2017. Real patterns in biological explanation. *Philosophy of Science* 84(5): 879–891.

Burnston, D. C. 2019. Review of Angela Potochnik's *Idealization and the Aims of Science*. *Philosophy of Science* 86(3): 577–583.

Bursten, J. R. 2018a. Conceptual strategies and inter-theory relations: The case of nanoscale cracks. *Studies in History and Philosophy of Modern Physics* 62: 158–165.

Bursten, J. R. 2018b. Smaller than a breadbox: Scale and natural kinds. *British Journal for the Philosophy of Science* 69(1): 1–23.

Byrne, H., and Drasdo, D. 2009. Individual-based and continuum models of growing cell populations: A comparison. *Journal of Mathematical Biology* 58(4–5): 657–687.

Carruthers, P. 2006. *The Architecture of the Mind: Massive Modularity and the Flexibility of Thought*. Oxford: Oxford University Press.

Cartwright, N. 1983. *How the Laws of Physics Lie*. Oxford: Oxford University Press.

Cartwright, N. 2004. Causation: One word, many things. *Philosophy of Science* 71(5): 805–819.

Castiglione, F., Pappalardo, F., Bianca, C., Russo, G., and Motta, S. 2014. Modeling biology spanning different scales: An open challenge. *Biomed Research International* 2014: 1–9.

Chakravartty, A. 2010. Perspectivism, inconsistent models, and contrastive explanation. *Studies in History and Philosophy of Science* 41(4): 405–412.

Charnov, E. L. 1982. *The Theory of Sex Allocation*. Princeton, NJ: Princeton University Press.

Charnov, E. L. 1989. Phenotypic evolution under Fisher's fundamental theorem of natural selection. *Heredity* 62: 113–116.

Chirimuuta, M. 2014. Minimal models and canonical neural computations: The distinctness of computational explanation in neuroscience. *Synthese* 191(2): 127–153.

Chirimuuta, M. 2018. Explanation in computational neuroscience: Causal and noncausal. *British Journal for the Philosophy of Science* 69(3): 849–880.

Chopard, B., Borgdorff, J., and Hoekstra, A. G. 2014. A framework for multi-scale modelling. *Philosophical Transactions of the Royal Society* 372: 20130378.

References

Churchland, P. 2013. *Touching a Nerve: Self as Brain*. New York: Norton.

Collins, P. H. 1986. Learning from the outsider within: The sociological significance of black feminist thought. *Social Problems 33*(6): S14–S32.

Collins, P. H. 2000. *Black Feminist Thought*. New York: Routledge.

Colyvan, M., Cusbert, J., and McQueen, J. 2018. Two flavours of mathematical explanation. In *Explanation beyond Causation: Philosophical Persepctives on Non-causal Explanations*, ed. A. Reutlinger and J. Saatsi, 206–227. Oxford: Oxford University Press.

Conee, E., and Feldman, R. 2011. Replies. In *Evidentialism and Its Discontents*, ed. T. Dougherty, 283–323. Oxford: Oxford University Press.

Cooper, M. M. 2015. Why ask why? *Journal of Chemical Education 92*: 1273–1279.

Corsano, G., Montagna, J. M., Iribarren, O., and Aguirre, P. 2009. *Mathematical Modeling Approaches for Optimization of Chemical Processes*. New York: Nova Science Publishers.

Corwin, I. 2016. Kardar-Parisi-Zhang universality. *Notices of the American Mathematical Society 63*(3): 230–239.

Craver, C. 2006. When mechanistic models explain. *Synthese 153*: 355–376.

Craver, C. 2007. *Explaining the Brain: Mechanisms and the Mosaic Unity of Neuroscience*. Oxford: Oxford University Press.

Craver, C. 2014. The ontic account of scientific explanation. In *Explanation in the Special Sciences: The Case of Biology and History*, ed. M. Kaiser, O. Scholz, D. Plenge, and A. Hüttemann, 27–52. Dordrecht, Netherlands: Springer.

Craver, C., and Darden, L. 2013. *In Search of Mechanisms: Discoveries across the Life Sciences*. Chicago: University of Chicago Press.

Craver, C., and Kaplan, D. 2020. Are more details better? On the norms of completeness for mechanistic explanations. *British Journal for the Philosophy of Science 71*(1): 287–319.

Dada, J., and Mendes, P. 2011. Multi-scale modelling and simulation in systems biology. *Integrative Biology 3*: 86–96.

Dallon, J. C. 2010. Multiscale modeling of cellular systems in biology. *Current Opinion in Colloid & Interface Science 15*: 24–31.

Daniel, K. J., Heggie, D. C., and Varri, A. L. 2017. An approximate analytic model of a star cluster with potential escapers. *Monthly Notices of the Royal Astronomical Society 468*(2): 1453–1473.

Davidson, L. A., von Dassow, M. ,and Zhou, J. 2009. Multi-scale mechanics from molecules to morphogenesis. *International Journal of Biochemistry & Cell Biology 41*(11): 2147–2162.

Deisboeck, T. S., Wang, Z., Macklin, P, and Cristini, V. 2011. Multiscale cancer modeling. *Annual Review of Biomedicine, 13*, doi:10.1146/annurev-bioeng-071910-124729.

de Mazancourt, C., and Dieckmann, U. 2004. Trade-off geometries and frequency-dependent selection. *The American Naturalist 164*(6): 765–778.

DePaul, M., and Grimm, S. 2007. Review essay on Jonathan Kvanvig's *The Value of Knowledge and the Pursuit of Understanding*. *Philosophy and Phenomenological Research* 74(2): 498–514.

de Regt, H. W. 2009a. The epistemic value of understanding. *Philosophy of Science 76*(5): 585–597.

de Regt, H. W. 2009b. Understanding and scientific explanation. In *Scientific Understanding: Philosophical Perspectives*, ed. H. W. de Regt, S. Leonelli, and K. Eigner, 21–42. Pittsburgh: University of Pittsburgh Press.

de Regt, H. W. 2017. *Understanding Scientific Understanding*. New York: Oxford University Press.

de Regt, H. W., Leonelli, S., and Eigner, K. (eds.). 2009. *Scientific Understanding: Philosophical Perspectives*. Pittsburgh: University of Pittsburgh Press.

D'Humiéres, D., and Lallemand, P. 1986. Lattice gas automata for fluid mechanics. *Physica* 140: 326–335.

Douglas, H. 2009. *Science, Policy, and the Value-Free Ideal*. Pittsburgh: University of Pittsburgh Press.

Eden, M. 1961. A two-dimensional growth process. In *Proceedings of the 4th Berkeley Symposium on Mathematical Statistics and Probability*, vol. 4: Contributions to Biology and Problems of Medicine, 223–239. Berkeley: University of California Press.

Elgin, C. Z. 2007. Understanding and the facts. *Philosophical Studies 132*: 33–42.

Elgin, C. Z. 2017. *True Enough*. Cambridge, MA: MIT Press.

Elgin, M., and Sober, E. 2002. Cartwright on explanation and idealization. *Erkenntnis* 57: 441–450.

Fillion, N., and Moir, R. H .C. 2018. Explanation and abstraction from a backward-error analytic perspective. *European Journal for Philosophy of Science 8*(3): 735–759.

Fisher, M. E. 1983. Scaling, universality and renormalization group theory. In *Critical Phenomena*, vol. 186 of *Lecture Notes in Physics*, ed. F. J. W. Hahne. Summer School held at the University of Stellenbosch, South Africa, January 18–29, 1–139. Berlin: Springer Verlag.

Fisher, R. A. 1922. On the Dominance Ratio. *Proceedings of the Royal Society of Edinburgh 42*: 321–341.

References

Fisher, R. A. 1930. *The Genetical Theory of Natural Selection*. Oxford: Clarendon Press.

Fodor, J. 1974. Special sciences; or, the disunity of science as a working hypothesis. *Synthese 28*(2): 97–115.

Forber, P. 2010. Confirmation and explaining how possible. *Studies in the History and Philosophy of Biological and Biomedical Sciences* 41: 32–40.

Frank, S. A. 2009. The common patterns of nature. *Journal of Evolutionary Biology 22*: 1563–1585.

Friedman, M. 1974. Explanation and scientific understanding. *Journal of Philosophy 71*: 5–19.

Galton, F. 1877. Typical laws of heredity. *Nature 15*: 492–495.

Galton, F. 1886. Regression towards mediocrity in hereditary stature. *Journal of the Anthropological Institute of Great Britain and Ireland 15*: 246–263.

Garfinkel, A. 1981. *Forms of Explanation: Rethinking the Questions in Social Theory*. New Haven, CT: Yale University Press.

Gerisch, A., Penta, R., and Lang, J. 2017. *Multiscale Models in Mechano and Tumor Biology*. Cham, Switzerland: Springer.

Giere, R. 1988. *Explaining Science: A Cognitive Approach*. Chicago: University of Chicago Press.

Giere, R. 2006. *Scientific Perspectivism*. Chicago: University of Chicago Press.

Giere, R. 2010. An agent-based conception of models and scientific representation. *Synthese 172*: 269–281.

Gillespie, J. H. 1974. Natural selection for within-generation variance in offspring number. *Genetics 76*: 601–606.

Gillespie, J. H. 1977. Natural selection for variances in offspring numbers: A new evolutionary principle. *American Naturalist 111*: 1010–1014.

Gisiger, T. 2001. Scale invariance in biology: Coincidence or evidence of a universal mechanism? *Biological Review 76*: 161–209.

Glennan, S. 2002. Rethinking mechanistic explanation. *Philosophy of Science 69*(S3): 342–353.

Glennan, S. 2017. *The New Mechanical Philosophy*. Oxford: Oxford University Press.

Godfrey-Smith, P. 2003. *Theory and Reality*. Chicago: University of Chicago Press.

Godfrey-Smith, P. 2009. Abstractions, idealizations, and evolutionary biology. In *Mapping the Future of Biology: Evolving Concepts and Theories*, ed. A. Barberousse, M. Morange, and T. Pradeu, 47–55. Boston: Springer.

Golden, K. M. 2014. Mathematics of sea ice. In *The Princeton Companion to Applied Mathematics*, ed. N. J. Higham, M. Dennis, P. Glendinning, F. Santosa, and J. Tanner, 694–705. Princeton, NJ: Princeton University Press.

Goldenfeld, N., and Kadanoff, L. P. 1999. Simple lessons from complexity. *Science* 284: 87–89.

Goss-Custard, J. D. 1977. Predator responses and prey mortality in Redshank, *Tringa tetanus* (L.), and a preferred prey, *Corophium volutator* (Pallas). *Journal of Animal Ecology* 46(1): 21–35.

Green, S. 2013. When one model is not enough: Combining epistemic tools in systems biology. *Studies in History and Philosophy of Biological and Biomedical Sciences* 44: 170–180.

Green, S., and Batterman, R. 2017. Biology meets physics: Reductionism and multiscale modeling of morphogenesis. *Studies in History and Philosophy of Biological and Biomedical Sciences* 61: 20–34.

Grene, M. 1961. Statistics and selection. *British Journal for the Philosophy of Science* 12(45): 25–42.

Grimm, S. 2006. Is understanding a species of knowledge? *British Journal for the Philosophy of Science* 57(3): 515–535.

Grimm, S. 2008. Explanatory inquiry and the need for explanation. *British Journal for the Philosophy of Science* 59(3): 481–497.

Grimm, S. 2011. Understanding. In *The Routledge Companion to Epistemology*, ed. Duncan Pritchard and Sven Berneker, 84–94. New York: Routledge.

Grimm, S. 2012. The value of understanding. *Philosophy Compass* 7(2): 103–117.

Gross, F. 2015. The relevance of irrelevance: Explanation in systems biology. In *Explanation in Biology: An Enquiry into the Diversity of Explanatory Patterns in the Life Sciences*, ed. P. Braillard and C. Malaterre, 175–198. New York: Springer.

Grüne-Yanoff, T. 2009. Learning from minimal economic models. *Erkenntnis* 70: 81–99.

Hacking, I. 1990. *The Taming of Chance*. Cambridge: Cambridge University Press.

Halliday, D., Resnick, R., and Walker, J. 2011. *Fundamentals of Physics*. 9th ed. Hoboken, NJ: Wiley.

Hamilton, W. D. 1967. Extraordinary sex ratios. *Science* 156(3774): 477–488.

Hammerstein, P., and Selten, R. 1994. Game theory and evolutionary biology. In *Handbook of Game Theory*, vol. 2, ed. R.J. Aumann and S. Hart, 929–987. Amsterdam: Elsevier Science B.V.

References

Harding, S. 2004. A socially relevant philosophy of science? Resources from standpoint theory's controversiality. *Hypatia 19*(1): 25–47.

Harding, S. 2015. *Objectivity and Diversity: Another Logic of Scientific Research*. Chicago: Chicago University Press.

Harte, J., Kinzig, A. P., and Green, J. 1999. Self-similarity in the distribution and abundance of speices. *Science 286*: 334–336.

Harte, J., Smith, A. B., and Storch, D. 2009. Biodiversity scales from plots to biomes with a universal species-area curve. *Ecology Letters 12*: 789–797.

Hartmann, A. K., and Rieger, H. 2002. *Optimization Algorithms in Physics*. Berlin: Wiley.

Hartmann, S. 1998. Idealization in quantum field theory. In *Idealization in Contemporary Physics*, ed. N. Shanks, 99–122. Amsterdam: Rodopi.

Hartwell, L. H., Hood, L., Goldberg, M. L., Silver, L. M., and Veres, R.C. 2000. *Genetics: From Genes to Genomes*. Boston: McGraw-Hill.

Hausman, D., and Woodward, J. 1999. Independence, invariance, and the causal Markov condition. *British Journal for the Philosophy of Science 50*(4): 521–583.

Hausman, D. M., and Woodward, J. 2004. Modularity and the causal Markov condition: A restatement. *British Journal for the Philosophy of Science 55*(1): 147–161.

Hempel, C. 1965. *Aspects of Scientific Explanation*. New York: Free Press.

Hempel, C., and Oppenheim, P. 1948. Studies in the logic of explanation. *Philosophy of Science 15*(2): 135–175.

Henriksen, R. N. 2015. *Scale Invariance: Self-Similarity of the Physical World*. Weinheim, Germany: Wiley-VCH.

Hillerbrand, R. 2015. Explanation via micro-reduction: On the role of scale separation for quantitative modelling. In *Why More is Different: The Frontiers Collection*, ed. B. Falkenburg and M. Morrison, 69–88. Berlin: Springer.

Hitchcock, C., and Woodward, J. 2003. Explanatory generalizations, part II: Plumbing explanatory depth. *Noûs 37*: 181–199.

Hochstein, E. 2017. Why one model is never enough: A defense of explanatory holism. *Biology and Philosophy 32*: 1105–1125.

Hohenegger, C., Alali, B., Steffen, K. R., Perovich, D. K., and Golden, K. M. 2012. Transition in the fractal geometry of Arctic melt ponds. *The Cryosphere 6*: 1157–1162.

Hrdy, S. 1981. *The Woman That Never Evolved*. Cambridge, MA: Harvard University Press.

Hrdy, S. 1986. Empathy, polyandry, and the myth of the coy female. In *Feminist Approaches to Science*, ed. Ruth Bleier, 119–146. New York: Pergamon.

Hume, D. 1739. *A Treatise of Human Nature*. London: Oxford University Press.

Hume, D. 1999. *An Enquiry Concerning Human Understanding*, ed. Tom L. Beauchamp. Oxford: Oxford University Press.

Huneman, P. 2010. Topological explanations and robustness in biological sciences. *Synthese 177*: 213–245.

Jackson, F., and Pettit, P. 1992. In defense of explanatory ecumenism. *Economics and Philosophy 8*(1): 1–21.

Jhun, J. 2019. Economics, equilibrium methods, and multi-scale modeling. *Erkenntnis*, https://doi.org/10.1007/s10670-019-00113-6.

Jones, M. 2005. Idealization and abstraction: A framework. In *Idealization XII: Correcting the Model; Idealization and Abstraction in the Sciences*, ed. M. Jones and N. Cartwright, 173–218. Warsaw: Rodopi.

Kadanoff, L. 1966. Scaling laws for Ising models near T_c. *Physics 2*: 263–272.

Kadanoff, L. 1990. Scaling and universality in statistical physics. *Physica A 163*: 1–14.

Kadanoff, L. P. 2000. *Statistical Physics: Statics, Dynamics, and Renormalization*. Singapore: World Scientific.

Kadanoff, L. P. 2013. Theories of matter: Infinities and renormalization. In *The Oxford Handbook of Philosophy of Physics*, ed. Robert Batterman, 141–188. Oxford: Oxford University Press.

Kaplan, D. M. 2011. Explanation and description in computational neuroscience. *Synthese 183*: 339–373.

Kaplan, D. M., and Craver, C. F. 2011. The explanatory force of dynamical and mathematical models in neuroscience: A mechanistic perspective. *Philosophy of Science 78*(4): 601–627.

Kardar, M., Parisi, G., and Zhang, Y.C. 1986. Dynamic scaling of growth interfaces. *Physics Review Letters 56*(9): 889–892.

Keller, E. F. 1982. Feminism and science. *Signs 7*(3): 589–602.

Keller, E. F. 1983. *A Feeling for the Organism: The Life and Work of Barbara McClintock*. San Francisco: W. H. Freeman.

Keller, E. F. 1985. *Reflections on Gender and Science*. New Haven, CT: Yale University Press.

Kennedy, A. G. 2012. A non representationalist view of model explanation. *Studies in History and Philosophy of Science 43*(2): 326–332.

Khalifa, K. 2012. Inaugurating understanding or repackaging explanation? *Philosophy of Science 79*(1): 15–37.

Khalifa, K. 2013. The role of explanation in understanding. *British Journal for the Philosophy of Science 64*(1): 161–187.

Khalifa, K. 2017. *Understanding, Explanation, and Scientific Knowledge.* Cambridge: Cambridge University Press.

Khalifa, K., Doble, G., and Millson, J. 2019. Counterfactuals and explanatory pluralism. *British Journal for Philosophy of Science*, https://doi:10.1093/bjps/axy048.

Khaluf, Y., Ferrante, E., Simoens, P, and Huepe, C. 2017. Scale invariance in natural and artificial collective systems: A review. *J. R. Society Interface 14*: 20170662. http://dx.doi.org/10.1098/rsif.2017.0662.

Kim, J. 1994. Explanatory knowledge and metaphysical dependence. *Philosophical Issues 5*: 51–69.

Kitcher, P. 1981. Explanatory unification. *Philosophy of Science 48*(4): 507–531.

Kitcher, P. 1989. Explanatory unification and the causal structure of the world. In *Scientific Explanation*, ed. Philip Kitcher and Wesley Salmon, 410–505. Minneapolis: University of Minnesota Press.

Kitcher, P. 1993. *The Advancement of Science.* New York: Oxford University Press.

Knuuttila, T., and Merz, M. 2009. An objectual approach to scientific understanding. In *Scientific Understanding: Philosophical Perspectives*, ed. H. W. de Regt, S. Leonelli, and K. Eigner, 146–168. Pittsburgh: University of Pittsburgh Press.

Krebs, J. R., and Davies, N. B. 1993. *An Introduction to Behavioral Ecology.* 3rd ed. Oxford: Blackwell Scientific Publications.

Kuorikoski, J., Lehtinen, A., and Marchionni, C. 2010. Economic modeling as robustness analysis. *British Journal for the Philosophy of Science 61*(3): 541–567.

Kvanvig, J. L. 2003. *The Value of Knowledge and the Pursuit of Understanding.* New York: Cambridge University Press.

Kvanvig, J. L. 2009. Responses to critics. In *Epistemic Value*, ed. A. Haddock, A. Millar, and D. Pritchard, 339–353. New York: Oxford University Press.

Lange, M. 2013a. Really statistical explanations and genetic drift. *Philosophy of Science 80*(2): 169–188.

Lange, M. 2013b. What makes a scientific explanation distinctively mathematical? *British Journal for the Philosophy of Science 64*(3): 485–511.

Lange, M. 2014. On "minimal model explanations": A reply to Batterman and Rice. *Philosophy of Science 82*(2): 292–305.

Lange, M. 2017. *Because without Cause: Non-causal Explanations in Science and Mathematics.* Oxford: Oxford University Press.

Lange, M., and Rosenberg, M. 2011. Can there be a priori causal models of natural selection? *Australasian Journal of Philosophy* 89(4): 591–599.

Le Bihan, S. 2017. Enlightening falsehoods: A modal view of scientific understanding. In *Explaining Understanding*, ed. S. Grimm, C. Baumberger, and S. Ammon, 111–135. New York: Routledge.

Leonelli, S. 2009. Understanding in Biology: The Impure Nature of Biological Kinds. In *Scientific Understanding: Philosophical Perspectives*, ed. H. W. de Regt, S. Leonelli, and K. Eigne, 189–209. Pittsburgh: University of Pittsburgh Press.

Leplin, J. 1997. *A Novel Defense of Scientific Realism*. New York: Oxford University Press.

Levins, R. 1966. The strategy of model building in population biology. *American Scientist* 54: 421–431.

Levy, A. 2011. Makes a difference: Review of Michael Strevens' Depth: An Account of Scientific Explanation. *Biology and Philosophy* 26: 459–467.

Levy, A. 2015. Modeling without models. *Philosophical Studies* 172: 781–798.

Levy, A., and Bechtel, W. 2012. Abstraction and the organization of mechanisms. *Philosophy of Science* 80(2): 241–261.

Lewis, D. 1986. Causal explanation. In *Philosophical Papers*, vol. 2, 214–240. Oxford: Oxford University Press.

Lewontin, R. C. 1961. Evolution and the theory of games. *Journal of Theoretical Biology* 1(3): 382–403.

Lipton, P. 1991. *Inference to the Best Explanation*. New York: Routledge.

Lipton, P. 2009. Understanding without explanation. In *Scientific Understanding: Philosophical Perspectives*, ed. Henk W. de Regt, Sabina Leonelli, and Kai Eigner, 43–63. Pittsburgh: University of Pittsburgh Press.

Lombrozo, T. 2011. The instrumental value of explanations. *Philosophy Compass* 6: 539–551.

Longino, H. 1990. *Science as Social Knowledge*. Princeton, NJ: Princeton University Press.

Longino, H. 2002. *The Fate of Knowledge*. Princeton, NJ: Princeton University Press.

Longino, H. 2013. *Studying Human Behavior: How Scientists Investigate Aggression and Sexuality*. Chicago: Chicago University Press.

Love, A., and Nathan, M. J. 2015. The idealization of causation in mechanistic explanation. *Philosophy of Science* 82(5): 761–774.

Machamer, P. K., Darden, L., and Craver, C. 2000. Thinking about mechanisms. *Philosophy of Science* 67(1): 1–25.

References

Majumdar, S. N., and Nechaev, S. 2004. Anisotropic ballistic deposition model with links to the Ulam problem and the Tracy-Widom distribution. *Physics Review E 69*: 011103.

Majumdar S. N., and Schehr, G. 2013. Top eigenvalue of a random matrix: Large deviations and third order phase transitions. *Journal of Statistical Mechanics: Theory and Experiment 2014*: 10–12.

Massimi, M. 2018. Perspectival modeling. *Philosophy of Science 85*(3): 335–359.

Matthen, M., and Ariew, A. 2009. Selection and causation. *Philosophy of Science 76*(2): 201–224.

Matthewson, J., and Calcott, B. 2011. Mechanistic models of population-level phenomena. *Biology and Philosophy 26*(5): 737–756.

Matthewson, J., and Weisberg, M. 2009. The structure of tradeoffs in model building. *Synthese 170*(1): 169–190.

Maynard Smith, J. 1978. Optimization theory in evolution. *Annual Review of Ecological Systems 9*: 31–56.

Maynard Smith, J. 1982. *Evolution and the Theory of Games*. Cambridge: Cambridge University Press.

Maynard Smith, J., and Price, G. A. 1973. The logic of animal conflict. *Nature 246*: 15–18.

McKenna, T. forthcoming. Marc Lagne and minimal model explanations: A defense of Batterman and Rice. *Philosophy of Science*.

McMullin, E. 1985. Galilean idealization. *Studies in History and Philosophy of Science 16*(3): 247–273.

Meier-Schellersheim, M., Fraser, I. D. C., and Klaushcen, F. 2009. Multi-scale modeling in cell biology. *Wiley Interdisciplinary Review of Systems Biology Medicine 1*(1): 4–14.

Millstein, R. L. 2006. Natural selection as a population-level causal process. *British Journal of the Philosophy of Science 57*(4): 627–653.

Milton, G. W. 2002. *Theory of Composites*. Cambridge: Cambridge University Press.

Mitchell, S. D. 2008. Exporting causal knowledge in evolutionary and developmental biology. *Philosophy of Science 75*(3): 697–706.

Mitchell, S. D. 2009. *Unsimple Truths: Science, Complexity, and Policy*. Chicago: University of Chicago Press.

Mitchell, S. D. 2012. Emergence: Logical, functional and dynamical. *Synthese 185*: 171–186.

Mizrahi, M. 2012. Idealizations and scientific understanding. *Philosophical Studies 160*: 237–252.

Moore, T. A. 2003. *Six Ideas That Shaped Physics*. 2nd ed. New York: McGraw-Hill.

Morrison, M. 1996. Physical models and biological contexts. *Philosophy of Science* 64(Suppl.): S315–S324.

Morrison, M. 2002. Modelling populations: Pearson and Fisher on Mendelism and biometry. *British Journal for the Philosophy of Science* 53(1): 39–68.

Morrison, M. 2004. Population genetics and population thinking: Mathematics and the role of the individual. *Philosophy of Science* 71(5): 1189–1200.

Morrison, M. 2009. Understanding in physics and biology. In *Scientific Understanding: Philosophical Perspectives*, ed. H. W. de Regt, S. Leonelli, and K. Eigner, 123–145. Pittsburgh: University of Pittsburgh Press.

Morrison, M. 2011. One phenomenon, many models: Inconsistency and complementarity. *Studies in History and Philosophy of Science Part A* 42(2): 342–351.

Morrison, M. 2014. Complex systems and renormalization group explanations. *Philosophy of Science* 81(5): 1144–1156.

Morrison, M. 2015. *Reconstructing Reality: Models, Mathematics, and Simulations*. New York: Oxford University Press.

Morrison, M. 2018a. The non-causal character of renormalization group explanations. In *Explanation beyond Causation: Philosophical Persepctives on Non-causal Explanations*, ed. A. Reutlinger and J. Saatsi, 206–227. Oxford: Oxford University Press.

Morrison, M. 2018b. Turbulence, emergence and multi-scale modeling. *Synthese*, https://doi.org/10.1007/s11229-018-1825-5.

Nagel, E. 1961. *The Structure of Science*. New York: Harcourt, Brace, and World.

Northcott, R. 2010. Walsh on causes and evolution. *Philosophy of Science* 77(3): 457–467.

Nozick, R. 1981. *Philosophical Explanations*. Cambridge, MA: Harvard University Press.

Oden, J. T. 2006. *Simulation-Based Engineering Science: Revolutionizing Engineering Science through Simulation*. Washington, DC: NSF Publications. www.nsf.gov/pubs/reports/sbes_final_report.pdf.

Odenbaugh, J. 2005. Idealized, inaccurate, but successful: A pragmatic approach to evaluating models in theoretical ecology. *Biology and Philosophy* 20: 231–255.

Odenbaugh, J. 2011. True lies: Realism, robustness, and models. *Philosophy of Science* 78(5): 1177–1188.

Okruhlik, K. 1994. Gender and the biological sciences. *Canadian Journal of Philosophy* 20(Suppl.): 21–42.

Orzack, S. H., and Sober, E. 2001. *Adaptationism and Optimality*. Cambridge: Cambridge University Press.

Ostling, A., and Harte, J. 2003. A community-level fractal property produces power-law species-area relationships. *Oikos 103*(1): 218–224.

Parker, G. A. 1978. Searching for mates. In *Behavioral Ecology: An Evolutionary Approach*, ed. J. Krebs and N. Davies, 214–255. Oxford: Blackwell Scientific Publications.

Parker, G. A., and Maynard Smith, J. 1990. Optimality theory in evolutionary biology. *Nature 348*(6296): 27–33.

Parker, W. S. 2006. Understanding pluralism in climate modeling. *Foundations of Science 11*(4): 349–368.

Parker, W. S. 2011. When climate models agree: The significance of robust model prediction. *Philosophy of Science 78*(4): 579–600.

Parunak, H. V. D., Brueckner, S., and Savit, R. 2004. Universality in multi-agent systems. In *Proceedings of the Third International Joint Conference on Autonomous Agents and Multi-Agent Systems*, ed. N. R. Jennings, 930–937. New York: Association for Computing Machinery.

Peters, D. 2014. What elements of successful scientific theories are the correct targets for "selective" scientific realism? *Philosophy of Science 81*(3): 377–397.

Pexton, M. 2014. How dimensional analysis can explain. *Synthese 191*: 2333–2351.

Pincock, C. 2007. A role for mathematics in the physical sciences. *Noûs 41*: 253–275.

Pincock, C. 2011. Modeling reality. *Synthese 180*: 19–32.

Pincock, C. 2012. *Mathematics and Scientific Representation*. Oxford: Oxford University Press.

Pincock, C. 2018. Accommodating explanatory pluralism. In *Explanation beyond Causation: Philosophical Persepctives on Non-causal Explanations*, ed. A. Reutlinger and J. Saatsi, 39–56. Oxford: Oxford University Press.

Pindyck, R. S., and Rubinfeld, D. L. 2009. *Microeconomics*. 7th ed. Upper Saddle River, NJ: Pearson Education.

Pitts-Taylor, V. 2017. *The Brain's Body: Neuroscience and Corporeal Politics*. Durham, NC: Duke University Press.

Plutynski, A. 2004. Explanation in classical population genetics. *Philosophy of Science 71*(5): 1201–1214.

Potochnik, A. 2007. Optimality modeling and explanatory generality. *Philosophy of Science 74*(5): 680–691.

Potochnik, A. 2009a. Levels of explanation reconceived. *Philosophy of Science 77*(1): 59–72.

Potochnik, A. 2009b. Optimality modeling in a suboptimal world. *Biology and Philosophy* 24(2): 183–197.

Potochnik, A. 2015. Causal patterns and adequate explanations. *Philosophical Studies* 172(5): 1163–1182.

Potochnik, A. 2017. *Idealization and the Aims of Science*. Chicago: University of Chicago Press.

Potochnik, A. 2018. Eight other questions about explanation. In *Explanation beyond Causation: Philosophical Persepctives on Non-causal Explanations*, ed. A. Reutlinger and J. Saatsi, 57–72. Oxford: Oxford University Press.

Povich, M. 2018. Minimal models and the generalized ontic conception of scientific explanation. *British Journal for the Philosophy of Science* 69(1): 117–137.

Pritchard, D. 2009. Knowledge, understanding, and epistemic value. In *Epistemology*, ed. A. O'Hear, 19–43. Cambridge: Cambridge University Press.

Pruessner, G. 2012. *Self-Organised Criticality*. Cambridge: Cambridge University Press.

Psillos, S. 2011. Living with the abstract: Realism and models. *Synthese* 180: 3–17.

Putnam, Hillary. 1975. Philosophy and our mental life. In *Mind, Language and Reality*, 291–303. Cambridge University Press.

Pyke, G. H. 1984. Optimal foraging theory: A critical review. *Annual Review of Ecology and Systematics* 15(1): 523–575.

Pyke, G. H. 2019. Optimal foraging theory: An introduction. In *Encyclopedia of Animal Behavior*, vol. 1., 2nd ed., ed. J. C. Choe, 111–117. Cambridge, MA: Elsevier.

Qu, Z., Garfinkel, A., Weiss, J. N., and Nivala, M. 2011. Multi-scale modeling in biology: How to bridge the gaps between scales? *Progress in Biophysics and Molecular Biology* 107(1): 21–31.

Railton, P. 1981. Probability, explanation, and information. *Synthese* 48: 233–256.

Raup, D. M. 1986. Biological extinction in earth history. *Science* 231: 1528–1533.

Reisman, K., and Forber, P. 2005. Manipulation and the causes of evolution. *Philosophy of Science* 72(5): 1113–1123.

Relethford, J. H. 2012. *Human Population Genetics*. Hoboken, NJ: Wiley-Blackwell.

Resnik, D. B. 1991. How-possibly explanations in biology. *Acta Biotheoretica* 39: 141–149.

Reutlinger, A. 2014. Why is there universal macrobehavior? Renormalization group explanations as non-causal explanation. *Philosophy of Science* 81(5): 1157–1170.

References

Reutlinger, A. 2016. Is there a monistic theory of causal and noncausal explanations? The counterfactual theory of scientific explanation. *Philosophy of Science* 83(5): 733–745.

Reutlinger, A. 2018. Extending the counterfactual theory of explanation. In *Explanation beyond Causation: Philosophical Persepctives on Non-causal Explanations*, ed. A. Reutlinger and J. Saatsi, 74–95. Oxford: Oxford University Press.

Reutlinger, A., and Saatsi, J. (eds.). 2018. *Explanation beyond Causation: Philosophical Persepctives on Non-causal Explanations*. Oxford: Oxford University Press.

Rice, C. 2012. Optimality explanations: A plea for an alternative approach. *Biology and Philosophy* 27(5): 685–703.

Rice, C. 2013. Moving beyond causes: Optimality models and scientific explanation. *Noûs* 49(3): 589–615.

Rice, C. 2016. Factive scientific understanding without accurate representation. *Biology and Philosophy* 31(1): 81–102.

Rice, C. 2017. Models don't decompose that way: A holistic view of idealized models. *British Journal for the Philosophy of Science* 70(1): 179–208.

Rice, C. 2018. Idealized models, holistic distortions and universality. *Synthese* 195(6): 2795–2819.

Rice, C. 2019a. Understanding realism. *Synthese*, doi: 10.1007/s11229-019-02331-5.

Rice, C. 2019b. Universality and the problem of inconsistent models. In *Understanding Perspectivism: Scientific Challenges and Methodological Prospects*, ed. Michela Massimi and C. D. McCoy, 85–108. New York: Taylor and Francis.

Rice, C. 2020. Universality and modeling limiting behaviors. *Philosophy of Science*, https://doi.org/10.1086/710623.

Rice, C. manuscript. Modeling multiscale patterns: Active matter, minimal models, and universality.

Rice, C., and Rohwer, Y. 2020. How to reconcile a unified account of explanation with explanatory diversity. *Foundations of Science*, doi: 10.1007/s10699-019-09647-y.

Rice, C., Rohwer, Y., and Ariew, A. 2018. Explanatory schema and the process of model building. *Synthese* 196: 4735–4757. https://doi.org/10.1007/s11229-018-1686-y.

Rohwer, Y., and Rice, C. 2013. Hypothetical pattern idealization and explanatory models. *Philosophy of Science* 80(3): 334–355.

Rohwer, Y., and Rice, C. 2016. How are models and explanations related? *Erkenntnis* 81(5): 1127–1148.

Rosindell, J., and Cornell, S. J. 2013. Universal scaling of species-abundance distribution across multiple scales. *Oikos 122*: 1101–1111.

Ross, L. 2015. Dynamical models and explanation in neuroscience. *Philosophy of Science 82*(1): 32–54.

Roughgarden, J., Berman, A., Shafir, S., and Taylor, C. 1996. Adaptive computation and ecology and evolution: A guide for future research. In *Adaptive Individuals in Evolving Populations: Models and Algorithms*, ed. R. K. Belew and M. Mitchell, 23–30. Reading, MA: Addison-Wesley.

Rubin, H., and O'Connor, C. 2018. Discrimination and collaboration in science. *Philosophy of Science 85*(3): 380–402.

Saatsi, J. 2019. Realism and explanatory perspectives. In *Understanding Perspectivism: Scientific Challenges and Methodological Prospects*, ed. Michela Massimi and C. D. McCoy, 65–84. New York: Taylor and Francis.

Saatsi, J., and Pexton, M. 2013. Reassessing Woodward's account of explanation: Regularities, counterfactuals, and non-causal explanations. *Philosophy of Science 80*(5): 613–624.

Salmon, W. 1978. Why ask "why"? An inquiry concerning scientific explanation. *Proceedings and Addresses of the American Philosophical Association 51*(6): 683–705.

Salmon, W. C. 1984. *Scientific Explanation and the Causal Structure of the World*. Princeton, NJ: Princeton University Press.

Salmon, W. C. 1989. *Four Decades of Scientific Explanation*. Pittsburgh: University of Pittsburgh Press.

Salmon, W. C. 1997. Causality and explanation: A reply to two critiques. *Philosophy of Science 64*(3): 461–477.

Schelling, T. 1978. *Micromotives and Macrobehavior*. New York: Norton.

Schurz, G., and Lambert, K. 1994. Outline of a theory of scientific understanding. *Synthese 101*(1): 65–120.

Seger, J., and Stubblefield, J. W. 1996. Optimization and adaptation. In *Adaptation*, ed. M. Rose and G. V. Lauder, 93–123. Cambridge: Cambridge University Press.

Shapiro, L., and Sober, E. 2007. Epiphenomenalism—the do's and the don'ts. In *Studies in Causality: Historical and Contemporary*, ed. G. Wolters and P. Machamer, 235–264. Pittsburgh: University of Pittsburgh Press.

Sklar, L. 1993. Idealization and explanation: A case study from statistical mechanics. *Midwest Studies in Philosophy 18*(1): 259–270.

Sledge, P., and Rice, C. manuscript. An epistemic argument for diversity in science.

Sober, E. 1980. Evolution, population thinking, and essentialism. *Philosophy of Science 47*(3): 350–383.

Sober, E. 1983. Equilibrium explanation. *Philosophical Studies 43*(2): 201–210.

Sober, E. 1984. *The Nature of Selection.* Cambridge, MA: MIT Press.

Sober, E. 1997. Two outbreaks of lawlessness in recent philosophy of biology. *Proceedings of the 1996 Biennial Meetings of the Philosophy of Science Association, Part II: Symposia Papers (December 1997) 64* (Suppl.): S458–S467.

Sober, E. 1999. The multiple realizability argument against reductionism. *Philosophy of Science 66*(4): 542–564.

Sober, E. 2000. *The Philosophy of Biology.* 2nd ed. Boulder, CO: Westview Press.

Sober, E. 2011. A priori causal models of natural selection. *Australasian Journal of Philosophy 89*(4): 571–589.

Sober, E. 2015. *Ockham's Razors: A User's Manual.* New York: Cambridge University Press.

Solé, R. V., and Bascompte, J. 2006. *Self-organization in Complex Ecosystems.* Princeton, NJ: Princeton University Press.

Solomon, M. 2001. *Social Empiricism.* Cambridge, MA: MIT Press.

Stanford, P. K. 2003. No refuge for realism: Selective confirmation and the history of science. *Philosophy of Science 70*(5): 913–925.

Stanford, P. K. 2006. *Exceeding Our Grasp: Science, History, and the Problem of Unconceived Alternatives.* Oxford: Oxford University Press.

Stanley, H. E., Amaral, L. A. N., Gopikrishnan, P., Ivanov, P. C., Keitt, T. H., and Plerou, V. 2000. Scale invariance and universality: Organizing principles in complex systems. *Physica A 281*(1): 60–68.

Stanley, H. E., Amaral, L. A. N., Gopikrishnan, P., Plerou, V., and Salinger, M. A. 2002. Scale invariance and universality in economic phenomena. *Journal of Physics: Condensed Matter 14*: 2121–2131.

Stephens, C. 2004. Selection, drift, and the "forces" of evolution. *Philosophy of Science 71*(4): 550–570.

Stephens, D. W., and Krebs, J. R. 1986. *Foraging Theory.* Princeton, NJ: Princeton University Press.

Stigler, S. M. 1990. *The History of Statistics: The Measurement of Uncertainty before 1900.* Reprint ed. Cambridge, MA: Belknap Press of Harvard University Press.

Stigler, S. M. 2010. Darwin, Galton and the Statistical Enlightenment. *Journal of the Royal Statistical Society 173*: 469–482.

Strevens, M. 2004. The causal and unification approaches to explanation unified-causally. *Noûs* 38(1): 154–176.

Strevens, M. 2008. *Depth: An Account of Scientific Explanation*. Cambridge, MA: Harvard University Press.

Strevens, M. 2013. No understanding without explanation. *Studies in History and Philosophy of Science* 44(3): 510–515.

Strevens, M. 2017. How idealizations provide understanding. In *Explaining Understanding: New Essays in Epistemology and the Philosophy of Science*, ed. S. R. Grimm, C. Baumberger, and S. Ammon, 37–49. New York: Routledge.

Suárez, M. 1999. The role of models in the application of scientific theories. In *Models as Mediators: Perspectives on Natural and Social Science*, ed. Mary S. Morgan and Margaret Morrison, 168–195. Cambridge: Cambridge University Press.

Sullivan, E., and Khalifa, K. 2019. Idealizations and understanding: Much ado about nothing? *Australasian Journal of Philosophy* 97(4): 673–689.

Tao, T. 2012. E pluribus unum: From complexity, universality. *Daedalus* 141(3): 23–34.

Torquato, S. 2002. *Random Heterogeneous Materials: Microstructure and Macroscopic Properties*. New York: Springer.

Tracy, C. A., and Widom, H. 1993. Level-spacing distribution and the Airy kernel. *Physics Letters B 305*: 115–118.

Tracy, C. A., and Widom, H. 1994. Level-spacing distributions and the Airy kernel. *Communications in Mathematical Physics 159*: 151–174.

Trout, J. D. 2002. Scientific explanation and the sense of understanding. *Philosophy of Science* 69(2): 212–233.

Trout, J. D. 2007. The psychology of explanation. *Philosophy Compass* 2: 564–596.

Vaihinger, H. 1952. *The Philosophy of "As If": A System of the Theoretical, Practical and Religious Fictions of Mankind*. 2nd ed. London: Routledge and Kegan Paul.

van Fraassen, B. C. 1980. *The Scientific Image*. Oxford: Oxford University Press.

Vickers, P. 2017. Understanding the selective realist defence against the PMI. *Synthese 194*: 3221–3232.

Walsh, D. M. 2007. The pomp of superfluous causes: The interpretation of evolutionary theory. *Philosophy of Science* 74(3): 281–303.

Walsh, D. M. 2010. Not a sure thing: Fitness, probability, and causation. *Philosophy of Science* 77(2): 147–171.

Walsh, D. M. 2015. Variance, invariance and statistical explanation. *Erkenntnis 80*: 469–489.

References

Walsh, D. M., Ariew, A., and Matthen, M. 2017. Four pillars of statisticalism. *Philosophy, Theory, and Practice in Biology* 9(1). http://dx.doi.org/10.3998/ptb.6959004.0009.001.

Walsh, D. M., Lewens, T., and Ariew, A. 2002. The trials of life: Natural selection and random drift. *Philosophy of Science* 69(3): 452–473.

Wayne, A. 2011. Expanding the scope of explanatory idealization. *Philosophy of Science* 78(5): 83–841.

Weisberg, M. 2007a. Three kinds of idealization. *Journal of Philosophy* 104(12): 639–659.

Weisberg, M. 2007b. Who is a modeler? *British Journal for the Philosophy of Science* 58(2): 207–233.

Weisberg, M. 2008. Challenges to the structure of chemical bonding. *Philosophy of Science* 75(5): 932–946.

Weisberg, M. 2013. *Simulation and Similarity: Using Models to Understand the World*. Oxford: Oxford University Press.

Weslake, B. 2010. Explanatory depth. *Philosophy of Science* 77(2): 273–294.

Wilson, M. 2017. *Physics Avoidance*. Oxford: Oxford University Press.

Wimsatt, W. C. 1976. Reductive explanation: A functional account. In *PSA 1974, Proceedings of the 1974 Biennial Meeting, Philosophy of Science Association*, ed. R. S. Cohen, C. A. Hooker, A. C. Michalos, and J. W. van Evra, 671–710. Dordrecht, Netherlands: Reidel.

Wimsatt, W. 1994. The ontology of complex systems: Levels of organization, perspectives, and causal thickets. *Canadian Journal of Philosophy* 20: 207–274.

Wimsatt, W. 2007. *Re-engineering Philosophy for Limited Beings: Piecewise Approximations of Reality*. Cambridge, MA: Harvard University Press.

Winsberg, E. 2006. Handshaking your way to the top: Inconsistency and falsification in intertheoretic reduction. *Philosophy of Science* 73: 582–594.

Winsberg, E. 2010. *Science in the Age of Computer Simulation*. Chicago: University of Chicago Press.

Wolchover, N. 2014. At the far ends of a new universal law. *Quanta*, October 15. https://www.quantamagazine.org/beyond-the-bell-curve-a-new-universal-law-20141015.

Wolfram, S. 1984a. Cellular automata as models of complexity. *Nature* 311: 419–424.

Wolfram, S. 1984b. Universality and complexity in cellular automata. *Physica D10*: 1–35.

Woodward, J. 1997. Explanation, invariance, and intervention. *Proceedings of the 1996 Biennial Meetings of the Philosophy of Science Association, Part II: Symposia Papers (December 1997) 64* (Supplement): S26–S41.

Woodward, J. 2003. *Making Things Happen: A Theory of Causal Explanation.* Oxford: Oxford University Press.

Woodward, J. 2010. Causation in biology: Stability, specificity, and the choice of levels of explanation. *Biology and Philosophy* 25: 287–318.

Woodward, J. 2018. Some varieties of non-causal explanation. In *Explanation beyond Causation: Philosophical Perspectives on Non-causal Explanation,* ed. J. Saatsi and A. Reutlinger, 117–137. Oxford: Oxford University Press.

Woody, A. 2015. Re-orienting discussions of scientific explanation: A functional perspective. *Studies in History and Philosophy of Science 52*: 79–87.

Worrall, J. 1989. Structural realism: The best of both worlds? *Dialectica 43*: 99–124.

Wylie, A. 2004. Why standpoint matters. In *The Feminist Standpoint Theory Reader,* ed. S. Harding, 339–352. New York: Routledge.

Yurk, B. P., and Cobbold, C. A. 2018. Homogenization techniques for population dynamics in strongly heterogeneous landscapes. *Journal of Biological Dynamics 12*(1): 171–193.

Zagzebski, L. 2001. Recovering understanding. In *Knowledge, Truth, and Duty: Essays on Epistemic Justification, Responsibility, and Virtue,* ed. M. Steup, 235–252. New York: Oxford University Press.

Zhang, J., Zhang, Y. C., Alstøm, P., and Levinsen, M. T. 1992. Modeling forest fire by a paper-burning experiment, a realization of the interface growth mechanism. *Physica A 189*(3–4): 383–389.

Index

Note: Page numbers in *italics* indicate figures.

ABMs. *See* Agent-based models (ABMs)
Abstraction. *See also* Holistically distorted models
  distortion introduced by, 11, 16
  epistemic contributions of, 279, 284–286, 290
  versus idealization, 3–4, 93, 120–121, 123–124
  in mechanistic modeling, 36
  in minimal model explanations, 85
  representational limitations of science and, 148–149
Accurate representation of causes
  decompositional strategy and, 12–13, 30, 34–37, 133–145, 260, 298, 303n9
  exceptions to, 302n9
  isomorphism, 130
  minimalist idealizations and, 9
  priority of, 279
  realist approach to, 1–2, 9, 260–261
  scientific reliance on, 7–8, 12
  standard view of, 2–3, 29–34, 196–197, 230, 296–297
  tension with widespread use of distortions, 4–5
  understanding and, 33–34, 266–267
Achinstein, P., 106, 276–277
Agent-based models (ABMs), 208
Antireductionist approach, 226, 280

Arctic melt pond modeling. *See* Melt pond modeling
Ariew, André, 67, 305n23
Asymmetry problem, 23–24
  causal approach to, 26–27, 56
  counterfactual approach to, 90, 96–97, 116
  minimal model approach to (explanatory asymmetry), 79–80
Atomic scale, 204
Autonomous statistical explanations
  Galton's work on, 60–63, 305nn21–22
  natural selection and, 63–69
Autonomy
  in multiscale modeling, 206, 220, 223, 227, 281
  universality and, 164, 182

Bacterial growth modeling, 169–173, *170*, 310nn5–7
Baker, Alan, 114
Barabási, A., 185
Baron, S., 115
Barzel, B., 185
Bascompte, J., 211, 212, 218
Batterman, R. W.
  account of minimal model explanations, 83–85
  on common feature accounts, 157

Batterman, R. W. (cont.)
  on homogenization, 219
  investigation of multiscale modeling techniques, 228, 282
  on multiple realizability, 180–181
  on renormalization group explanations, 80, 306n28
Bechtel, Bill, 28, 30, 48–49
Beckner, M., 240
Benford's law, 310n8
Bidirectional interactions among models, 227–228
Biological optimality models. *See* Optimality explanations
Biological population modeling. *See* Population genetics
Bishop, R. C., 208
Block spin transformation, 78–79
Bohr, Niels, 3
Bokulich, Alisa, 86, 94, 128–131, 183, 265
Boltzmann, Ludwig, 174
  Maxwell-Boltzmann distribution, 136–138
Borgdorff, J., 209
Bottom-up multiscale modeling, 203, 226–227
Boyle's law, 31–32, 136
Brandon, R., 246
Bridges of Königsberg, 112–113, *113–114*
Bromberger, S., 276
Brueckner, S., 208
Bursten, J. R., 282, 312n18

Cancer development modeling, scales in, 204–205
Canonical explanations, 31–32
Caricatures, idealized models as, 288
Cartwright, Nancy, 4, 57, 148, 301n7
Causal explanation, 1–3, 119, 297. *See also* Accurate representation of causes; Mechanistic accounts of explanation
  asymmetry problem in, 24
  canonical explanations, 31–32
  "capturing" causes of interest in, 32–33
  causal-mechanical (CM) model, 3, 24–25, 294, 313n4
  causal pattern explanations, 27
  diachronic nature of, 56
  dominance of, 7, 22–23
  generality of, 24, 50, 293–294
  idealizations in, 30–32
  interventionist account, 25–26, 56–57, 94–96, 108–109, 302n3, 307n11
  invariance and, 127, 183–184
  link between explanation and understanding in, 33–34
  modularity of causes in, 56–57
  natural selection explanations, 67–68
  role of irrelevance information in, 29, 31, 126–128
  Strevens's kairetic account, 26, 108–109
  unification of causal and noncausal explanations, 111–117, 298
Causal influences, 25
Causal-mechanical (CM) model, 3, 24–25, 294, 313n4
Causal pattern explanations, 27
Causal properties, 25–26
Cell biology models, scales in, 200–203, *201*
Cellular automata models, 185
Central limit theorem (CLT), 63, 70–71, 136, 139, 142, 173–174, 289
Central propositions, 34, 250
"Change-relating" relationships, 127
Characteristic scale, 30, 187, 310n1
Charles's law, 136
Charnov, Eric, 47, 55
Chebyshev, Pafnuty, 174

Index

Checkerboard model (Schelling), 235–237, 244, 246–247
Chopard, B., 209
Classes, universality. *See* Universality classes
Climate modeling
  melt pond modeling, 165–169, *166–167*, 212–213, *213*, 223
  problem of inconsistent models in, 11, 191
CLT. *See* Central limit theorem (CLT)
Cobbold, C. A., 221
Colyvan, M., 115
Common feature accounts, 157–158
Communicative approach to explanation, 106–108, 276–277
Complex behaviors, emergence of, 223, 312n16
Conceptual strategies account, 312n18
Conflicting models. *See* Problem of inconsistent models
Constraints
  counterfactual account, 108, 115
  multiscale modeling, 224–231
  optimality explanations, 42–48, 51–56, 60, 147, 160
Continuum (finite element) model, 225
Control variables, optimality models, 42
Copernicus, Nicolaus, 265
Corwin, Ivan, 171, 173
Cost-benefit analysis, in optimality models
  dung fly example, 45–46
  equilibrium sex ratios example, 47–48
  lapwing foraging example, 43–44, *43–44*
Counterfactual account of explanation, 15–16, 89–90, 298–299. *See also* Holistically distorted models; Holistic distortion view; Universality
  asymmetry problem resolved by, 96–97, 116
  Bokulich's account compared to, 128–131
  communicative and ontological aspects of explanation in, 106–108, 276–277
  constraints, 108, 115
  counterexamples to, 89, 99–100, 307n13
  departures from standard view, 108–109, 298–299
  equilibrium explanations in, 98, 115, 128
  explanatory context in, 102–106, 108, 183
  idealizations in, 120–124
  linking of explanation and understanding in, 117–120, 256–259, 277–278
  motivation for, 38
  objections to, 109–111
  psychology of explanation in, 278–279
  Reutlinger's account compared to, 124–128, 308n24
  role of explanatory context in, 102–106
  role of irrelevance information in, 124–128
  set of counterfactual dependence and independence relations required in, 97–104, 109–110, 116–117, 126–127, 243–247, 297, 307n12
  statement of, 93–94, 108
  traditional counterexamples to, 99–101, 307n13
  understanding produced by modal information, 243–244
  unification of causal and noncausal explanations in, 111–117, 298
  view of counterfactual dependence relations in, 94–97, 110–111
  Woodward's account compared to, 94–96, 108–109

Covering-law models, 112, 293. *See also* Deductive-nomological (DN) model
Craver, Carl, 3, 27, 29–30, 36, 41, 48–49, 126, 163
Currency, in optimality models, 42, 44, 56, 303n3

Daniel, Katherine J., 240
Darwin, Charles, 266
Darwinian selection, 69
Decompositional strategy, 260
  assumptions of, 30, 34–35, 133–134, 303n9
  centrality to philosophical accounts, 12–13, 35–37
  challenges to, 135–145, 298
Deductive-nomological (DN) model, 22–24, 71. *See also* Covering-law models
  central ideas of, 22–23
  counterexamples to, 23–24, 71–72, 93, 96–97
Deep-water wave dispersion model, 122
Degrees of understanding, 257–258, 264, 313n8
Deidealization, 8–9, 283, 284
Deisboeck, T. S., 204
Dependency condition, in Reutlinger's account, 125
de Regt, H. W., 234, 258, 313n10
Descartes, René, 313n11
Design variables, optimality model, 42
Diachronic nature of causal representations, 56
Diamagnetic Rydberg spectra, 265
Distortion. *See also* Holistically distorted models; Holistic distortions, leveraging of; Holistic distortion view; Idealizations
  centrality to science, 8–12
  responses to, 12–15
  tension with accurate representation requirements, 4–5
  widespread and essential use of, 4–8

Distributions
  Gaussian, 70, 173–175, 310n8
  limiting, 70
  Tracy-Widom, 176–180
Diversity, value of, 272, 282–283
Drift, genetic. *See also* Population genetics
  in biological optimality models, 46, 48, 53, 304n16
  in Eden Growth Model, 171
  selection and, 304n16
  in statistical models, 64, 66, 71, 139–142
Dung fly model, 45–46, 51, 56, 129, 304n9

Ecological diffusion, 220–221
Ecological models, 311n10
  causal mechanisms in, 304n11
  homogenization approach to, 220–221
  spatial ecology models, 220–221, 311n15
  species abundance distribution (SAD), 215
Eden, M., 169. *See also* Eden Growth Model
Eden Growth Model, 169–173, *170*, 310nn5–7
Elgin, C. Z.
  accurate representation requirements, 5–6
  exemplification, 68, 289
  on idealizations, 121–122, 251, 288–289, 312n5
  on problem of inconsistent models, 195, 199
  on understanding, 117–118, 248, 250, 254
Elgin, Mehmet, 30–31, 51
Emergence of complex behaviors, 223, 312n16
Emergentist approach to multiscale modeling, 226

… Index 341

Energy costs, in optimality models, 43
Engineering simulation, scales in, 205–206
Environmental pressures, idealization of, 4, 53, 64
Epistemic contributions of idealizations, 120–124. *See also* Understanding without explanation
  application of theoretical modeling tools, 71, 82, 123, 135–140, 147, 283–286
  better explanations, 92–93, 286–288
  focus on universal patterns, 293–296
  improved understanding, 3–8, 33–34, 288–291, 303n6
  intertwined idealizations and holistic distortion, 123, 291–293
  realism about epistemic products, 296–297
Equilibrium explanations, 303n7
  in counterfactual account, 98, 115, 128
  Hardy-Weinberg equilibrium model, 140–141
  ideal gas law, 105, 112, 136–138, 159–160, 255
  in optimality explanations, 44, 48–50, 304n11
Equilibrium sex ratios, Fisher's model of, 47–48, 50, 58, 304n10
Essential idealizations. *See also* Holistic distortion view; Idealizations; Idealized models
  hypothetical pattern idealizations, 10
  in minimal model explanations, 81–82, 143
  in optimality explanations, 51–55, 141–143, 147
  reliance on, 81–82, 107, 263
  required by mathematical modeling techniques, 135–140, 147, 283–286
  in statistical explanations, 69–71
Evolution, Darwin's theory of, 266. *See also* Natural selection

Evolutionarily stable strategy (ESS), 239
Exemplification, 68, 289
Explanation, 106–108, 234–235. *See also* Causal explanation; Counterfactual account of explanation; Noncausal explanations; Understanding
  as cluster concept, 89
  communicative versus ontological features in, 106–108, 276–277
  construction over time, 281–282
  contributions of idealizations to, 92–93, 120–124, 286–288
  explanatory dependence relations in, 91–92, 96, 108
  generality in, 182–183
  importance to science, 21–22, 302n1
  levels of explanation, 163, 280–281
  manipulationist, 25–26, 279–280
  potential explanation, 312n1
  psychology of, 278–279
  questions for future research, 281–283, 299–300
  relationship with other aims of scientific inquiry, 279–280
  representational aims of, 279
  requirements for an account of, 90–93
Explanatory asymmetry, 79–80. *See also* Asymmetry problem
Explanatory context, in counterfactual account, 102–106, 108, 183
Explanatory dependence relations, 91–92, 96, 108
Explanatory knowledge, 22
Explanatory relevance relation, 26
Explicit appeals to universality, examples of, 164–165. *See also* Universality classes
  Eden Growth Model, 169–173, *170*, 310nn5–7
  Gaussian universality, 173–175
  melt pond model, 165–169, *166–167*, 212–213, *213*, 223
  Tracy-Widom distribution, 176–180

Exploration of possibility space, 240–243, *241–242*
Extinction events, power laws in, 214–215, *216*

Factive account of understanding, 17–18, 118, 261–264, 296–297, 312n4
"Felicitous falsehoods," idealizations as, 5
Feminist epistemology, 270
Fillion, Nicolas, 295
Fisher, Michael, 83
Fisher, R. A., 174
  idealizing assumptions, 63–64, 69–71, 138–140
  model for equilibrium sex ratios, 47–48, 50, 58, 304n10
Fitness, 303n3
  in dung fly model, 46
  in equilibrium sex ratios model, 47
  in evolving biological systems, 57–58
  in foraging strategy models, 58–59, 305n19
  in natural selection explanations, 67–68, 72, 304n16
  trait, 64–67, *65*
Fixed points, *78*, 79–80
Flagpole's shadow case, 23, 96
Foraging models
  idealizing assumptions in, 52, 138–140
  lapwing, 43–44, *43–44*
  mathematical representation of noncausal features in, 58–59, 305n19
  prey and patch choice models, 45
  redshank sandpiper, 58–59, 305n19
  redshank sandpiper foraging strategies, 58–59, 305n19
Fossil record, power laws in, 214–215, *216*

Fractal structure
  definition of, 311n8
  in melt pond modeling, 166, *167*, 213–215, 311n8
Frequency-dependent optimality models, 44, 303n1
Frequency-independent optimality models, 44, 45–46
Fresnel, Augustin-Jean, 313n11
Friedman, M., 91

Galilean idealization, 8–9
Galton, Francis, 174, 305n21
  autonomous statistical explanations, 60–63, 305nn21–22
Game-theoretic models, 44, 51, 303n1
Garden-variety idealizations, 10–11
Gases, laws of. *See* Ideal gas law
Gaussian distribution, 70, 136, 173–176, *176*, 310n8
Generality of explanation, 182–183. *See also* Universality
  in causal accounts, 24, 50
  in optimality models, 47–48, 50, 54
Genetical selection, 69
Genetic drift. *See also* Population genetics
  in biological optimality models, 46, 48, 53, 304n16
  in Eden Growth Model, 171
  selection and, 304n16
  in statistical models, 64, 66, 71, 139–142
Giere, R., 4, 195
Gillespie, J. H., 64–65, *65*
Gillespie Point (GP), 65, *65*
Gisiger, T., 156, 214, 215
Glennan, Stuart, 28, 92
Godfrey-Smith, P., 1
Golden, Kenneth M., 168
Goldenfeld, N., 79, 159
Goss-Custard, J. D., 59

Grasping of systematic relationships, 248–249
Green, S., 282
Grene, Marjorie, 69
Grimm, Stephen, 118, 119, 256, 312n4

Hamiltonians, 78, *78*, 80, 125–126
Handshaking account of multiscale modeling, 225–226
Hardy-Weinberg equilibrium model, 140–141
Harte, J., 215
Hawk-Dove game, 10, 238–240, 244–245, 290
Heggie, D. C., 240
Hempel, Carl, 93, 246
  deductive-nomological (DN) model, 22–24, 71–72, 93, 96–97
  inductive-statistical (IS) model, 302n2
Henriksen, R. N., 211
Heredity. *See* Population genetics
Hexed-salt case, 23, 96–97
Hoekstra, A. G., 209
Holism, 145–146
Holistically distorted models, 37–39, 153–154. *See also* Multiscale modeling; Universality; Universality classes
  Eden Growth Model, 169–173, *170*, 310nn5–7
  Gaussian universality, 173–175, 310n8
  justifying through appeal to universality classes, 154–162, 279, 298
  melt pond model, 165–169, *166–167*, 212–213, *213*, 223
  parameters of universality in, 162–164
  Tracy-Widom distribution, 176–180
Holistic distortions, leveraging of, 120–124, 298. *See also* Understanding without explanation
  application of theoretical modeling tools, 71, 82, 123, 135–140, 147, 283–286
  better explanations, 92–93, 286–288
  focus on universal patterns, 293–296
  improved understanding, 3–8, 33–34, 288–291, 303n6
  intertwined idealizations and holistic distortion, 123, 291–293
  realism about epistemic products, 296–297
Holistic distortion view, 134–135, 298, 309n6. *See also* Holistically distorted models; Holistic distortions, leveraging of
  argument for, 16–17, 38–39, 145–150
  as methodological prescription, 146
Homogenization, 166–169, 219–223
How-possibly explanations, 246
Hume, David, 293
Huneman, P., 305n20
Hypothetical pattern idealizations, 10
Hypothetical scenarios, modeling of, 237–240

Ice ponds, modeling melting of. *See* Melt pond modeling
Ideal gas law
  idealizing assumptions in, 105, 112, 137–138, 255
  statement of, 136–137
  universality and, 159–160
Ideal intervention, 25–26, 302n3
Idealizations. *See also* Holistic distortion view; Idealized models
  abstraction versus, 3–4, 93, 120–121, 123–124
  in causal explanations, 30–32
  in counterfactual explanations, 92–93, 120–124
  Galilean, 8–9
  garden-variety, 10–11
  "harmless," 30–31
  hypothetical pattern, 10
  in ideal gas law, 136–138
  minimalist, 8–10, 31–32

Idealizations (cont.)
  in minimal model explanations, 81–82, 122, 143
  multiple models, 8, 11–12
  in optimality explanations, 51–55, 147
  positive contributions of (*see* Epistemic contributions of idealizations)
  problems posed by, 8–12
  reliance on, 3–4, 81–82, 107, 263, 297
  responses to, 12–15
  in statistical models, 60, 62–63, 69–71
Idealized models. *See also* Holistically distorted models; Mathematical modeling techniques; Multiscale modeling
  decompositional strategy for, 12–13, 30, 34–37, 133–145, 260, 298, 303n9
  extraction of modal information from, 128–131, 147–148, 158, 199–200, 261–262, 286, 291, 313n10
  holistic distortion view of, 16–17, 38–39, 134–135, 145–150, 291–293, 298
  multiple conflicting (*see* Problem of inconsistent models)
  representational accuracy in (*see* Accurate representation of causes)
Implication condition, 125
Inaccessibility, 1, 301n1
Inclusive fitness, 303n3
Inconsistent models. *See* Problem of inconsistent models
Induction, problem of, 293
Inductive-statistical (IS) explanation, 302n2
Infinite population size, idealization of, 46, 53, 139–143, 304n16
Inheritance
  Galton's statistical explanations of, 60–63, 305nn21–22
  phenotypic strategy inheritance, 52–53, 139–143, 147, 304n14

Instrumentalism, 229, 262
Interactional complexity, 217
Interscale modeling techniques, 281
  homogenization, 166–169, 219–223
  renormalization, 77–81, 84, 124–128, 156, 217–219, 306n28
  simultaneous use of, 223
Intervention, ideal, 25–26, 302n3
Interventionist account of explanation, 25–26, 56–57, 94–96, 108–109, 302n3, 307n11
Invariance, scale, 211–217, *212*, 311n11
Invariance, Woodward's concept of, 127, 183–184
Investigation of necessity claims, leveraging holistic distortions for, 235–237
Irrelevance information, role of. *See also* Decompositional strategy
  in causal and mechanistic accounts, 29, 31, 126–128
  in counterfactual account, 97–104, 109–110, 124–128, 297, 307n12
  in objections to deductive-nomological (DN) model, 96–97
  in equilibrium explanations, 128
  in holistically distorted models, 286–288
  influence of context on, 102–106
  in minimal model explanations, 73, 80–86
  in Reutlinger's account, 124–128, 308n24
  in statistical explanations, 60, 71–73, 128
IS explanation. *See* Inductive-statistical (IS) explanation
Isomorphism, 130

Kadanoff, L. P., 79, 81, 105, 146, 159
Kaplan, David, 27, 29–30, 36, 41, 163
Kaplan, D. M., 3
Kardar, M., 171

Kardar-Parisi-Zhang (KPZ) universality class, 171–172, 177
Keizer's paradox, 311n5
Khalifa, K., 94, 235, 255
Khaluf, Y., 211
Kim, Jaegwon, 117
Kinetic theory of gases, 136. *See also* Ideal gas law
Kitcher, P., 91
Knuuttila, T., 79–80
Königsberg, bridges of, 112–113, *113–114*
KPZ (Kardar-Parisi-Zhang) growth equation, 171
KPZ (Kardar-Parisi-Zhang) universality class, 171–172, 177
Kvanvig, J. L., 34, 248, 250, 312n4

Lange, Marc, 71, 83, 157, 163
LaPlacean demon, 181
Lapwing foraging model, 43–44, *43–44*
Large numbers, law of, 309n2
Lattice gas automaton (LGA) model, 73–81, 306n27
  computational algorithm for, 73–75, *74–75*
  differences between model and target system in, 73, 76–81
  renormalization techniques in applications of, 77–81, *78*, 84, 306n28
  universality classes in, 79–81, 84
Law of large numbers, 309n2
Laws of nature, 24
Levels of explanation, 163, 280–281
Levels of reality, 163
Levins, R., 193
Levy, Arnon, 37, 48–49
LGA model. *See* Lattice gas automaton (LGA) model
Limiting distributions, 70
Lipton, P., 94, 246, 312n1
Liquid-drop model, 191, *192*
Localization, idealizations in, 30

Longino, H., 191, 270
Lyapunov, Aleksandr, 174

Macroscopic patterns. *See also* Universality
  ideal gas law, 136–138
  independence from microscopic counterparts, 158, 174–175
  lattice gas automaton (LGA) model, 75
  multiscale modeling and, 227
  self-similarity in, 211–213
  top-down multiscale modeling of, 203–205
Macroscopic scale, in cancer development models, 204–205
Majumadar, Satya, 177
Manipulationist accounts of explanation, 25–26, 279–280
Mapping assumption, 35, 133, 140–145, 303n9
Markov, Andrey, 174
Massimi, M., 195
Mathematical explanations, 86, 112–115
Mathematical modeling techniques, 39. *See also* Holistically distorted models; Multiscale modeling
  characteristic scale in, 30, 187, 310n1
  homogenization, 166–169, 219–223
  idealizations essential to, 71, 82, 123, 135–140, 147, 283–286
  minimal model explanations and, 80
  in optimality models, 55–60
Matthen, M., 67, 305n23
Maxwell-Boltzmann distribution, 136–138
Maynard Smith, J., 238, 239
McMullin, Ernan, 8
Mechanistic accounts of explanation
  counterfactual information in, 111
  decompositional strategy and, 36
  description of, 27–28
  diachronic nature of, 56

Mechanistic accounts of explanation (cont.)
   explanation of general patterns in, 294
   factive requirements for, 2–3
   modularity of causes in, 56–57
   relationship with mechanisms, 302n4
   representational accuracy in (*see* Accurate representation of causes)
   role of irrelevance information in, 29, 31, 126–128
   truth in, 2–3
Melt pond modeling, 223
   appeal to universality in, 165–169, *166–167*
   self-similarity in, 212–213, *213*
Mendelism genetics, Hardy-Weinberg equilibrium model of, 140–141
Merz, M., 79–80
Metaphysics. *See also* Problem of inconsistent models
   metaphysical aspects of explanation, 55–56, 92, 94, 111, 277
   metaphysical dependence relation, 26
   metaphysical intuitions about causes, 60
   modeling and idealization as independent from, 135, 229, 308n1
   multiscale modeling and, 229
   universality and, 181
   Wilson on, 285
Microscopic scale, in cancer development modeling, 204
Middle-out multiscale modeling, 203–204
Minimalist idealization, 8, 9–10, 31–32
Minimality condition, 126
Minimal model explanations, 15, 310n7. *See also* Renormalization
   captured by counterfactual account, 116
   differences between model and target system in, 73, 76–81
   essential idealizations in, 81–82, 122, 143
   explanatory questions in, 82–83
   lattice gas automaton (LGA) model example, 73–81, *74–75*, *78*, 306nn27–28
   macroscale limiting behaviors modeled in, 73–76
   problem of asymmetry in, 79–80
   role of irrelevance information in, 73, 80–86
   universality classes in, 79–81
Mirroring of causal relations, 32–33
Mizrahi, M., 250
Modal account of understanding. *See also* Modal information, extraction of
   factive understanding, 17–18, 118, 247–259, 261–264, 296–297
   focus on modal information in, 256–259
   objections to, 109–111
   role of diversity in, 269–270
   scientific progress and, 264–269
   Understanding Realism, 259–264
   understanding without explanation, 234–245
Modal information, extraction of, 128–131, 147–148, 158, 199–200, 261–262, 286, 291, 313n10
Model decomposition assumption, 35, 133, 135–140, 303n9
Model systems, 7, 116, 129–130, 155, 301n5
Model-to-mechanism-mapping (3M) requirement, 29–30, 41
Modularity of causes, 56–57
Moir, Robert, 295
Molecular scale, in cancer development models, 204
Morrison, M., 4, 63, 139, 146, 148, 191, 195–197, 219, 261
Multimodel ensemble studies, 194–195

Multiple models idealization, 8, 11–12
Multiple realizability, 180–182
Multiscale modeling, 17, 187–190, 281, 298. *See also* Tyranny of scales
  antireductionist approach to, 226
  bidirectional interactions in, 227–228
  bottom-up, 203, 226–227
  Bursten's conceptual strategies account, 225
  emergence of complex behaviors in, 223, 312n16
  emergentist approach to, 226
  homogenization, 166–169, 219–223
  Keizer's paradox, 311n5
  middle-out, 203–204
  pluralism in, 227–230
  practical constraints on, 224–231
  problem of inconsistent models, 12, 17, 189–203, 310n4
  relative autonomy in, 206, 220, 223, 227, 281
  renormalization, 77–81, 84, 124–128, 156, 217–219, 306n28
  scale dependence, 207–210, *210*, 223, 311n7
  scale invariance, 211–217, *212*, 311n11
  scale separation, 207–210
  top-down, 203, 226–227
  tyranny of scales, 17, 189–190, 203–223
  Winsberg's handshaking account of, 225–226

Nagel, Ernest, 22
National Science Foundation, 205
Natural selection
  autonomous statistical explanations and, 68–69
  causal interpretations of, 67–68
  genetical versus Darwinian selection, 69
  manipulationist criteria for causation, 67
  measure of trait fitness in, 64–67, *65*
Necessity claims, investigation of, 235–237
Neural systems, canonical model of, 85
Newton, Isaac, 313n11
Newton's second rule of scientific reasoning, 163
Noise, 171–172, 311n5
"No miracles" argument, 260
Noncausal explanations, 15–16
  common features of, 41, 87, 297–298
  definition of, 86
  mathematical explanations, 86, 112–115
  minimal model explanations, 73–86
  optimality explanations, 42–60
  statistical explanations, 60–73
  unifying with causal explanations, 111–117, 298
Nonfactive understanding, view of science as, 13, 296, 301n2
Nonrealists, 13
Nozick, R., 256
Nuclear phenomena
  conflicting idealized models of, *192*
  idealized models of, 191
Nucleus, Bohr's collective model of, 3

Ontic accounts of explanation, 106–108, 276–277
Ontological sense of explanation, 106–111, 276–277
Optimality explanations
  components of, 42–44
  constraints and trade-offs, 42–48, 51–56, 60, 147, 160
  currency, 42, 44, 56, 303n3
  dung fly example, 45–46, 51, 56, 304n9
  equilibrium explanations, 44, 48–50, 304n11

Optimality explanations (cont.)
  equilibrium sex ratios example, 47–48, 58, 304n10
  essential idealizations in, 51–55, 141–143, 147
  frequency-dependent, 44
  frequency-independent, 44, 45–46
  generality of explanation in, 47–48, 50, 54
  "harmless" idealizations in, 30–31
  lapwing foraging example, 43–44, *43–44*
  mathematical representation of non-causal features in, 55–60
  optimization theory in, 42
  prevalence of, 42
  universality and, 160
Optimization criterion, 42
Optimization theory, 42
Orbit models
  planets, 265
  stars, 240–243, *241–242*, 245
Ostling, A., 215
Overlapping universality classes, 197–198, *198*, 201–203

Parisi, G., 171
Parker, G. A.
  dung fly model, 45–46, 51, 56, 129, 304n9
  lapwing foraging model, 43–44
Parker, W. S., 190, 194
Partially accurate representations, appeal to. *See* Decompositional strategy
Parunak, H. V. D., 208
Patterns, stability of. *See also* Universality
  generality in explanations and, 182–183
  invariance and, 183–184
  multiple realizability, 180–182
Percolation theory, 218

Peripheral propositions, 34, 250, 253–254
Perspectivalism, 196–197
Pexton, M., 147
Phenotypic strategy inheritance, 52–53, 139–143, 147, 304n14
Pincock, Christopher, 37, 109–113, 122, 229
Planetary motion, Copernicus's theory of, 265
Pluralism, 89, 110, 226–227, 272
Plutynski, A., 70, 247
Population genetics
  autonomous statistical explanations of, 63–69, 305n24
  causal interpretations of, 67–68
  essential idealizations in, 70–71, 138–140, 247
  genetical versus Darwinian selection, 69
  Hardy-Weinberg equilibrium model, 140–141
  Hawk-Dove game, 10, 238–240, 244–245, 290
  manipulationist criteria for causation, 67
  measure of trait fitness in, 64–65, *65*
  optimality explanations for, 47–48, 50
  trait fitness and, 64–67, *65*
Population size, idealization of, 46, 53, 139–143, 304n16
Possibility space, exploration of, 240–243, *241–242*
Potochnik, Angela, 94
  accurate representation requirements, 3, 5–6, 302n10
  causal account of explanation, 5–6, 27, 32, 41, 108–109, 163
  communicative sense of explanation, 106–108, 276–277
  on generality in explanation, 182
  on idealizations, 309n5

on metaphysical inferences from models, 229
on optimality explanations, 51, 58–59, 304n13
on problem of inconsistent models, 195
on stability of patterns, 183
view of irrelevance information, 103
view of scientific understanding, 254, 267–268, 278, 302n10
Povich, M., 83
Power laws, 213–215, 311n10
Pragmatic sense of explanation, 276–277
Prediction, explanation and, 279–280
Preemption cases, 100–101, 307n13
Prey and patch choice models, 45
Price, G. A., 239
Pritchard, D., 312n4
Problem of asymmetry
  causal approach to, 26–27, 56
  counterfactual approach to, 90, 96–97, 116
  description of, 23–24
  Lange's explanatory asymmetry objection, 79–80
  minimal model approach to, 79–80
Problem of inconsistent models, 8, 11–12, 17, 282
  accurate representation requirements as cause of, 191–193, 196–197
  definition of, 189
  examples of, 190–191, *192*, *201*
  methodological challenges generated by, 310n4
  models exemplifying different aspects of phenomena, 195–196
  multimodel ensemble studies, 194–195
  perspectivalism and, 195
  realism and, 193, 195, 262
  robustness analysis and, 193–195, 302n5
  universality and, 197–203

Problem of induction, 293
Problem of relevance, 23, 26
Production of structure, 25
Propagation of structure, 25
Propositions, central/peripheral, 34, 250, 253–254
Proto-understanding, 256–257
Pseudoprocesses, 24–25
Psychology of explanation, 278–279
Pyke, Graham, 58

Quantum mechanics, 266
"Quasi-factive" accounts of understanding, 250
Quenched KPZ class (qKPZ), 172

Raup, D. M., 214
Real fluids. *See* Lattice gas automaton (LGA) model
Realism, 1–2, 296–297, 308n1. *See also* Decompositional strategy
  compatibility with idealization, 39, 262, 313n9
  defenses of, 36–37
  emphasis on accurate representation in, 260–261
  "no miracles" argument, 260
  problem of inconsistent models and, 193, 195, 262
  Understanding Realism, 17–18, 261–264
Redshank sandpiper foraging strategies, 58–59, 305n19
Reductionism, 163–164, 226
Relevance, problem of, 23, 26
Renormalization, 77–81, 84, 124–128, 156, 217–219, 306n28
Representational accuracy. *See* Accurate representation of causes
Reutlinger, A., 89, 94, 124–128, 308n24
Ripley, D., 115
Robustness analysis, 193–195, 302n5

Rohwer, Yasha, 10
Ross, Lauren, 85–86
Roughgarden, Joan, 237

Saatsi, Juha, 313n9, 313n11
SAD. *See* Species abundance distribution (SAD)
Salmon, Wesley, 24–25, 91–92
Savit, R., 208
Scale. *See* Multiscale modeling; Tyranny of scales
Scatophaga stercoraria (dung fly) model, 45–46, 51, 56
Schehr, Grégory, 177
Schelling, T., 235–237, 244, 246–247
Schrödinger's equation, 83
Scientific understanding. *See* Understanding
Segregation, Schelling's checkerboard model of, 235–237, 244, 246–247
"Selective confirmation" defense of realism, 36–37, 260. *See also* Decompositional strategy
Self-organized criticality (SOC), 185
Self-similarity, 211–212
Separation of scales, 207–210
Sex ratios, Fisher's optimality model of, 47–48, 58, 304n10
Shapiro, L., 67
Shell model, 191, *192*
Sober, Elliott, 30–31, 49, 51, 68, 226, 246, 304n8
SOC. *See* Self-organized criticality (SOC)
Solé, R. V., 211, 212, 218
Space-time independent noise, 171
Spatial ecology models, 220–221, 311n15
Spatial scales. *See* Multiscale modeling; Tyranny of scales
Species abundance distribution (SAD), 214–215
Stability of patterns. *See also* Universality
generality in explanations and, 182–183
invariance and, 183–184
multiple realizability, 180–182
Standard approach, 1–3. *See also* Causal explanation; Decompositional strategy
counterfactual account compared to, 108–109, 298–299
definition of, 3
differences among views in, 3, 301nn3–4
need for alternative to, 37–39
prevalence of, 3
problems with, 41–42
representational accuracy requirements (*see* Accurate representation of causes)
Stanford, P. K., 13
Star orbit modeling, 240–243, *241–242*, 245
Statistical explanations, 15
essential idealizations in, 69–71
Galton's autonomous statistical explanations, 60–63, 305nn21–22
Lange's account of, 71
natural selection, 63–69
prevalence of, 60
role of irrelevance in, 60, 71–73, 128
statistical dependence in, 71–73, 305n24
subsumed under counterfactual account of explanation, 115–116
Strategy sets, 42, 52
Strevens, Michael, 3, 41
accurate representation requirements, 2, 31, 36, 143
causal-mechanical approach, 163, 294, 313n4
on equilibrium explanations, 48–49
on generality in explanations, 182

kairetic account of event explanation, 26, 108–109
ontological sense of explanation, 106–108, 276–277
on scientific understanding, 234–235, 246, 250–251, 266, 278
view of irrelevance information, 103, 127–128, 138
Substitution cost, 47, 50
Sullivan, E., 255
Supervenience, 181

Target decomposition assumption, 34, 133, 135, 303n9
Tao, Terence, 158, 174
Temporal scales. *See* Multiscale modeling; Tyranny of scales
Theoretical modeling tools, idealizations essential for, 71, 82, 123, 135–140, 147, 283–286
Theories, models versus, 18
Thermodynamic limit, 81–82, 122, 146–147
3M (model-to-mechanism-mapping) requirement, 29–30
Top-down multiscale modeling, 203, 226–227
Topological explanations, 305n20
Topologies, scale-invariant, 211, *212*
Tracy-Widom distribution, 176–180
Trade-offs, in optimality models, 42–48, 51–52, 55–56
Trait fitness, 64–67, *65*
Trout, J. D., 234, 246, 258
"True enough," 252, 289
Truth. *See* Accurate representation of causes
Tyranny of scales, 17, 189–190, 298
characteristic scale, 30, 187, 310n1
concept of, 203–204
in engineering simulation techniques, 205
examples of, 204–206

homogenization, 166–169, 219–223
renormalization, 77–81, 84, 124–128, 156, 217–219, 306n28
scale dependence, 207–210, *210*, 223, 311n7
scale invariance, 211–217, *212*, 311n11
scale separation, 207–210

Understanding, 2, 233–234, 308n19. *See also* Understanding without explanation
as body of information, 117–118, 248–249, 308n20
caveats for account of, 247–248
contributions of idealizations to, 3–8, 33–34, 288–291, 303n6
degrees of, 257–258, 264, 313n8
explanation and, 33–34, 91–92, 117–120, 277–278, 288–291
factive requirement for, 118, 247–255, 261–264, 312n4
grasping of systematic relationships in, 118–119, 248–249
holistic distortions and, 288–291
proto-understanding, 256–257
"quasi-factive" accounts of, 250
role of diversity in promoting, 269–273, 282–283
and scientific progress, 264–269
Understanding Realism, 17–18, 261–264
Understanding without explanation, 234–235, 277–278, 298
exploration of possibility space, 240–243, *241–242*
investigation of necessity claims, 235–237
modeling of hypothetical scenarios, 237–240
objections to, 245–247
role of modal information in, 243–247, 256–259

Unificationist theories of explanation, 111–112
Universality, 7–8, 17, 109, 308n25. *See also* Holistically distorted models; Problem of inconsistent models; Tyranny of scales
 advantages of, 184–186
 autonomy and, 164, 182
 definition of, 155
 Gaussian, 173–175, 310n8
 generality of, 162–164, 182–183
 importance to science, 7–8, 281, 293–296
 invariance and, 183–184
 minimal model explanations of, 73–86, 98, 116, 122
 multiple realizability, 180–182
 parameters of, 162–164
 reductionism, 163–164
 renormalization, 77–81, 84, 124–128, 156, 217–219, 306n28
 role of explanatory context in, 102–106
 scale dependence and, 207–210
 scale invariance and, 211–217, *212*, 311n11
 scope of, 162
 self-organized criticality (SOC) example, 185
 Tracy-Widom distribution, 176–180
Universality classes, 16–17, 39, 164–165, 176–180, 298. *See also* Holistically distorted models; Multiscale modeling
 advantages of, 134, 158, 184–186
 in cellular automata models, 185
 differentiation of, 197–198
 Eden Growth Model, 169–173, *170*, 310nn5–7
 Gaussian distribution, 173–175, *176*, 310n8
 independence of universal behaviors, 164
 justifying use of idealized models with, 154–162, 298
 Kardar-Parisi-Zhang (KPZ), 171–172, 177
 lattice gas automaton (LGA) model, *78*, 79–80, 84
 melt pond modeling, 165–169, *166–167*, 212–213, *213*, 223
 overlapping, 177, 197–198, *198*, 201–203
 parameters of, 162–164

van Fraassen, Bas, 13, 106, 108, 125, 276–277
Varri, A. L., 240
Veridical, deterministic, and atomic causal model, 26–27
Veridicality condition, 125
Veritism, 254

Walsh, D. M., 92
Wavefunction scarring, 265
Weisberg, Michael
 accurate representation requirements, 143
 causal account of explanation, 3, 41
 on idealizations, 8–9, 11, 31–32, 36, 138
 model-world relation, 130, 301n5
 on problem of inconsistent models, 11, 193
 view of similarity, 32
"What if things had been different?" questions, 25, 59, 94–95, 102, 111–112, 119, 162, 265, 313n11
Wilson, M., 70, 191, 282, 285
Wimsatt, W. C., 190, 193–194, 197, 217, 228
Winsberg, E., 225–226, 227, 282
Wolfram, S., 185

Woodward, James, 3, 41. *See also* "What if things had been different?" questions
  concept of invariance, 127, 183–184
  on covering-law accounts, 112
  on explanatory context, 102, 104–105
  interventionist account of explanation, 25–26, 56–57, 94–96, 108–109, 302n3, 307n11
  on role of irrelevance information, 84–85, 98–99, 103, 127, 307n11

Young Earth creationists example, 249
Yurk, B. P., 221

Zhang, Y.C., 171
Zipf's law, 310n8